T0136843

Impact Analysis of Total Productive Maintenance

José Roberto Díaz-Reza · Jorge Luis García-Alcaraz
Valeria Martínez-Loya

Impact Analysis of Total Productive Maintenance

Critical Success Factors and Benefits

José Roberto Díaz-Reza
Universidad Autónoma de Ciudad Juárez
Ciudad Juárez, Chihuahua, Mexico

Valeria Martínez-Loya
Universidad Autónoma de Ciudad Juárez
Ciudad Juárez, Chihuahua, Mexico

Jorge Luis García-Alcaraz
Universidad Autónoma de Ciudad Juárez
Ciudad Juárez, Chihuahua, Mexico

ISBN 978-3-030-13190-6 ISBN 978-3-030-01725-5 (eBook)
https://doi.org/10.1007/978-3-030-01725-5

Softcover re-print of the Hardcover 1st edition 2019

This Springer imprint is published by the registered company Springer Nature Switzerland AG
The registered company address is: Gewerbestrasse 11, 6330 Cham, Switzerland

This work is dedicated, to my mother Socorro Díaz, who since the first day has given me her unconditional love and has always encouraged me to be a better person.
To my aunt Gela for being another pillar of the family.
To my brother (Demetrio), who has always been an unconditional support.
To my sister-in-law Marisol, for being the one who has always been a great support.
To my sisters (Laura and Juani) for accompanying my mother, Aunt, and nephews (Zaid, Yair, Iram, Yael and Sofia), who are the joy of the family ... Thank you very much for everything that each one has contributed to my life.
RP!

José Roberto Díaz-Reza

Humans take inspiration when they set a goal. To me, my family is my inspiration, which is why I dedicate this book to:
God. I thank Him for everything.
To my parents; my life teachers.
To my children (Jorge Andres and Mariana Odette), they are the reason to be of my life,

my greatest pillars, and strengths.
To my wife, Ana Blanca Rodríguez-Rendon, for her unconditional support in all the projects that I've undertaken.
To my brothers and sisters, who have taught me the best lessons at home.
Those brothers who recognize me and accept me as such.

Jorge Luis García-Alcaraz

To God for being my guidance allowing me to reach this stage of my life, providing me light and strength every time I need it.
To my parents (Rosa and Rafael) for their unconditional love, trust, and because they've always have supporting me on my decisions, ideas, goals, dreams, and mistakes.
To my sister (Dilia) for being my accomplice and friend, besides all the support and unconditional motivation to never give myself up.

Valeria Martínez-Loya

Foreword

It is always a pleasure to have the scoop of reading a book before it is published, which is an advantage that only the ones who are invited to write a foreword have. On this occasion, I have read the book titled Impact Analysis of Total Productive Maintenance—Critical success factors and Benefits that is written by José Roberto Díaz-Reza, Jorge Luis García-Alcaraz, and Valeria Martínez-Loya, all of them from the Autonomous University of Ciudad Juarez in Mexico. In general terms, this book arises from the need that exists in the industrial field to link the different critical factors or activities associated with the Total Productive Maintenance (TPM) implementation programs along with the benefits obtained from itself.

The authors divide their book into five parts, which are divided into chapters, according to an addressed theme. In addition, these parts are briefly described below in order to motivate the reader to look at its content:

Part I is called *Concepts and Evolution of TPM*, where the authors have carried out a literature review to understand what has been achieved by that lean manufacturing tool over time, as well as the different approaches and applications that have been given in different industrial sectors. In addition, the analysis that is reported regarding the magazines that publish about TPM, years, and industrial sectors is significant, since they are required to keep their machines and equipment on their productive systems in optimal conditions.

Part II is called *Activities and Benefits Associated with TPM*, which lists 75 activities reported in the literature review that are required to achieve the success of this tool in a productive system, which are human and operational. In the same way, 22 benefits are listed that can be associated with the proper TPM implementation, which are divided into those related with the organization, the productivity indexes in the company, and the employees' safeness. In addition, that section presents an idea about what should be performed to incorporate TPM, as well as what a manager may expect in response to the actions he or she takes. Also, it is relevant to mention that each of the activities and benefits is widely analyzed through a literature review, and its study and importance in this book are justified.

Part III has been called *Research Problem, Objectives, and Methodology*, where the authors define the problems that exist in their environment regarding the TPM implementation. Moreover, it is argued that there are many activities associated with TPM where 75 have been identified, as well as benefits; however, there are no studies that seek to relate them directly and indirectly, also there is no quantification about the effect that these activities have on the benefits. Thus, the objective of this book is to present through structural equation models, the relationships between those activities that are required to implement TPM and the benefits that are obtained. In addition, a series of sequential activities that the authors carry out in order to fulfill this goal are defined, where the data gathered from the sector, its validation, and the generation of the causal models can be mentioned.

Part IV is called *Validation and Analysis of Data*, where the authors already applied the methodology to generate validity indexes from data obtained about the industrial sector. Additionally, in this process, some of the activities and benefits are deleted to improve those indexes, and as well as to generate stronger structural equation models. Also, a descriptive analysis of the TPM tasks and benefits is reported, where central tendency measures and dispersion are described, which allows a discussion from a univariable perspective.

Finally, Part V is called *Estructural Equation Models* and it has been considered that in this section, questions that researchers asked at the beginning of their book are answered, because it is where the required tasks for TPM are completely related to the benefits obtained. As a matter of fact, there are two types of structural equation models: the simple ones, in which an activity is linked to a benefit, therefore they are presented on purpose, since they are easy to understand, as the authors are introducing more and more difficult models to the second ones that they call complexes, where four variables already intervene. In addition, it is important to see how a sensitivity analysis is performed for each complex to determine the probability of occurrence from the variables when there are at their low and high values.

I hope that readers on this book will find the usage that is sought in it, since the outcomes will allow identifying the main activities associated with the TPM implementation and the benefits that are obtained. In addition, I know that many readers may mention that relationships are sometimes logical and have common sense, but a real contribution of this book is that it quantifies that relationship and will make it easier for those interested to focus their attention on activities that are crucial for their business in a specific way.

Tudela, Spain

Juan Ignacio Latorre-Biel
Department of Engineering
Public University of Navarra—Campus Tudela

Cali, Colombia

Diego Fernando Manotas Duque
Department of Industrial Engineering
Universidad del Valle

Preface

Nowadays, production systems must be highly competitive, so the resources administration is for the company survival in order to obtain as much benefits as they can, therefore, many lean manufacturing tools are applied, and one aims that the machinery and equipment have a high availability level to be able to attend production orders at any time, which is called total productive maintenance (TPM). In addition, the main reason to study TPM is that the damaged machines can represent unused resources and undelivered production orders in time or with quality out from certain specifications, which logically affect the image and economic profitability in the company.

Furthermore, TPM has proven its efficiency throughout history, that is why managers from many companies pursue to implement it into their productive systems, and often they do not have enough information about how to carry out the implementation process as well as about the coordination of the different aspects that intervene on it. In addition, there's a book entitled Impact Analysis of Total Productive Maintenance—Critical Success Factors and Benefits, where based on literature review TPM is proposed; it is identical to the main activities required to guarantee the success of TPM, as well as its benefits.

However, it is important to understand why the impact of the activities is considered in TPM programs to achieve its benefits. In this book, for its study, the activities and benefits are divided into different groups called latent variables and analyzed through structural equation models that allow them to be related. Additionally, there is a high trust level that this type of models will help managers focus on activities that are significantly depending on their company's needs, as they can identify those activities that affect the benefits they want to acquire.

Moreover, the book is divided into 13 chapters, which are integrated into five parts, which are briefly discussed in the part below:

Part I is about two introductory chapters and its content is the following:

Chapter 1 is titled *TPM Background*, which lists some concepts of TPM over time and from different contexts, the approaches that have been assigned to this tool, the objectives that are aimed to achieve, the losses that may be obtained if

TPM it is not applied properly, as well as a sequence of activities that must be carried out to implement it.

In Chap. 2, which is titled *TPM Literature Review*, the importance of TPM in the industry is described and a literature review is presented, where some graphs are illustrated to represent the articles that are published per year in that TPM area, the type of publication, the names of the scientific journals, the sectors where the case studies are reported, and the main editorial houses that disseminate this type of research. Likewise, with the objective to motivate the reader, some success stories from some companies that have applied TPM in diverse industrial sectors are presented.

Part II has two chapters, which are described below:

Chapter 3 is titled *Activities Associated with the Success of TPM*, which is one of the most important and it is the basis for the definition of all the other chapters, since it describes a total of 75 activities, which are grouped into categories, associated with human and operational factors. Also, a description for each of these activities is mentioned, as well as a literature review about their importance in the TPM implementation process, which is justified.

Chapter 4 is titled *Benefits Associated with the TPM Implementation in the Industry* and as its name suggests, it reports 22 benefits from TPM, which are divided according to their focus and repercussion, therefore, they are divided into three categories: the benefits for the organization, productivity, and security. In addition, as in the previous chapter, a justification for each of these benefits is established based on a literature review.

Part III is also composed of two chapters, which are described below:

Chapter 5 is entitled *Definition of the Problem and Objective of the Research*, which presents the need to relate the 75 activities that are included in Chap. 3 along with the 22 benefits from Chap. 4. Also, the importance to perform each of these activities in order to obtain the benefits that TPM can offer, since, although it is known that this relationship exists, there is no quantification of it. Thus, the objective of this book is to present a set of models that allow relating the activities required for TPM with the benefits obtained.

Chapter 6 is entitled *Methodology*, which displays a list of eight activities that are carried out in chronological order to achieve the objective previously discussed in Chap. 5. In addition, it is vital to mention that the methodology includes fieldwork to gather information from the industry, integrating the experience from several experts in the maintenance area. Also, the data is validated statistically, purified, and analyzed throughout the means of causal models allowing to use the structural equation modeling technique, which shows the relationships between activities and benefits.

Part IV includes two chapters, where the data analysis is described:

Chapter 7 is entitled *Validation of Variables*, which are integrated into structural equation models. In this case, the activities associated with human factors are divided into Work culture, Suppliers, Managerial commitment, and Clients, while the operational factors are divided into PM Implementation, TPM implementation,

Technological status, Layout, and Warehouse management and the benefits are divided into organizational, safety, and productivity. In addition, for each one of these variables, efficiency indexes are obtained in order to know if the activities and benefits actually contribute to its explanation.

Chapter 8 is titled *Descriptive Analysis*, which reports a series of indexes and parameters associated with the characteristics of the sample and the data obtained. In addition, 397 cases from the surveys applied to the maquila sector are reported, as well as the sectors involved in the study, the gender, and years of experience from the participants responding to the survey, as well as the position within it are analyzed. Additionally, regarding activities and benefits, since they are estimated on a Likert scale, the median is reported as a central tendency measure, and the interquartile range as a measure of dispersion.

In Part V, which includes five chapters, the structural equation models are described, where the activities required for TPM are related to the benefits; these chapters are described below:

Chapter 9 is titled *Simple Models*, where three simple structural equation models are reported, which integrate only one activity with a benefit from TPM, which are duly justified. Similarly, a list of all the possible relationships that may exist between the variables associated with the activities and the benefits is addressed, but they are no longer fully explained, since only the dependent indexes are indicated. In addition, in this chapter, the quantifications that determine the relationship between the variables (activities and benefits) start to show up.

Chapter 10, which is titled *Structural Equation Models: Human Factor—Part I*, presents two complex structural equation models, where the activities associated with human factors are related to other activities, since it is assumed that there is dependency between them. Also, a sensitivity analysis is reported at the end of each model, which indicates the probabilities of occurrence for each of the variables together and independently.

Chapter 11 is titled *Structural Equation Models-Human Factor—Part II*, which presents two complex structural equation models, where the activities associated with the human factor in the TPM implementation are already linked with the benefits obtained from the tool. Also, the analyzes are interesting, since here the first causal models appear, which allow managers to identify the activities where they should focus their efforts on.

Chapter 12 is called *Structural Equation Models-Technical Factors*, as its name indicates, it relates the technical factors associated with the activities required to implement TPM with the benefits, and two complex models come out. In addition, as in other models, the hypotheses are presented and justified with a literature review; they are validated, and a conclusion is written from themselves.

Finally, Chap. 13 is titled *Structural Equation Models—Methodological Factors*, which presents a complex structural equation model that integrates the methodological factors and a second-order integrative model that coordinates all the associated variables in a single model along with the activities and benefits, which is incorporated by three latent variables. Also, this is the most significant model, because it is the one allowing to observe from a general perspective the importance that human resources have in the TPM implementation.

Authors hope that the content of this book will contribute to understand managers and decision-makers in the TPM implementation, the relevance of this tool, the appropriate management of the machinery and equipment available in their production systems, and identify the tasks that are essential to acquire a specific benefit focusing their efforts on it. In addition, we hope that enthusiast, academics, and researchers can find this as a useful book for understanding the TPM in industry with real applications.

Ciudad Juárez, Mexico

José Roberto Díaz-Reza
Jorge Luis García-Alcaraz
Valeria Martínez-Loya

Acknowledgements

The present project would not have been completed without the support of many people who were directly and indirectly involved, and therefore, the authors want to thank the following people and institutions:

- To all the maintenance managers and associated areas that answered our survey to gather data from their industry, who commented on it, and allowed the access to their companies.
- To all those students and coworker teachers who helped during the gathering data process from the companies.
- To the National Council of Science and Technology (CONACYT), which through the project called Thematic Network of Optimization Industrial Processes has supported this project.
- To the Secretariat of Public Education of Mexico, who through the Program for Professional Teacher Development (PRODEP) have supported this project.
- Special thanks to the Autonomous University of Ciudad Juárez (UACJ), our work and study place, for their financial support during the book translation and editing process.
- Also, our gratefulness to Alejandra Bautista, who is responsible for translating the book, and her advice and patience has contributed to this project.

Contents

Part V Estructural Equation Models

List of Figures

List of Tables

Part I
Concepts and Evolution of TPM

Chapter 1
TPM Background

Abstract Total Productive Maintenance is considered a lean manufacturing tool increasingly used within the current industrial environments. Therefore, the objective of this chapter is to present a series of concepts and definitions described throughout the literature review, discuss them, and show its evolution as well.

1.1 TPM Definition

According to the literature review, there are some definitions about the TPM concept, therefore, with the purpose of having a broader knowledge and approach, which, although it has changed over time, still has the same basis. In addition, the following is a list of definitions about TPM.

- TPM is productive maintenance carried out by all the employees through small groups of activities (Nakajima 1989).
- TPM is an innovative maintenance approach that optimizes the equipment effectiveness, eliminates breakdowns, and promotes the operators' autonomous maintenance (Nakajima 1988).
- TPM is an association between the maintenance and productive functions in the organization to improve product quality, reduce waste, reduce manufacturing costs, increase equipment availability, and improve the company's maintenance status (Rhyne 1990).
- TPM is a production-driven improvement methodology that is designed to optimize equipment reliability and ensure the efficient management in plant assets (Ginder et al. 1995).
- TPM is reduced to understanding the following strategies (Suzuki 1996):
  - Maximizing the overall efficiency of the team,
  - Establishing a comprehensive preventive maintenance (PM) system that covers the entire life of the team,
  - Involving all the departments that plan, use, and maintain equipment,

J. R. Díaz-Reza et al., *Impact Analysis of Total Productive Maintenance*,
https://doi.org/10.1007/978-3-030-01725-5_1

  - Involve all employees from the senior management along with the machine operators,
  - Promote the PM through an organized management, that is, through a small group of autonomous activities.

- TPM is not a maintenance program itself, it is a team management program that combines and promotes the concepts of continuous improvement, and total quality as well as the employees' empowerment to achieve zero stoppages and defects (Stephens 2004).
- TPM is a procedure to introduce maintenance considerations into the organization's activities. It involves an operative and maintenance team that work together to reduce waste, minimize downtimes, and improve the final product quality (Eti et al. 2004).
- TPM is a methodology that aims to increase the availability/effectiveness of the existing equipment in a certain situation, through the effort of minimizing entries (improving and maintaining the equipment at an optimum level to reduce the life cycle cost), and the investment in human resources that results in a better hardware usage (Chan et al. 2005).
- TPM is a larger part of a Lean initiative, and its goal is generally to improve uptime and equipment reliability (Press 2005).
- TPM is a philosophy that involves the entire organization, which increases the knowledge levels, performance, efficiency, and teamwork in each area (Sun et al. 2003).
- TPM is a maintenance program, which implies a recently defined concept for plants and equipment maintenance; it can be considered as the "medical science" for industrial machines (Mâinea et al. 2010).
- TPM represents a system for the effective technology process usage (Friedli et al. 2010).
- TPM is a set of techniques to ensure that each machine in a production process can always perform the required tasks (Anvari et al. 2014).
- TPM is a continuous improvement process focused on structured teams that seeks to optimize production effectiveness by identifying and eliminating equipment losses as well as production efficiency, throughout the production system life cycle through the employees' active participation at all levels of the operational hierarchy (Attri et al. 2013a).
- TPM is a maintenance management approach focused on involving all the employees of an organization in the equipment improvement. It consists of a variety of methods, known for the maintenance management experience that are effective in improving reliability, quality, and production (Das et al. 2014).
- TPM includes maintenance prevention, maintenance improvement, and preventive maintenance (Benjamin 1997; Filho and Utiyama 2016).
- TPM is the combination of preventive maintenance activities and the Total Quality Management philosophy to create a TPM culture by providing

integration with maintenance, engineering, and management units to ensure that employees protect the equipment and the machinery that they use and ensure that machines work always correctly (Arslankaya and Atay 2015).
- TPM is a proactive approach that aims to identify problems as soon as possible, and plans to avoid any problems before they happen; its slogan is zero errors, zero accidents, and zero losses (Kiran 2016).
- TPM is widely used to improve the effective equipment usage and to obtain a world-class manufacturing system in quality and cost terms (Shinde and Prasad 2017).
- The TPM concept can be summarized based on three meanings represented by the total word (Wang and Lee 2001):
  - Total effectiveness (includes productivity, costs, quality, inter-gas, safety, environment, welfare, and morale).
  - Total maintenance system (including maintenance prevention (MP) and maintenance improvement (MI)).
  - Employee participation.
- TPM can be analyzed based on the three terms (Kiran 2016; Mwanza and Mbohwa 2015):
  - Total: represents the total employees' involvement, that is, from the senior managers to the line operators. Similarly, it also represents the total team effectiveness.
  - Productive: indicates the production of goods and services to meet or exceed customers' expectations, which is possible if the machinery and equipment maintain a high reliability level.
  - Maintenance: symbolizes keeping the equipment and the plant running all the time, in other words, in a condition as good or better than the original.

Although the first accepted TPM definition was provided by the Japan Institute of Plan Engineers in Nakajima (1989), through time it has evolved adding other key elements.

In general, it can be summarized that TPM is a methodology that works as a tool, in which the purpose is to maintain the equipment and machinery used in the production of goods and services in optimal conditions to be able to provide products and/or services that achieve and even exceed customers' expectations.

In addition, it also seeks to reduce waste, minimize equipment inactivity, and improve quality, but mainly focus on equipment maintenance programs to optimize efficiency and performance through activities to improve maintenance, preventive and predictive maintenance. Therefore, the TPM basis is: "zero errors, accidents, and losses".

However, in order to function properly, TPM is considered as a philosophy, because it requires total commitment from all the hierarchical levels of the organization, which symbolizes the teamwork and high coordination of the activities between the administration, production, and maintenance areas.

Finally, it could be summarized that TPM is the result of effort, teamwork, communication, technology, knowledge, and continuous improvement within a company.

1.2 TPM Evolution and Origin

From the second world war, companies began to worry about offering products with a higher quality. In fact, Japanese industries were the pioneers, but based on the ideas taken from the USA and then transforming them into successful practices within those industries (Nakajima 1988). In such a way, at that time, the main focus was on the preventive machinery and equipment maintenance, and Seiichi Nakajima, considered as the father of the TPM, belonging to the Japan Institute of Plant Maintenance (JIPM) (Kiran 2016), established the bases for the development of such maintenance approach, based on the preventive maintenance and by considering the following aspects (Kunio 2017).

1 Breakdown maintenance (BM): broken equipment repairs.
2 Preventive maintenance (PM): planned inspections program, replacements, and planned repairs to avoid failures and control deterioration.
3 Maintainability improvement (MI), sometimes referred to as corrective maintenance (CM): includes broader activities as part of preventive maintenance by adding modifications and procedures to prevent less downtimes.
4 Maintenance prevention (MP): includes the machinery and equipment designers to develop better equipment and more efficiency to maintain them.
5 Productive maintenance (PM): it is the PM, CM, and MP consolidation, that is, the activities required to keep the equipment in optimal conditions, in which the main objective is to maximize productivity, synonymous to profitability.

According to Nakajima (1988), preventive maintenance was introduced during the 1950s, while productive maintenance was fully established during the 1960s. Also, it was not until the 1970s that the indications about the TPM concept emerged as such, and elements beyond simple routines and maintenance activities were already included.

In this sense, the evolutionary MPR process is presented in a broader way in the section below, where it is appreciated how the approach has shown a broader turn through the time as it can be seen in Table 1.1.

Table 1.1 TPM evolution through time (Nakajima 1989; Peng 2012)

Year	Period	Approach	Target	Concepts
Preindustrial revolution	Worker is responsible for repairing their work area	Mechanical expert workers	Bifunctional workers	
Prior-1950	Breakdown maintenance	Machinery repair only if damaged	Equipment failures repair in a reasonable time	"Repair only if it is broken"
1950	Preventive maintenance	Maintenance functions establishment Maintenance based on time (time-based)	Increase the equipment useful life Reduce dead time due to work stoppages or defects	PM (preventive maintenance) PM (productive maintenance) MI (maintainability improvement)
1960	Productive maintenance	Reliability Maintainability Costs	Reduce dead time due to work stoppages or defects and increase maintenance efficiency	Reliability engineering Maintainability engineering Engineering economy Reliability-centered maintenance (RCM) Behavioral sciences Systems engineering
1970	TPM (total productive maintenance)	Predictive maintenance with TQC, commitment and total involvement of employees	Zero breakdowns as well as zero defects	Ecology Maintenance prevention Just in time (JIT) TQC and TQM Terotechnology
1980–1990	TPM with predictive maintenance	TPM praxis Maintenance based on the environment conditions	Zero breakdowns and zero defects Availability optimization	Computerized management maintenance
2000s	Post-maintenance era	CMMS application TPM and the factory of the future	Zero defects, zero breakdowns, zero accidents, zero pollution, zero inventory	Artificial intelligence and expert systems

1.3 TPM Objectives

In the first instance, the overall TPM objective is to increase the company and the team productivity through a small set of activities and autonomous maintenance developed by the team operator (Nakajima 1988; Wang and Lee 2001). In other words, it is to improve the equipment efficiency and maximize the outputs (Baglee et al. 2008), as well as making the most of its usefulness, availability, and avoiding its degradation (Mwanza and Mbohwa 2015).

In addition, TPM seeks the reduction of the following aspects (Cudney et al. 2013):

- Stops (breakdowns)
- Barriers between departments
- Quality issues
- Safety or environmental accidents
- Costs
- Emergency or unplanned maintenance.

On the other hand, Agustiady and Cudney (2016) claim that the TPM main objectives are to: (1) Avoid waste within changing environments, (2) reduce manufacturing costs, (3) produce low batches quantities as soon as possible, and finally (4) provide free defects goods.

Similarly, TPM pursues to achieve the five zeros; (1) zero stoppages (breakdowns) (2) zero defects, (3) zero accidents, (4) zero contamination, and (5) zero inventories (Kiran 2017).

According to Smith and Mobley (2011), the TPM objectives are broader because they focus more on improving performance, employee interaction, and positive effort than on maintenance technology. Therefore, it means that there must be a balance between the human and technical factor.

1.4 TPM and Big Losses

In order to achieve the main TPM objective previously commented, the companies seek to maximize the equipment effectiveness as a whole, and at the same time minimize costs; however, sometimes the companies' efforts to achieve it are not enough to eliminate certain barriers, which are called big losses that cause the machines performance and veracity to get affected.

According to Nakajima (1988), there are six big losses, which can be classified into three categories:

- Downtime: Stops due to equipment failures, setup, and adjustments.
- Speed losses: Inactivity and minor stoppages (abnormal operations), and speed reduction.

- Defects: Process and rework defects reduced the performance between the start of the machine and stable production.

Likewise, Agustiady and Cudney (2016) classify the losses by focusing on different levels:

- Availability: breakdowns and setup.
- Operational: idling and minor stoppages.
- Quality: quality factors and rework.

However, for Smith and Hawkins (2004), the great losses are not only six but have increased to the same ones that are classified in the following four categories.

- Planned losses
  - No production, breaks and/or shift changes.
  - Planned maintenance.
- Inactivity time losses
  - Equipment failures or faults.
  - Configurations and changes (setup and changeover).
  - Parts or tools changes.
  - Start and adjustments.
- Efficient development losses
  - Minor stops (less than six minutes).
  - Reduction in speed or cycle time.
- Quality losses
  - Product or raw material waste.
  - Defects or rework.
  - Transition process performance or losses.

1.5 TPM Origin

Since the first TPM official definition appeared in 1971 by JIPM and Nakajima (1989), five fundamental principles were established to allow TPM to be successful, which are (McCarthy and Rich 2004; Nakajima 1989):

- Adopt improvement activities designed to increase the total effectiveness of the equipment by attacking the losses.
- Improve current planned and predictive maintenance systems to extend the equipment.
- Establish a self-maintenance and cleaning level carried out by highly trained operators.

- Increase operators and engineers' skills and motivation through individual and group development.
- Apply early management techniques to design a low life cycle by creating reliable, safe equipment, and processes that are easy to operate and maintain.

1.6 TPM Implementation Steps

According to Shen (2015), the TPM implementation and adaptation to any company take an average of 2.5–3 years, but its success and duration may vary since it depends on the focus and corporate state from each of them. Also, the fact is that it is a process that involves multiple activities, in addition to a broad commitment on all the areas in general not only in the maintenance area as it might seem at first sight.

For an appropriate TPM implementation, it must be supported with the implementation of 12 appropriate activities in 4 categories as it is shown below (Shen 2015):

- Preliminary stage
  - TPM introduction to the organization announcement.
  - TPM initial education and promotion.
  - TPM departmental structure establishment.
  - TPM policies and objectives definition.
  - TPM master plan design for its implementation.
- Beginning and introduction TPM stage
  - TPM implementation.
- TPM implementation stage
  - Efficiency system for the production department establishment.
  - Management systems for new products and new equipment establishment.
  - Quality maintenance system establishment.
  - An efficient system for the administration and indirect departments establishment.
  - Safety, health, and environment management system establishment.
- Performance stage
  - Complete TPM implementation.

1.7 TPM Pillars

The TPM implementation is not a simple task since there is no series of rigorous steps to follow to ensure its successful implementation. However, it must be based on certain concepts that allow it to establish the base to guarantee its success, in this sense, multiple "pillars" represented by various tools and practices that serve as a guide to achieve the proposed objectives have been generated, which are illustrated in Fig. 1.1.

In addition, within this figure, there is a house consisting of eight pillars where there is a logical sequence that must be followed to implement TPM. Therefore, the pillars must be established before the "house" can be built. It should be mentioned that the definition of the pillars depends too much on the philosophy and internal structure in each company, in such a way that it is personalized and adapted according to the existing culture and the approaches in the company itself. For instance, Kunio (2017) reports five pillars, as well as Ahuja (2009) who considers five pillars. On the one hand, Levitt (2010) contemplates ten concepts, where six are classified as pillars while the four remaining are considered as rungs or base steps. On the other hand, Morales Méndez and Rodriguez (2017) acknowledge that the important pillars are nine.

In this research, eight significant pillars are considered, which are listed below:

1. Autonomous Maintenance (Jishu Hozen)
2. Focused improvement (Kobetsu Kaizen)
3. Planned Maintenance

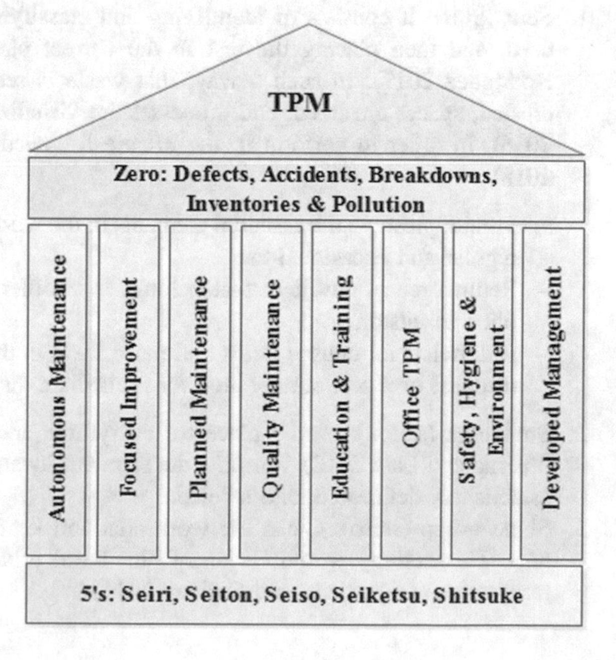

Fig. 1.1 TPM pillars

4. Education and Training
5. Quality Maintenance (Hinshitsu Hozen)
6. Office TPM (OTPM)
7. Safety, Hygiene, and Environment (SHE)
8. Development Management

1.7.1 Pillar 1: Autonomous Maintenance (Jishu Hozen)

TPM starts with the 5S establishment, in other words, this tool is considered the cornerstone of the TPM implementation (Singh et al. 2013), that is the reason why some authors consider it as another independent pillar (Morales Méndez and Rodriguez 2017).

In addition, the 5S represent five disciplines to keep the visual workspace, and it is a universal tool that can be applied in any situation and anywhere, from the floor machines to the finance department and the central office (Mohan Sharma and Lata 2018).

Also, the 5S achieve a serene workplace environment by engaging employees with a commitment to implement and practice systematic cleaning (Levitt 2010). On the other hand, if the 5S are not given the necessary importance, they will lead the company towards the 5Ds: Delays, Defects, Dissatisfied customers, Declining profits, and Demoralized employees (Singh et al. 2013).

These five principles to create an efficient and effective company are as follows:

1. Seiri (Sort): It consists of identifying and classifying which elements will be used, and then placing the rest in the correct places (Morales Méndez and Rodriguez 2017). In such a way, that waste is reduced, a safe work area is created, spaces are freed, and processes are visualized (Agustiady and Cudney 2016). In order to perform it, the 3R are followed (Mohan Sharma and Lata 2018).

   – Retain: preserve the essential elements in the work area to function, either for regular and occasional use.
   – Return: return any item that belongs to another department, location, supplier, or customer.
   – Rid: delete all unused items and place them in the recycling or trash bin for disposal or a preliminary area for immediate disposal.

2. Straighten (seiton): Find a place for everything and set it in its place (Mohan Sharma and Lata 2018). For this, the standard location for each article must be established, delineated, and labeled.
3. Shine/sweep (seiso): Clean the work area and keep everything organized and neat. This activity would eliminate dirt, build pride in work areas, and build team values (Agustiady and Cudney 2016).

4. Standardize (Seiketsu): It consists in documenting the procedures so that they can be repeated in an easy, continuous, and effective way, and therefore, the new personnel can be trained correctly. In addition, the appropiate resources, personnel definition, documents, and standard times are required to perform each task (Morales Méndez and Rodriguez 2017).
5. Sustain (Shitsuke): it is about forming a continuous improvement of procedures habit in addition to seeking to train and discipline people about the 5S.

As a matter of fact, the 5S software is considered by multiple authors as the TPM support because they are the direct personnel involved responsibility, which requires standardization and discipline (Morales Méndez and Rodriguez 2017) crucial elements from the main TPM, that is, autonomous maintenance (Shinde and Prasad 2017).

Furthermore, Autonomous Maintenance (Jishu Hozen) aims to achieve the operator belonging sense (Shinde and Prasad 2017). This contributes to the significant losses reduction with the operator being responsible for their processes since they have been trained and qualified on thesc functions (Morales Méndez and Rodriguez 2017). Some of the activities that must be developed in this stage are: cleaning, lubrication, adjustments, visual inspections, and readjustments production equipment (Singh et al. 2013).

According to Ahuja and Khamba (2008), some of the key performance indicators (KPIs) on autonomous maintenance to increase profitability include the following aspects:

- Failures or stoppages due to poor self-maintenance
- Spill of products
- Dry machines (zero leaks)
- Autonomous maintenance activities carried out
- Corrected defects
- Meetings held
- Quality defects
- Accidents and unsafe places
- Training sessions
- Kaizens Autonomous maintenance registered and implemented

In order to achieve proper self-maintenance, seven main steps must be followed (Panneerselvam 2012):

- Employees training
- Cleaning and initial machinery and equipment arrangement
- Countermeasures for areas with difficult access establishment
- Repair standards that include the preparation of programming for cleaning, inspection, and lubrication, and the establishment of all the details
- General inspection training for employees in areas such as pneumatics, electrical, hydraulic, lubricant, coolant, bolts, screws, and safety
- Autonomous inspection
- Standardization

1.7.2 Pillar 2: Focused Improvement (Kobetsu Kaizen)

This pillar is about all the activities that maximize the equipment effectiveness, processes, and organization through the waste elimination as well as allowing to improve the performance (Vilarinho et al. 2017).

Kaizen means "continuous improvement", with the 5S being one of the most common elements implemented in the pursuit for continuous improvement (Kiran 2017).

The principle behind Kaizen is a large number of small improvements that are more effective than some great value improvements (Singh et al. 2013). In general, Kaizen philosophy is based on reducing cycle times and delivery schedules, which in turn increases the productivity, reduces work in process (WIP), reduces defects, increases capacity, increases flexibility, and designs improvement through visual management techniques (Agustiady and Cudney 2016). However, these activities are not limited only to the production areas but can also be implemented within the administrative areas (Levitt 2010).

Mainly through the waste identification and minimization, increasement in quality and losses decreasement by using the "why-why" structure, and the FMEA analysis development (Abhishek et al. 2014) eliminating losses helps to improve OEE (Shinde and Prasad 2017).

1.7.2.1 Overall Equipment Effectiveness

Overall equipment effectiveness (OEE) is the main TPM measure and it is used to determine how efficient a machine is working (Ahmad et al. 2018). In addition, OEE has been defined as the relationship between the time spent to the approved quality products production and the scheduled time (Kumar and Soni 2015). Also, its main objective is to identify the losses categorized in: availability, performance rate, and quality (Ahmad et al. 2018). The formula to estimate it is the following:

$$\text{OEE} = \text{Availability} \times \text{Performance Rate} \times \text{Quality Rate} \tag{1.1}$$

- Availability: It is a comparison between the amount of time the machine is producing and the amount of time it was programmed to produce (Ahmad et al. 2018). Availability considers the loss of time due to inactivity, includes any event that stops planned production for an appreciable period (usually several minutes, enough time to be recorded as a traceable event). Examples include equipment failures, material shortages, and time to change (Mwanza and Mbohwa 2015). The formula 1.2 is needed to estimate availability:

$$\text{Availability} = \frac{\text{Required availability} - \text{Downtime}}{\text{Required availability}} \times 100 \tag{1.2}$$

- Performance: The performance considers the speed loss, that is, any factor that causes the process to operate at less than the maximum possible speed. For example, machines, poor quality materials (Mwanza and Mbohwa 2015). The formula 1.3 illustrates how to obtain performance:

$$\text{Performance} = \frac{\text{Theoretical cycle time} \times \text{units output}}{\text{Actual cycle time}} \times 100 \tag{1.3}$$

- Quality: It can be expressed as the production output in the process or equipment minus the volume or the number of quality defects that exist, then divided by the production output. In addition, it considers the quality loss, which includes produced parts that do not meet the quality standards, including those that require some type of rework (Ahmad et al. 2018).

$$\text{Quality} = \frac{\text{Production output} - \text{defects}}{\text{Production output}} \tag{1.4}$$

In general, OEE is relevant because it helps to establish priorities among the several improvement projects, consequently, it reflects results appropriately.

1.7.3 Pillar 3: Planned Maintenance (PM)

The planned maintenance goal (keikaku hozen) is to achieve and maintain the machines availability, the optimal maintenance cost, improve machines reliability and maintainability, zero failures and breakdowns, and ensure always the spare parts availability (Singh et al. 2013). Normally, planned maintenance involves work led by highly trained maintenance technicians (McKone et al. 1999).

Also, planned maintenance is about maintenance practices and approaches such as preventive maintenance, time-based maintenance (TBM), condition-based maintenance (CBM), and corrective maintenance (CM) (Jasiulewicz-Kaczmarek 2016). The key to effective planned maintenance is to have a PM plan for each tool, which includes the following main activities (Jasiulewicz-Kaczmarek 2016):

- Guidance and support for autonomous maintenance occupations
- Planned maintenance (stabilize the MTBF, extend the useful life of the equipment, identify when to use the different maintenance tasks by predictive maintenance technology means)
- Lubrication management
- Customize the planned maintenance structure
- Spare parts management
- Reduction of maintenance cost activities
- Improvement and updating of maintenance skills
- Success in the use of predictive maintenance tools

– In general, compliance with the PM plan is a successful implementation measure from the maintenance tools and the execution of the plans (McKone et al. 2001).

1.7.4 *Pillar 4: Quality Maintenance (Hinshitsu Hozen)*

The quality maintenance seeks to keep the equipment in good working conditions, as the highest quality products are delivered to customers through faultless manufacturing (Panneerselvam 2012). In other words, it focuses on the monitoring of activities that affect the variability in product quality (Shinde and Prasad 2017). In this pillar, it is about the transition from reactive to proactive, and this is to evolve from quality control to quality assurance (Singh et al. 2013).

The QM activities consist in establishing the equipment conditions that prevent quality defects based on the basic concept of maintenance on the perfect equipment to maintain the perfect product quality (Attri et al. 2013b). In addition, these conditions are verified and measured in time series to verify that the measured values are within the standard to avoid defects. Also, the transition of the measured values is observed through graphs to predict the defects possibilities in order to take the appropriate measures before they are presented (Singh et al. 2013).

At this point, it is the prevention of defects perspective in origin and focuses on the use of poka-yoke as well as an error-proof system (foolproof system), as well as, online defects detection and segregation, and the effective operator quality assurance implementation are contemplated (Ngadiman et al. 2012).

1.7.5 *Pillar 5: Education and Training*

This pillar is fundamental because here the initial understanding of the TPM importance is produced, followed by the understanding of the correct operation processes, the operation machines, and the standards rigor (Morales Méndez and Rodriguez 2017).

The element purpose is to increase the operators and the people involved morale and experience by providing skills and technical training (Shinde and Prasad 2017). In such a way, they are eager to work and perform all the required functions effectively and independently (Attri et al. 2013b). Since it is not enough to "know-how" but also to "know-why" in a way, a factory full of experts is achieved (Levitt 2010). Similarly, this pillar is fundamental to another pillar, that is, the same continuous improvement that is only possible through in the people continuous improvement knowledge and ability at different levels (Singh et al. 2013).

Panneerselvam (2012) considers that an adequate training process should consider the following fundamental aspects:

- Focus on improving knowledge, skills, and techniques.
- Create a training environment for self-learning based on needs.
- Create a training curriculum, training tools, and training evaluation for the employees' revitalization.
- Train to remove fatigue from employees and make work more enjoyable.

1.7.6 *Pillar 6: Office TPM (OTPM)*

This pillar is the next step for the other four pillars (JII, Kaizen, QM y PM), which must be followed to increase the administrative functions productivity and efficiency (Singh et al. 2013) through the identification and elimination of losses.

The OTPM objectives should include functional loss, the organization of highly efficient offices, service provision, and support to production departments focusing on the workplace and standardized work procedures effective organization (Ahuja 2009).

Moreover, this is where the technical specialists in the process specify the type of machine required, the maintenance technicians establish the most efficient maintenance strategies, the machine operators strictly comply with the equipment optimal use, the most effective strategy for the management is established by critical parts, safeness, and healthy work is guaranteed (Shinde and Prasad 2017).

As it was previously mentioned, the OTPM purpose is to eliminate losses, which includes the processes and procedures analysis to increase the automation of the office (Levitt 2010). In this sense, according to Panneerselvam (2012), some of the losses that are looked to be solved and avoided are as follows:

- Processing losses
- Losses in costs including the accounting, purchasing, market technology, and sales areas by increasing inventory levels
- Loss of communication
- Loss due to inactivity
- Losses due to adjustments and precision
- Office equipment breakdowns
- Interruptions in the communication channel, telephone, or internet
- Time spent on information retrieval
- No online connection
- Customer complains due to logistics
- Loss due to low quality
- Expenses due to dispatches or emergency purchases
- Loss due to inactive operators

1.7.7 Pillar 7: Safety, Hygiene, and Environment (SHE)

The main purpose of the SHE is to ensure a workplace where there are zero accidents, zero occupational diseases, and zero environmental accidents (Morales Méndez and Rodriguez 2017). Likewise, it is pursuing that if there are areas at health risk, they are identified, improved, and at the same time, activities that preserve the environment are carried out (Kiran 2017). Also, organizations should treat people respectfully as well as the environment (Levitt 2010). For instance, the promotional security activities as: Kaizen activities, competition on security, safety poster creation, month of the security, celebration of the week, and to inform on the best initiatives against loss and accesses can contribute a lot to improve the security in the workplace (Ahuja 2009).

1.7.8 Pillar 8: Developed Management

This TPM component is responsible for incorporating knowledge and manufacturing skills acquired by maintaining the existing equipment to be applied in new equipment designs (Ahuja 2009). The layout, commissioning, and adequate testing will ensure that the team can develop reliable products and along with the granted specifications (Levitt 2010).

Furthermore, some of the aspects that are aimed to cover at this point are the minimization of problems, implementation on time, and improvement in the maintenance of new equipment from experiences obtained according to the previous team (Abhishek et al. 2014). In fact, considering several data, such as equipment performance, life cycle costs, reliability, maintenance objectives, equipment test plans, operational documentation, and training is essential for this development (Ahuja 2009).

According to Ahuja and Khamba (2008), some key activities in this pillar are as follows:

- Total number of preventive maintenance records (sheets)
- Number of required days for the new machinery to reach 85% of OEE
- Number of prevented defects
- Amount of energy/fuel consumed
- Number of required days for product development
- LCC implementation on new machinery
- Number of design standards
- Amount of automatic machines
- Number of registered maintenance initiatives

References

Abhishek J, Rajbir B, Harwinder S (2014) Total productive maintenance (TPM) implementation practice: a literature review and directions. Int J Lean Six Sigma 5(3):293–323. https://doi.org/10.1108/IJLSS-06-2013-0032

Agustiady TK, Cudney EA (2016) Total productive maintenance: strategies and implementation guide. CRC Press

Ahmad N, Hossen J, Ali SM (2018) Improvement of overall equipment efficiency of ring frame through total productive maintenance: a textile case. Int J Adv Manuf Technol 94(1):239–256. https://doi.org/10.1007/s00170-017-0783-2

Ahuja IPS (2009) Total productive maintenance. In: Ben-Daya M, Duffuaa SO, Raouf A, Knezevic J, Ait-Kadi D (eds) Handbook of maintenance management and engineering. Springer, London, pp 417–459

Ahuja IPS, Khamba JS (2008) Strategies and success factors for overcoming challenges in TPM implementation in Indian manufacturing industry. J Qual Maintenance Eng 14(2):123–147. https://doi.org/10.1108/13552510810877647

Anvari A, Zulkifli N, Sorooshian S, Boyerhassani O (2014) An integrated design methodology based on the use of group AHP-DEA approach for measuring lean tools efficiency with undesirable output. Int J Adv Manuf Technol 70(9):2169–2186. https://doi.org/10.1007/s00170-013-5369-z

Arslankaya S, Atay H (2015) Maintenance management and lean manufacturing practices in a firm which produces dairy products. Procedia Soc Behav Sci 207:214–224. https://doi.org/10.1016/j.sbspro.2015.10.090

Attri R, Grover S, Dev N, Kumar D (2013a) Analysis of barriers of total productive maintenance (TPM). Int J Syst Assur Eng Manag 4(4):365–377. https://doi.org/10.1007/s13198-012-0122-9

Attri R, Grover S, Dev N, Kumar D (2013b) An ISM approach for modelling the enablers in the implementation of total productive maintenance (TPM). Int J Syst Assur Eng Manag 4(4):313–326. https://doi.org/10.1007/s13198-012-0088-7

Baglee D, Trimble R, MacIntyre J (2008) Maintenance strategy development within SME's: the development of an integrated approach. IFAC Proc Volumes 41(3):222–227. https://doi.org/10.3182/20081205-2-CL-4009.00040

Benjamin SB (1997) An enhanced approach for implementing total productive maintenance in the manufacturing environment. J Qual Maintenance Eng 3(2):69–80. https://doi.org/10.1108/13552519710167692

Cudney EA, Furterer S, Dietrich D (2013) Lean systems: applications and case studies in manufacturing, service, and healthcare. CRC Press

Chan FTS, Lau HCW, Ip RWL, Chan HK, Kong S (2005) Implementation of total productive maintenance: a case study. Int J Prod Econ 95(1):71–94. https://doi.org/10.1016/j.ijpe.2003.10.021

Das B, Venkatadri U, Pandey P (2014) Applying lean manufacturing system to improving productivity of airconditioning coil manufacturing. Int J Adv Manuf Technol 71(1):307–323. https://doi.org/10.1007/s00170-013-5407-x

Eti MC, Ogaji SOT, Probert SD (2004) Implementing total productive maintenance in Nigerian manufacturing industries. Appl Energy 79(4):385–401. https://doi.org/10.1016/j.apenergy.2004.01.007

Filho MG, Utiyama MHR (2016) Comparing the effect of different strategies of continuous improvement programmes on repair time to reduce lead time. Int J Adv Manuf Technol 87(1):315–327. https://doi.org/10.1007/s00170-016-8483-x

Friedli T, Goetzfried M, Basu P (2010) Analysis of the implementation of total productive maintenance, total quality management, and just-in-time in pharmaceutical manufacturing. J Pharm Innov 5(4):181–192. https://doi.org/10.1007/s12247-010-9095-x

Ginder A, Robinson A, Robinson CJ (1995) Implementing TPM: The North American experience. Taylor & Francis

Jasiulewicz-Kaczmarek M (2016) SWOT analysis for planned maintenance strategy—a case study. IFAC-Papers OnLine 49(12):674–679. https://doi.org/10.1016/j.ifacol.2016.07.788

Kiran DR (2016) Total quality management: key concepts and case studies. Elsevier Science

Kiran DR (2017) Chapter 13—total productive maintenance. In Kiran DR (ed) Total quality management. Butterworth-Heinemann, pp 177–192

Kumar J, Soni VK (2015) An exploratory study of OEE implementation in Indian manufacturing companies. J Inst Eng (India) Ser C 96(2):205–214. https://doi.org/10.1007/s40032-014-0153-x

Kunio S (2017) TPM for workshop leaders. Taylor & Francis

Levitt J (2010) TPM reloaded: total productive maintenance. Industrial Press

Mâinea M, Duţă L, Patic PC, Căciulă I (2010) A method to optimize the overall equipment effectiveness. IFAC Proc Volumes 43(17):237–241. https://doi.org/10.3182/20100908-3-PT-3007.00046

McCarthy D, Rich N (2004) Lean TPM: a blueprint for change. Elsevier Science

McKone KE, Schroeder RG, Cua KO (1999) Total productive maintenance: a contextual view. J Oper Manag 17(2):123–144. https://doi.org/10.1016/S0272-6963(98)00039-4

McKone KE, Schroeder RG, Cua KO (2001) The impact of total productive maintenance practices on manufacturing performance. J Oper Manag 19(1):39–58. https://doi.org/10.1016/S0272-6963(00)00030-9

Mohan Sharma K, Lata S (2018) Effectuation of lean tool “5S” on materials and work space efficiency in a copper wire drawing micro-scale industry in India. Mat Today Proc 5(2, Part 1):4678–4683. https://doi.org/10.1016/j.matpr.2017.12.039

Morales Méndez JD, Rodriguez RS (2017) Total productive maintenance (TPM) as a tool for improving productivity: a case study of application in the bottleneck of an auto-parts machining line. Int J Adv Manuf Technol 92(1):1013–1026. https://doi.org/10.1007/s00170-017-0052-4

Mwanza BG, Mbohwa C (2015) Design of a total productive maintenance model for effective implementation: case study of a chemical manufacturing company. Procedia Manuf 4:461–470. https://doi.org/10.1016/j.promfg.2015.11.063

Nakajima S (1988) Introduction to TPM: total productive maintenance. Productivity Press

Nakajima S (1989) TPM development program: implementing total productive maintenance. Productivity Press

Ngadiman Y, Hussin B, Abdul Majid I (2012) A study of total productive maintenance implementation in manufacturing industry

Panneerselvam R (2012) Production and operations management. PHI Learning

Peng K (2012) Equipment management in the post-maintenance era: a new alternative to total productive maintenance (TPM). Taylor & Francis

Press P (2005) TPM: collected practices and cases. Taylor & Francis

Rhyne D (1990) Total plant performance advantages through total productive maintenance. In: Conference proceedings APICS, pp 683–686

Shen CC (2015) Discussion on key successful factors of TPM in enterprises. J Appl Res Technol 13(3):425–427. https://doi.org/10.1016/j.jart.2015.05.002

Shinde DD, Prasad R (2017) Application of AHP for ranking of total productive maintenance pillars. Wireless Pers Commun. https://doi.org/10.1007/s11277-017-5084-4

Singh R, Gohil AM, Shah DB, Desai S (2013) Total productive maintenance (TPM) implementation in a machine shop: a case study. Procedia Eng 51:592–599. https://doi.org/10.1016/j.proeng.2013.01.084

Smith R, Hawkins B (2004) Lean maintenance: reduce costs, improve quality, and increase market share. Elsevier Science

Smith R, Mobley RK (2011) Rules of thumb for maintenance and reliability engineers. Elsevier Science

Stephens MP (2004) Productivity and reliability-based maintenance management. Prentice Hall

Sun H, Yam R, Wai-Keung N (2003) The implementation and evaluation of total productive maintenance (TPM)—an action case study in a Hong Kong manufacturing company. Int J Adv Manuf Technol 22(3):224–228. https://doi.org/10.1007/s00170-002-1463-3

Suzuki T (1996) TPM en industrias de proceso. Taylor & Francis

Vilarinho S, Lopes I, Sousa S (2017) Design procedure to develop dashboards aimed at improving the performance of productive equipment and processes. Procedia Manuf 11:1634–1641. https://doi.org/10.1016/j.promfg.2017.07.314

Wang FK, Lee W (2001) Learning curve analysis in total productive maintenance. Omega 29 (6):491–499. https://doi.org/10.1016/S0305-0483(01)00039-1

Chapter 2
TPM Literature Review

Abstract This chapter presents some data related to the evolution of total productive maintenance in recent decades, specifically from 1994. As a matter of fact, the data is organized according to the type of publication, the published, number as well as the TPM use trend, information is presented on the number of published articles by each journal, the authors that publish information about TPM, and the industrial sectors that implement this tool. In addition, the data was obtained from a literature review in different databases, where keywords were used such as total productive maintenance (TPM). Also, the results show that in 2015, it was the year where more papers were published about TPM, the most common type of publication is research articles with 714, the journal with the most reported publications is International Journal of Production Economics, the sector that uses it more is the construction of industrial buildings, and the editorial that has published the most is Taylor & Francis LTD. Similarly, a series of case studies are included that demonstrate the success of the TPM implementation, both in the manufacturing industry and in-service companies.

2.1 TPM Importance

Although according to the literature, the emergence and TPM implementation were originated mainly within the automotive industries, more specifically within Toyota, from its origin, its popularity and spread have increased exponentially in all sectors.

First, according to Suzuki (1996), TPM has increased its popularity and expansion due to the following reasons:

1. It guarantees drastic and remarkable results: this includes a decreasement in breakdowns, errors, and accidents as well as the increasement in quality and employee participation.
2. It improves the work environment in such a way that clean, order, and, consequently, safe work areas are achieved.

J. R. Díaz-Reza et al., *Impact Analysis of Total Productive Maintenance*,
https://doi.org/10.1007/978-3-030-01725-5_2

3. It promotes employee growth: it means helping to raise the employees' knowledge and capacity level, mainly in the production and maintenance areas.
4. Achieving a higher participation and empowerment level, and also increasing the commitment level to their work area and to the company in general.

According to Pinto et al. (2016), the TPM importance not only improves the availability and trust in the team but also promotes progress and improvement in production terms, generation, and quality products development as well as the individual capacities increasement in generating a teamwork spirit and a better attitude (Siong and Ahmed 2007).

In fact, improving production and guaranteeing quality products, efficiency and time are the fundamental elements to ensure satisfaction with the customers' demand, therefore, the company must be competent in providing its products with the highest quality at the shortest time possible (Borkowski et al. 2014).

Due to the previous information, in recent years, multiple companies have increased their interest level in TPM, as proof, it is that not only that this tool is known in the oriental environment but its application and study have extended its horizons to the western region. Nowadays, many industries in North and South America as well as in Europe are in constantly searching and training to adapt and implement TPM successfully within their companies, having the main objective to achieve favorable results.

Therefore, by consulting frequently different media, it is possible to find several researches and studies about TPM at an international level in industries belonging to various sectors such as manufacturing, chemical, health, textile, among others.

2.2 TPM Publishing

In order to present a concept and a broader picture regarding TPM, a literature review was carried out through the consultation of various electronic information media, mainly academic databases that could be presented in figures the TPM use in some areas.

As a result, a report is described, which is organized based on the number of TPM publications per year, by type of publication, industry sector, journal, and publisher.

2.2.1 Publications Per Year

After the first successful TPM implementation report in Japan in the 70s, data that shows how the application and internationalization of this tool was has been found, and the first reports are from the beginning of 1994. In this sense, Fig. 2.1 details a total of 714 publications from 1994 to the first 2 months in 2018, as well as a

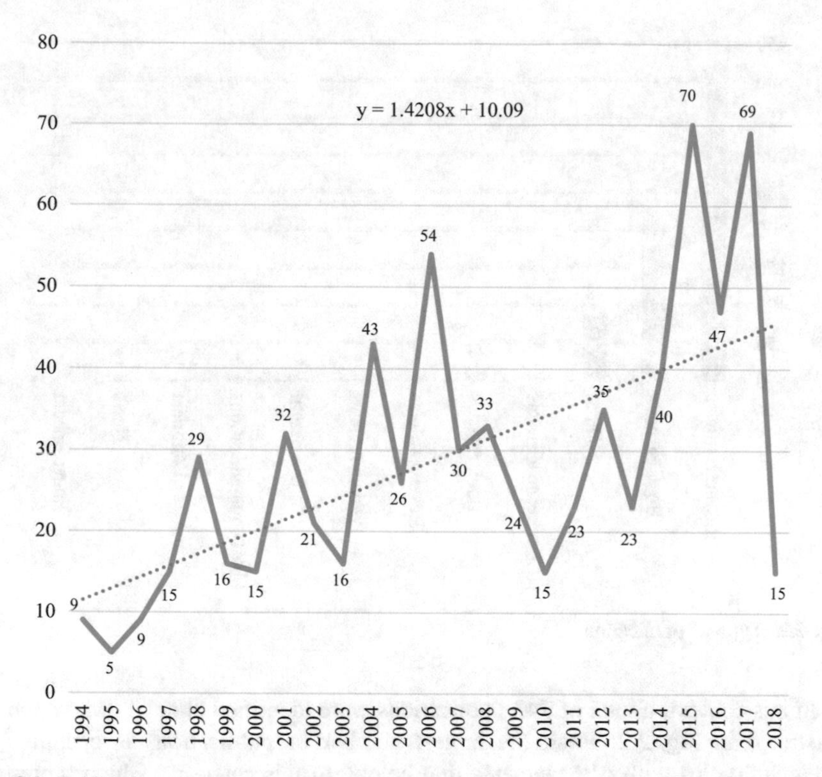

Fig. 2.1 Publications per year

positive trend line, which shows a notable growth regarding the TPM usage within different industries.

Furthermore, throughout 1994 and 2001, the main TPM deployment was observed as well as a total of 130 publications, it could be said that on average almost 19 publications per year. On the other hand, from 2002 to 2009, a total of 221 publications were reported, representing an increasement of 70% in publications during the same number of years. Also, between 2010 and 2018 first bimester, an increasement of over 30% is reported as related to the number of publications about TPM in the previous period, reporting a total of 290.

2.2.2 *Type of Publication*

Another classifications in the literature review consider the number of publications according to the following categories: articles, book chapters, congress publications, encyclopedias, news, abstracts, among others.

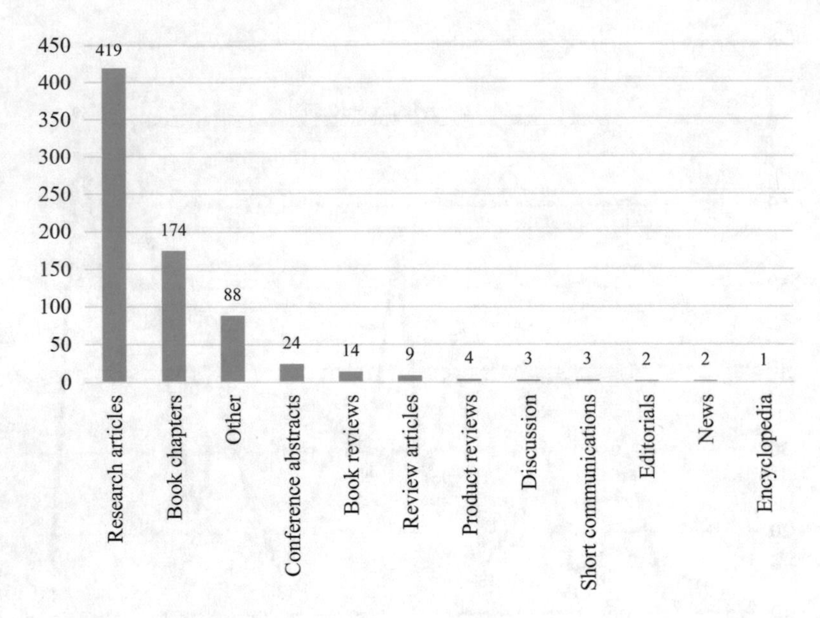

Fig. 2.2 Type of publication

In this context, a total of 743 publications were identified about TPM, which are illustrated in Fig. 2.2. First, the largest number of publications is occupied by research articles with 419 elements that belong to this category, which represents 56.39% of all publications, right before chapters in books with a contribution of 174 items, which represents 23.41% from the total.

As it can be perceived, the number of published articles and chapters together represents about 80% of the total publications about TPM.

2.2.3 *Scientific Magazines Publications*

Figure 2.3 shows the ten main journals that include dissemination topics about TPM, which have been organized in a descending order according to the number of publications that appear in them. For instance, a total of 292 publications stand out, being the magazine that presents the largest number of publications; the International Journal of Production Economics which presents a total of 49 publications.

Second, IFAC Proceedings Volumes present 46 elements, followed by the Journal of Manufacturing Systems with 44 publications and finally, Procedia CIRP, which presents a total of 35 publications. It should be noted that these four accounts represent 59.59% of the publications about TPM that could have been detected by the authors.

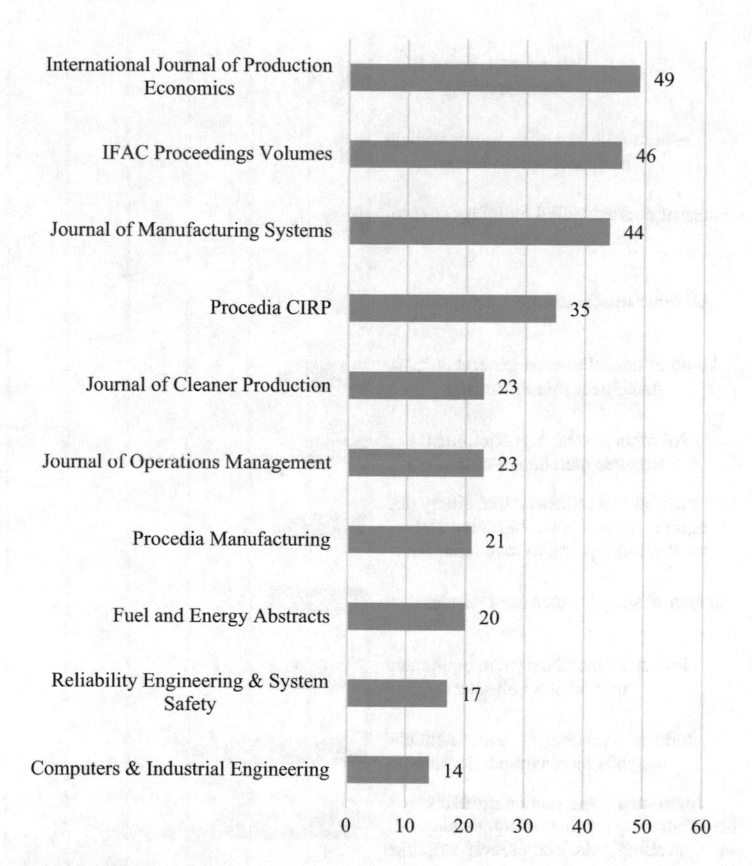

Fig. 2.3 Magazine publications

2.2.4 *Industrial Sectors*

Due to the TPM introduction and success, its versatility has been demonstrated to be implemented in different productive sectors and it is not exclusive to the automotive sector as it was initially considered. In addition, Fig. 2.4 illustrates the information on the industrial sectors where TPM has been implemented and, in the first instance, a total of 15 sectors were mentioned in 619 publications.

As it is observed, the sector related to the industrial construction is the one that reports a greater amount of application and TPM implementation with a total of 106 publications. Second, there are issues related to landscaping services including a total of 102 publications.

On the other hand, the consulting, physical distribution, and logistics process services are reported in 62 publications. While both the Industrial and manufacturing sector, the instrumental sector, and sectors related to manufacturing products for measuring, displaying, and controlling industrial process are reported in 52 publications.

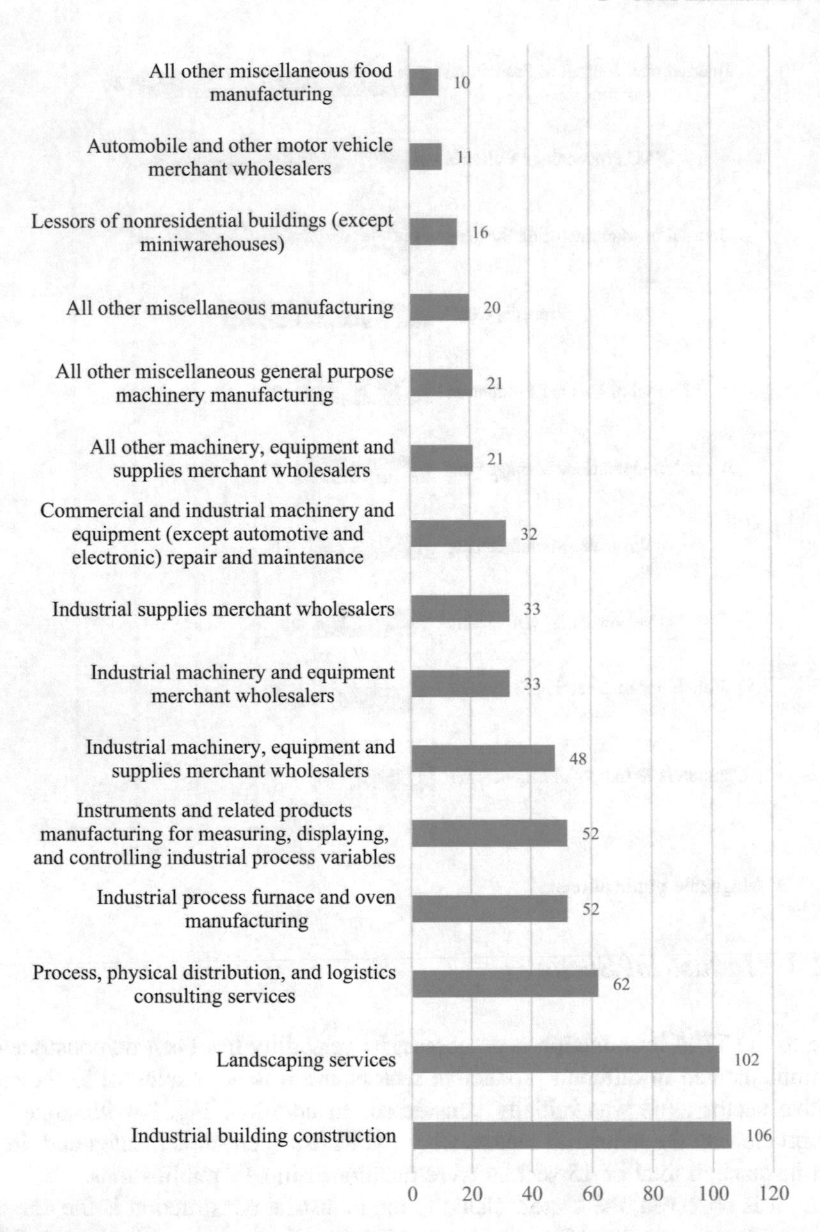

Fig. 2.4 Publications in the industrial sector

Finally, it is important to highlight that the five sectors previously mentioned are the principal where TPM is reported as an improvement tool and as a result of the publications included, these sectors represent 60.4% from the total publications that are shown in this category.

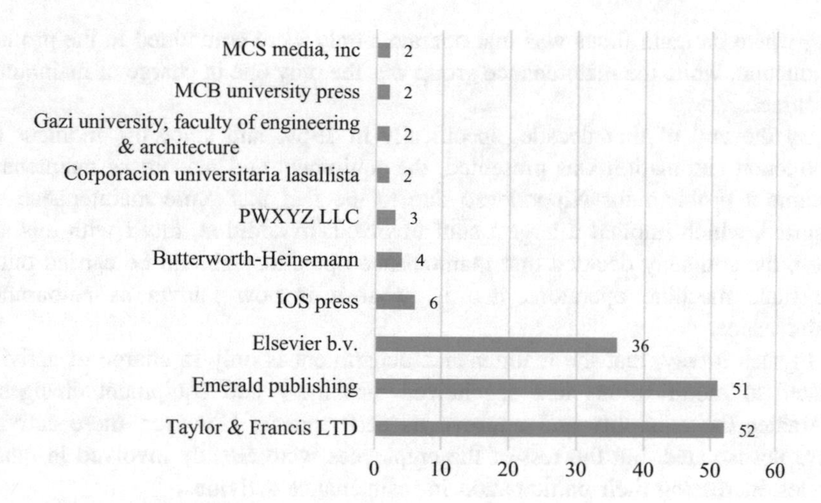

Fig. 2.5 Editorial publications

2.2.5 *Editorial Publications*

Finally, the editors that publish a larger amount of information regarding TPM are presented. In this case, the only information provided by different publishers that include a total of 160 publications is considered. Also, the most prominent publishers are: Taylor & Francis LTD with 52 publications, Emerald Publishing with 5, and Elsevier with 36. As a conclusion, the total among the three publishers is a total of 109 publications, which is equal to 86.8% of the information presented in Fig. 2.5. to each one provides the correspondin

2.3 TPM Successful Cases Background

Nowadays, Toyota is one of the most accepted models within the industrial processes, which is mainly based on the waste elimination, defects, and inventories reduction, which is supported in the TPM besides promoting those aspects includes zero stoppages and automation. Consequently, without TPM, the success that has been reported about TPS would not have worked, which has been demonstrated since Toyota suppliers with the purpose of not losing them as potential customers, quickly began with its implementation (Nakajima 1989).

In particular, a remarkable example is Nippodenso Co, which has been reported by Nakajima (1989) and has been taken as a base to work on by Levitt (2010), who was one of the main suppliers in Toyota, responsible for providing electrical parts and considered the pioneer in implementing preventive maintenance in the early

60s, where its main focus was that operators only were committed to the products production, while the maintenance group was the only one in charge of maintaining machines.

By the end of that decade, specifically in 1969, and since the moment that production automation was presented, the equipment and machinery maintenance became a problem for Nippodenso due to the fact that more maintenance was required, which implied a largest staff involved. In addition, faced with this situation, the company decided that maintenance operations should be carried out by the same machine operators, that is, what it is now known as autonomous maintenance.

In such a way, that the maintenance department is only in charge of activities related to modifications and specialized machinery and equipment changes to guarantee the reliability and improve its performance. However, these activities were not isolated, but the rest of the employees were already involved in quality circles, increasing their participation in maintenance activities.

Since Nippodenso was well known for having started with preventive maintenance, it worked to improve and prevent maintenance as well as make improvements on it, thus giving rise to the TPM productive maintenance emergence.

In 1971, the demanding work carried out by Nippodenso was reflected upon receiving the recognition called "Distinguished Plant Prize" granted by the Japanese Institute of Plant Engineers (JIPE) for the TPM development and implementation, becoming the first company receiving such certification. Since then, it is known as a PM award and is awarded annually, and it recognizes those companies that achieve a successful TPM implementation.

2.4 TPM Successful Cases in the Industry

TPM from its inception has achieved an acceptance by companies because even though the results obtained are not immediately noticeable, the results are highly significant and have great benefits for the company that decides to implement it. Although at a global level, the PM award is an award for recognizing the excellence of TPM implementation and successful monitoring, there are also other types of recognitions and success stories that companies experience when deciding to adopt TPM as a tool and improvement technique in their productive systems.

Throughout history, TPM has stood out for being more than a simple methodology or promising tool but it has retrieved reliable results in several industries. In this context, many cases are presented where different companies have obtained satisfactory results after the TPM adoption; therefore, it has allowed them to achieve an improvement in their performance, in general.

2.4.1 *TPM in the International Manufacturing Industry*

First, McKone et al. (2001) conduct a study to measure the TPM effect and impact on the manufacturing industry performance in four countries: the United States, Italy, Germany, and Japan. Some of the meaningful results that were obtained in the study are as follows:

- TPM helps to improve the team performance, which supports JIT efforts to reduce the storage, shorten delivery times, and eliminate other waste.
- TPM is mainly based on three self-maintenance activities, that is, cleaning, cross-training, and teamwork.
- TPM is also mainly based on planned maintenance: information tracking and disciplined planning.
- The TPM impact should not be considered in isolation, but it should be considered along with other manufacturing practices.
- TPM cannot only be used to control costs; it can improve cost, dimensions in quality and delivery.
- TPM can be a great contributor to the organization strength and it can improve preventive maintenance.

2.4.2 *TPM in the Electronic Industry*

In Chan et al. (2005), the TPM implementation is reported within an electronic manufacturing industry where it is mentioned that the following benefits were obtained:

- Effective equipment management.
- Equipment productivity improvement after the implementation with approximately 83% improvement.
- Reduction rate of equipment stops with a decreasement from 517 to 89 times. This improvement was the most significant in the team because it improved effectiveness and quality in the product.
- Employees empowerment. Empowering the workforce led to the development of a brighter, happier, and more relational workplace for people in the production area.
- Development of technical skills, work habits, growth, and multifunctional team promotion, which created an enthusiastic workforce to improve both company power competitive and image.

Logically, at the beginning of the TPM implementation, some difficulties were encountered. In the first instance, there is a lack of communication and adequate guidance on TPM (one of the steps considered as the main steps in the implementation of such a program), and this lack of information led the production workers to think that they are not selected to be part of the team that were going to work more than they did. In addition, the TPM team initially did not work together, since the habits from different teams in different shifts varied because each one had their own way of working. Also, they did not have a long-term vision due to some participating members educational level. Consequently, the TPM implementation was slow.

In order to face this problem and achieve an adequate implementation, the strategies shown below were followed:

- A specific guide/training was generated to realize the benefits in the production and maintenance department.
- Supervisors and/or management were required to convince their subordinates to accept the concept of TPM by providing adequate training.
- A meticulous selection of team members was necessary. Some of them had worked at the company for over 10 years and were not eager to learn or accept the new cultural change as well as the shift paradigm. Therefore, the selected team members should have a positive attitude and be willing to accept latest changes.
- The maintenance team skill was required to be transferred to the production operators, in fact, a well-developed maintenance training system is one of the key factors for the TPM implementation. In addition, the system is required to be frequently updated as the technology changes rapidly.
- The management support in the TPM implementation was very important since its commitment sustained and improved the production operators and the maintenance personnel morale.

2.4.3 TPM in the Pharmaceutical Industry

In the research by Friedli et al. (2010), improvements in the pharmaceutical industry related to the effectiveness and operational efficiency when using TPM in their production systems are reported. In addition, this study includes the collected data from pharmaceutical production sites based on surveys application between 2004 and 2009. The analysis is divided into four subsystems: TPM, TQM, JIT, and Management System.

In this case, only the results that the TPM implementation has provided to the pharmaceutical industry are addressed. Also, regarding TPM, companies have reported the following:

- There are formal programs to maintain machines and equipment.
- Maintenance plans and checklists are placed near the machines.
- All maintenance work is documented.

- Good maintenance stands out as a strategy to increase quality and plan compliance with production orders.
- Potential bottleneck machines are identified and supplied with additional spare parts.
- Maintenance programs are continuously optimized based on a dedicated fault analysis.
- The maintenance department focuses on helping machine operators perform their own preventive maintenance.
- Machine operators are actively involved in the decision-making process before buying new machines.
- The machines receive maintenance mainly internally. It is about avoiding the external maintenance service as far as possible since this represented additional costs.
- The statistical processes control is used to reduce the variations in the processes and it is considered in the TPM administration since the lack of quality is associated to the lack of calibrations.
- For the root cause analysis, there are standardized tools to obtain a deeper understanding of the factors that may influence.
- It operates with a high implementation level of technology analytical process for the supervision and processes control in real time.

Finally, the results obtained can be classified in three more ways that are presented; based on the interfunctional products development from the employee involvement, and based on the supplier quality management, which is shown in the section below:

- Interfunctional product development
  - Manufacturing engineers are heavily involved in a new medicine formulation as well as in necessary production development.
  - In companies, the product and processes development are closely related to each other.
  - Due to the close collaboration between the I + D and the manufacturing department, the product launch time can be significantly shortened.
  - For the products and processes transfers between different units or sites, there are standardized procedures that ensure a fast, stable, and fulfilled knowledge transference.
- Client involvement
  - Close contact with clients is often maintained.
  - Clients often provide their opinion on the delivery quality and performance.
  - Client requirements are regularly surveyed.
  - Clients satisfaction surveys are regularly conducted.
  - The just in time philosophy is assessed.
  - There are improvement programs along with clients to increase performance.

- Finally, supplier's quality management results are reported
  - Quality is the main criteria to select suppliers.
  - Classification of suppliers, and therefore, audits and qualification are carried out.
  - The evaluated and validated suppliers are used mainly in the supply system.
  - For a significant suppliers' percentage, there is no inspection in the incoming parts/materials because their confidence and quality are demonstrated over time.
  - Incoming material inspections are made in proportion to the previous quality performance or the type of supplier.
  - Basically, 100% of incoming shipments are inspected.
  - There are improvement programs along with suppliers to increase performance.

2.4.4 TPM in the Automotive Industry

The automotive industry has been considered the first to report the usage and TPM application. In this case, it is not the exception, and in a study carried out by Singh et al. (2013), the TPM success is described after its implementation within a machine that belongs to the automotive industry. Also, the implementation of eight pillars in TPM is reported and the benefits obtained can be translated into favorable results for the company. In addition, the results are presented focusing on each pillar.

- Pilar: 5S
  - A more organized workplace was obtained.
  - Delimitation of areas to store the tooling and spaces for storage on the shelves were labeled.
  - A cleaner place was achieved, and some space was released.
  - The existence of a greater visual communication.
- Pilar: Autonomous Maintenance
  - Machine operators are trained in order to each one provides the corresponding maintenance.
  - Sheets and checklists development.
- Pilar: Planned Maintenance
  - The planned maintenance, in this case, the preventive maintenance is carried out through the weekly training, which the maintenance manager oversees.
  - Predictive maintenance is carried out in the same way, and operators take care of the identified machines areas that have the highest failure probability, and thus the maintenance department is warned accordingly.

- Pilar: Continuous Maintenance, Kaizen
  - Poka-yokes introduction to avoid wrongly placing the keys in the machines.
  - Corrective execution actions to prevent oil leakage in machines.
  - A new layout implementation for two machines' purchase to fulfill the production requirements.
- Pilar: Quality Maintenance
  - Machines review parameters to avoid defects in the products.
  - Periodic verification parameters established.
- Pilar: Training
  - Training for workers in visual inspection.
  - Training for the gauges usage of "Go and no Go".
  - Training on the usage of instruments for machines measuring and calibrating.
  - Weekly training about quality tools.
- Pilar: TPM Office
  - Reduction in the required time to obtain data from the daily rejection of the Quality Department, which is described by the operators in their reports.
  - A new computer system proposal for the Maintenance Department.
  - The employees' details are shown on the bulletin board.
- Pilar: Health and Environment
  - Enough fire extinguishers that are provided throughout workshops.
  - Training is provided every 6 months to each person about the usage of extinguishers in case of an emergency.
  - Management receives suggestions for training the employee on the measures to be taken in case of an emergency.
  - Management also receives suggestions for drills at least once a year.

Finally, with the TPM application and the following pillars, the overall equipment efficiency presented a considerable increasement from 63 to 79%, which indicates an improvement in product quality and productivity.

2.4.5 TPM in the Industry and Psychology

Psychology plays a crucial role not only in life in general but also in the workplace. In that context, Pinto et al. (2016) conducted a study to evaluate the psychological sense of operators ownership in a particular plant section, in order to propose and encourage ways to make them capable of promoting high productivity levels.

For this, a descriptive study was carried out considering a sample of 30 participants who were living in Portugal. The information was collected through the application of the psychological property questionnaire developed by Avey James et al. (2009).

The results obtained with the TPM implementation allowed the development of a culture that promotes a sense of psychological ownership, which guarantees the workers' commitment and, at the same time, generates higher productivity levels.

Multiple elements and TPM tools, especially those related to autonomous maintenance were implemented in the industrial context, granting the following benefits:

- A more solid relationship among the operators and the company administration, the operational management in charge of the areas that were developed, and those with greater autonomy in the daily operators' management as well as the maintenance standards application.
- The implementation of autonomous maintenance promoted the interaction between operators and teams. In addition, by creating a plan, it was possible for the operators and the company to develop monitoring and interaction with the teams. It also provided a detection and anticipation of continuous abnormalities that resulted, logically, in fewer failures and quality defects.
- The strategic maintenance management activity is to make the team act to avoid possible failures. Also, the new organization way of some areas provided functional improvements to the work.
- It was found that operators have a sense of belonging to the lines and production areas of the company, which promoted a climate of continuous effort towards the objectives of the company.
- The operators' interest towards their work and performance was stimulated, providing organizational advantages in all the levels.
- In addition to the level of operational performance for the planning work, there was an improvement in the workers' attitudes towards the implementation of techniques such as TPM.

2.4.6 *TPM in the Textile Industry*

The textile industry represents a significant sector for the economy in different areas, in this sense, the work developed by Ahmad et al. (2017) details the implementation of one TPM pillars (Kaizen or continuous improvement) to increase the overall equipment efficiency in a company belonging to the textile industrial sector.

After the implementation, multiple improvements and favorable results for the company were reported, such as:

- The overall team efficiency increased from 75.09 to 86.02% after the training program.
- Operators started to pay more attention to production and small maintenance activities, as a result, the reduction of stop losses as well as the OEE improvement and the product quality; the production rate increased by 23.93%.
- The opportunity to improve their skills in the fundamental maintenance operations background for the TPM application was given to operators.
- Machine downtime was reduced from 375.5 to 217.5 min.
- The machine stoppage time loss in the initial stage was 37.5 min per shift, and after the TPM implementation, it was reduced to 21.75 min per shift.

2.4.7 *TPM in Automotive Industry*

In the research by Morales Méndez and Rodriguez (2017), a case study is presented detailing the TPM implementation in the bottleneck in auto parts machine line. In addition, the results show that after the TPM implementation in a period of 6 months, certain benefits were obtained, such as the evident hours assigned reduction, unplanned maintenance as well as its percentage impact on the total number of technical stoppages during the first 6 months. The most significant effects are presented in the section below:

- An improvement of 33.21% in lost parts was obtained, which represents a 10.7% increasement in production capacity terms.
- Regarding maintenance indicators, an improvement of 108% in average time between failures was obtained.
- Regarding the average repair time, an improvement of 30.2% was made.
- According to the general equipment effectiveness, an improvement of 18.75% was achieved, which exceeds what was planned for the period of 6 months.
- Improvements in equipment availability are reported as 8.9%.
- The equipment efficiency reports an increasement of 8.9%, and consequently, the quality was improved by 4.2%.

2.5 Conclusions

TPM is a technique that supports the objectives that have been raised in the implementation of lean manufacturing and can be considered as a tool that helps to maintain the materials flow, but in turn, it helps to ensure products processed quality in the machines. In addition, after the literature review, it can be observed that from an academic point of view, there are many references about this technique, and that it is no longer exclusive to the automotive sector where it had its origins, but its application has been expanded to several sectors.

Moreover, some cases have been reported where TPM has allowed many benefits to be obtained from various industrial sectors and it is observed that the results may vary; however, in some sectors, its application is more critical than in others, as in the case of the pharmaceutical industry, where maintenance errors are associated with the human health while in some sectors, it is associated with the operator safety and integrity. It is concluded that regardless of the sector where it is applied, TPM will always offer satisfactory results, although with differences among them, since human resources and work culture have a profound impact and influence. Therefore, the TPM implementation process should always be launched along with training and education about the topic.

References

Ahmad N, Hossen J, Ali SM (2017) Improvement of overall equipment efficiency of ring frame through total productive maintenance: a textile case. Int J Adv Manuf Technol. https://doi.org/10.1007/s00170-017-0783-2

Avey James B, Avolio Bruce J, Crossley Craig D, Luthans F (2009) Psychological ownership: theoretical extensions, measurement and relation to work outcomes. J Organ Behav 30(2):173–191. https://doi.org/10.1002/job.583

Borkowski S, Czajkowska A, Stasiak-Betlejewska R, Borade AB (2014) Application of TPM indicators for analyzing work time of machines used in the pressure die casting. J Ind Eng Int 10(2):55. https://doi.org/10.1007/s40092-014-0055-9

Chan FTS, Lau HCW, Ip RWL, Chan HK, Kong S (2005) Implementation of total productive maintenance: a case study. Int J Prod Econ 95(1):71–94. https://doi.org/10.1016/j.ijpe.2003.10.021

Friedli T, Goetzfried M, Basu P (2010) analysis of the implementation of total productive maintenance, total quality management, and just-in-time in pharmaceutical manufacturing. J Pharm Innov 5(4):181–192. https://doi.org/10.1007/s12247-010-9095-x

Levitt J (2010) TPM reloaded: total productive maintenance. Industrial Press

McKone KE, Schroeder RG, Cua KO (2001) The impact of total productive maintenance practices on manufacturing performance. J Oper Manag 19(1):39–58. https://doi.org/10.1016/S0272-6963(00)00030-9

Morales Méndez JD, Rodriguez RS (2017) Total productive maintenance (TPM) as a tool for improving productivity: a case study of application in the bottleneck of an auto-parts machining line. Int J Adv Manuf Technol 92(1):1013–1026. https://doi.org/10.1007/s00170-017-0052-4

Nakajima S (1989) TPM development program: implementing total productive maintenance. Productivity Press, New York

Pinto H, Pimentel C, Cunha M (2016) Implications of total productive maintenance in psychological sense of ownership. Procedia Soc Behav Sci 217:1076–1082. https://doi.org/10.1016/j.sbspro.2016.02.114

Singh R, Gohil AM, Shah DB, Desai S (2013) Total productive maintenance (TPM) implementation in a machine shop: a case study. Procedia Eng 51:592–599. https://doi.org/10.1016/j.proeng.2013.01.084

Siong SS, Ahmed S (2007) TPM implementation can promote development of TQM culture: experience from a case study in a Malaysian manufacturing plant. In: Proceedings of international conference on mechanical engineering

Suzuki T (1996) TPM en industrias de proceso. Taylor & Francis

Part II
Activities and Benefits Associated with TPM

Chapter 3
Activities Associated with the Success of TPM

Abstract In this chapter, a description from a total of 75 critical success factors for the total productive maintenance (TPM) implementation program within the industry is reported, these critical success factors were divided into 2 categories: activities related to the human factor and activities related to the operative factor, which is briefly described.

3.1 Introduction

As it has been mentioned in the previous chapters, TPM is a continuous improvement process focused on teams, however, it is not an easy task to carryout because it implies several activities that may vary from one company to another, although in the end their objective is shared, that is to say, increase production and at the same time, increase the morale of employees as well as job satisfaction (Attri et al. 2013a).

In order to ensure a successful development, TPM, like any other improvement program, it is based on a series of critical factors (Mohamed Sabry and Ali 2014). Normally, the critical success factors (CSFs) are the set of factors, components, constituent elements, or critical areas where the organization must focus on to achieve its mission and objectives, through the evaluation and categorization of its impacts (Jamali et al. 2010). In such a way that the TPM CSF contribute to overcome the challenges imposed by the global competition (Abhishek et al. 2014).

Moreover, CSF has been the research subject; therefore, multiple experts and academics have addressed the issue by doing various works to find the CSF that is considered crucial within the companies. In addition, results have somehow varied. Khanlari et al. (2008) mention eight main aspects such as understanding the TPM and committing employees. On the other hand, Bamber et al. (1999) consider nine main aspects such as organization, performance indexes, mission alignment, staff participation, TPM implementation plan, knowledge and beliefs, allocation of implementation time, management commitment, and the workforce motivation.

J. R. Díaz-Reza et al., *Impact Analysis of Total Productive Maintenance*,
https://doi.org/10.1007/978-3-030-01725-5_3

According to Ahuja and Khamba (2008a), there are only eight aspects that guarantee the success of TPM, and among them, there are the top managers contributions, the cultural transformations, the employee participation, the traditional and proactive maintenance policies, training and education, maintenance prevention, and production system improvements. Additionally, Mohamed Sabry and Ali (2014) consider these factors as organized variables in three stages highlighting leadership, management commitment, strategic vision, motivation, financial resources availability, knowledge and experience, organizational culture, and benchmarking and consulting.

As it is observed, there are multiple elements that influence the success of TPM so taking the information from above as a reference, the CSF have been classified as it is presented below, and emphasizing in two main aspects; the human and the operational factor, which are explained by other variables, which are described in the next section.

- Human Factor
  - Work Culture
  - Suppliers
  - Managerial Commitment
  - Clients
- Operational Factor
  - TPM Implementation
  - TPM Implementation Style
  - Technological Status
  - Design (Layout)
  - Warehouse Management.

3.2 Activities Associated with Human Factor

3.2.1 Work Culture

Undoubtedly, the culture promotion through values and principles within a company allow to strengthen the employees' performance at all levels and, consequently, the general business turnaround. In addition, since culture is developed by individuals within the organization, who want knowledge and skills, it must be guaranteed that these aspects are going from one generation to the next ones.

Indeed, with the development of a work culture, there is access to the creation of a general panorama that allows the administration to carryout in an easier way the appropriate decisions in the distinct aspects of the company, finding the critical points that must be improved. In terms of TPM, the work culture can be favored considering the following elements:

- *Clean and organized work areas*: TPM begins with 5S; one technique/tool from 5S is called the implementation of TPM cornerstone, since it is a Japanese cleaning (Singh et al. 2013) in order to keep the work area in optimum conditions, because issues cannot be recognized if the workplace is not organized, in other words, the work environment needs to be clean and properly organized to identify the problems and quickly solve them (Singh et al. 2013).
- *Tools and accessories organization and placement in designated places*: This activity is one of the basic principles on the 5S (seiton), it is very important that each element, tool or accessory has its designated place, because if it is required it can be found in an accessible and visible place. In addition, as soon as tools are no longer used , they must be returned to a previously designated area, therefore a better discipline among workers will be developed (Morales Méndez and Rodriguez 2017) as they are the ones who take care of these tools, through cleaning and area organization, therefore, defects may be detected on time (Attri et al. 2013b). Following this simple rule or principle is essential as it provides order and cleanliness in the workplace.
- *Departments and company cleaning*: TPM starts with cleaning because the hygiene and cleanliness work are transformed into high-quality standards for a company, since, cleaning provides the opportunity to inspect a machine that neatness may reveal anomalies facilitating its identification as well as allowing the rectification steps execution that results in improvements. The improvements are responsible for providing positive results, and at the same time, positive results increase the high-quality standards, which bring pride to the work environment (Kiran 2017). Normally, there are several committees that are in charge of verifying, modifying, and guiding the TPM implementation and progress, in addition, they are in charge of coordinating some of the activities (maintenance, training, cleaning, etc.) between the different departments, providing information on the various functional areas in the company (Mwanza and Mbohwa 2015).
- *Order and cleanliness promotion by senior management*: As any other, the 5S philosophy must start with the senior management commitment in order to systematically achieve the general cleanliness and standardization of the organization in the workplace, providing a motivated and pleasant environment for all employees in the organization (Jugraj Singh and Inderpreet Singh 2017). In fact, managers and senior managers are the organization, therefore, they are the ones who must set the example to the rest of the labor team.
- *Training for employees to carryout multiple activities*: An extremely important aspect is ensuring the most efficient use of human resources, supporting personal growth, and maximizing their skills and abilities through appropriate training and take advantage of their specialization in multiple tasks (Abhishek et al. 2014). Maintenance personnel with multiple skills are increasingly valuable in modern manufacturing plants, for example, where it implemented PLC, PC-based equipment and process control, automated tests, remote process control and supervision, and/or similar modern production systems (Smith and Hawkins 2004b). Maintenance technicians who can test and operate these

systems as well as perform mechanical adjustments, calibrations, and parts replacement, avoid the need to hire a greater number of personnel to perform multiple jobs in many other maintenance tasks (Smith and Hawkins 2004b). In this way, strategically trained employees know how to solve problems in the workplace to reduce losses in the productive sector and also contribute to the productivity improvement (Morales Méndez and Rodriguez 2017). In addition, plant processes should determine the need and benefits of including multi-skill training in the general training plan, since it has been demonstrated that operators and maintenance personnel cross-training can produce substantial efficiencies in the manufacturing environments (Smith and Hawkins 2004b).

- *Initial maintenance to the machinery and equipment by the operators*: The machines and production equipment cleaning gives the operators an idea about the current operation state of their machines, therefore, they can use all the available means, that is, their eyes, ears, nose, mouth, and hands to help their maintenance colleagues as an "early warning system" (Willmott and McCarthy 2001a). By working together as a team, they can ensure effective asset care and free maintenance people for tasks that require a higher training and skill level (Willmott and McCarthy 2001a).
- *The operators' opinions and ideas are considered before taking decisions related to maintenance*: While in most manufacturing environments, the operator is not seen as a member of the maintenance team, regarding TPM, the machine operator is trained to perform large number of daily tasks in simple maintenance and fault location, in other words, the basics of autonomous maintenance. Normally, in such environments, teams are created and organized, including a technical expert (often an engineer or maintenance technician) and production operators (Kiran 2017).

  In this configuration, the operator is enabled to understand the machinery and identify potential problems, correct them before they can affect production, and reduce downtimes as well as production costs (Kiran 2017). In these situations, their opinions are very valuable because they can avoid incurring unnecessary investments or actions.
- *Registration in work logs and maintenance for each machine*: The recording of the maintenance activities carried out becomes an essential task in the establishment of autonomous maintenance teams since a better communication and efficient teamwork should be promoted. For this reason, it is crucial that the company is responsible for designing an efficient system about recording data, so that the information is updated, accurate, and available for management anytime (Chand and Shirvani 2000).
- *Coordination of the maintenance schedule with the productive department*: Preventive maintenance (PM) and production are in conflict due to two main reasons; first, because the time spent in carrying out the activities related to the PM could be used for production. Second, re-translating the PM for production can increase the probability of failures, while maintenance managers try to achieve high equipment availability (Hnaien et al. 2016).

However, spending a large amount of time performing PM will reduce unscheduled interruptions, but investing very little time in PM will result in an increasement in unplanned interruptions in computer usage (Peng 2012). Generally, in praxis, the production planning and maintenance activities are carried out independently, that is, it cannot be guaranteed that the acquired plans are optimal according to the objective that minimizes the total maintenance and production cost. It can obtain better solutions when planning maintenance is integrated with manufacturing activities (Hnaien et al. 2016). For this, there must be an optimal balance and time where the effective measures to carryout a PM schedule are considered, which must be reviewed continuously considering the number of interruptions, whether they are scheduled or unscheduled (Peng 2012). Finally, the PM integration in decisions and production can reduce not only the interruption time but also the total expected cost (Hnaien et al. 2016). Normally, maintenance is scheduled during the no production hours (Ni et al. 2015).

- *Consider operators' opinions when programming the equipment maintenance*: The operator's cooperation on individual machines is extremely important, the fact is that they are the ones who are in the work area, therefore, they are more capable to detect unusual events that may indicate equipment malfunction, thus allowing taking the appropriate measures in certain situations and avoiding major problems. Also, once a general cooperative attitude is established, coordination should be refined monthly, weekly, daily, and possibly even in schedules (Mobley 2004).
- *Include operators' opinions in the reports drafting in logs*: It is necessary to make a formal entry for each inspection and preventive maintenance work, in a way that an exact control is saved, likewise, the information available in the case is required. At this point, the operator opinion plays a vital role, since they are the ones who carryout certain activities that only they know how to perform, therefore, they must be reported. On the other hand, if the work is very specific, the use of a checklist is necessary which upon completion of work and after being completed, it must be returned to the maintenance office (Mobley 2004). Any open preventive maintenance work order must be maintained in the report until the supervisor has verified the results to ensure quality and when they have signed the endorsement (Mobley 2004).
- *The maintenance can be requested by the operator when failures are presenting*: Within the TPM implementation, it is necessary to focus on one of the eight pillars, in other words, it is autonomous maintenance. In other words, it is defined as a set of activities that are implemented to preventive and predictive maintenance, which are carried out by the operators who are involved in machine manufacturing functions, therefore, they are in charge of its maintenance and operation (Rosimah et al. 2015). By assuming this responsibility, the operator is completely autonomous, this means that the worker can request the support of a maintenance team when intervention on the machine is needed or when help is required, without interfering in the productive process (Michał 2016).

- *Stopping the production line if the operator detects maintenance problems in the equipment*: When the employees are authorized to carryout maintenance work, they will commit themselves to solve the different problems that may arise without waiting for senior management instructions or approval, this situation will make the maintenance execution work easier, more effective and faster (Sani et al. 2012). In fact, authorization also contributes to the maintenance culture development, which process delegating decision-making authority to lower levels by encouraging employees to take the initiative and expand its reach (Sani et al. 2012).

3.2.2 Suppliers

The parts and services suppliers play a significant role in maintenance activities. In addition, the management related with suppliers plays a decisive role for the company's competition strength and profitability, which will directly impact on quality, costs, technologies, and time for new products (Yan et al. 2015).

In this way, the solid relationships establishment with sellers and suppliers can reduce the waiting time for support requests in the equipment and machinery, or accelerate the partial spare parts or new acquisitions deliveries as well as provide additional services beyond contractual obligations (Peng 2012). Also, some of the functions, activities, and responsibilities that must be respected and fulfilled by suppliers when offering their products or services, which are as follows:

- *Providing maintenance manuals for sold machinery and equipment*: A technical manual is a document that explains the use, maintenance, and handling of a product from delivery to disposal. In addition, it provides any technical information that a user may need during the product life (BS 1992), for example, product troubleshooting, updating, and elimination. Also, technical descriptions, specifications, parts lists, and maintenance programs are often needed to understand the instructions provided in the technical manual and thus plan maintenance activities (Setchi and White 2003). On the other hand, technical manuals are necessary since they provide adequate information and guidance for products that are operated and maintained by qualified users who work in close cooperation with the manufacturer and the product supplier (Setchi and White 2003).
- *Provide technical advice on the sold equipment*: The supplier's role goes beyond simply providing the product and the corresponding information on the overall equipment functionality. However, they must provide more information when there is a greater concentration, effort, and additional activities are required in terms of maintenance; they must also provide information on the available technology to guarantee their efficiency and effectiveness (Smith and Mobley 2011). In addition, at the end of the day, an exchange of information is expected, long-term commitment to quality, mutually shared benefits, and aligns

participation in the layout and products specification (Ross 2015a). For this, working together is required between the user (the manufacturing company) and the supplier (machines and production equipment) in the different activities of each phase in the production system design process (Bruch and Bellgran 2013).

– *Sign contracts with the buyer about the equipment maintenance*: When acquiring some equipment or machinery, the signature of some type of agreement or contract must be established, where the responsibilities are mentioned as obligations for both the seller and the buyer. Also, an effective maintenance contract represents an instrument to ensure that manufacturers and hire consultants have the common efficiency objective in the system, in terms of performance and costs (Manzini et al. 2010). Additionally, the responsibility for maintenance within such contracts relies on the supplier and not the buyer, and a major factor that influences the financial success of this agreement will be the efficiency and effectiveness, which repairs can be carried out (Zhang et al. 2017).
– *Establishment of the conditions in the guarantees for the sold equipment*: The products eventually fail, therefore, they are not reliable; a failure of an item may occur at the beginning of its life due to manufacturing defects or at the end of its life due to the nature of degradation in the item (Kim et al. 2004). In addition, most products are sold with a warranty that offers protection to buyers against early failures during the warranty period, as well as extended warranties, where an item is covered for an important part of its useful life (Kim et al. 2004). Therefore, the supplier must coordinate four important sets of client activities: warranty services, support and maintenance, system extensions, and consulting and optimization services (Helander and Möller 2008). Similarly, the buyer must pay attention to the conditions and terms where the guarantees have been extended.
– *Installation and start-up from the acquired equipment*: When the signing of the contracts about the equipment purchase or acquisition is established, the buyer must understand the type of negotiation clauses that are being described, in order to know whether those supplier services clauses that are offered are convenient or not. Typically, these services include equipment installations, upgrades, factory change orders (FCOs), engineering change orders (ECOs), as well as guard field engineering and supports (Peng 2012), but on some other cases, they have additional cost.
– *Training programs between the supplier and the buyer*: Currently, the purchasing companies depend *Training programs between the supplier and the buyer* more and more on their suppliers to offer technologically advanced and defect free products in a timely and profitable way (Chidambaranathan et al. 2009). If the supplier lacks some of these areas, the purchasing companies face the decision to look for an alternative source to work with the same supplier and fix their deficiencies (Chidambaranathan et al. 2009). Due to the uncertainty of finding a better source and the high cost of searching and evaluating new suppliers, purchasing companies can participate in the supplier's development (Krause and Ellram 1997). Also, the supplier's development requires the

purchasing and supplying companies to allocate financial, capital, and personnel resources for work; to share sensitive information; and create an effective means to measure performance (Chidambaranathan et al. 2009). Therefore, this strategy is a challenge for both the executives and buyer employees, who must be convinced that investing the company resources in a supplier is a worthy risk (Chidambaranathan et al. 2009).

- *Provide training at the supplier's facilities*: In terms of maintenance and equipment services, information on the layout and products manufacturing phases (e.g., the equipment) such as drawings and equipment maintenance instructions provided by manufacturers can be used in maintenance decisions as a starting point (Wan et al. 2018). Notably, it represents an opportunity to increase the knowledge and skills development levels related to the maintenance team who participate in training, being an advantage for the buying company. Finally, maintenance information and knowledge will not only be used to improve the manufacturing performance in the company that is receiving the purchase, but some type of feedback can also be provided to equipment manufacturers to improve reliability and quality in the future (Wan et al. 2018).
- *Provide training in the buyer's facilities*: Sometimes it is necessary for the supplier to go directly to facilities, since it may require training on certain machinery aspects in these types of activities, the buyer assistance and participation may be required to improve the supplier capabilities (Kadir et al. 2011). In addition, this is a supplier development program with reseller training by the buyer (Joshi et al. 2016), it means proper training. Also, the correct training selection is more beneficial to improve the technical knowledge and the supplier capabilities (Kadir et al. 2011), which is reflected in the increasement in the supplier's performance, consequently, it makes the buyer more competent.

3.2.3 Managerial Commitment

The senior management and manager's support, commitment, and participation are essential for the successful TPM implementation. However, for this commitment and involvement to be reflected in the company itself, it is necessary that several activities are carried out by the heads and managers, among them the following are listed:

- *The TPM responsibility from heads department*: In TPM, maintenance goes from being the maintenance department responsibility to being everyone's responsibility (Attri et al. 2013a). However, often both production and maintenance personnel are reluctant to accept such a change. In addition, the fact is that production employees and managers refuse to accept responsibility for maintenance activities due to concerns about whether production employees have the skills and/or enough time to perform maintenance tasks. Also, maintenance employees refuse to grant maintenance responsibilities to production employees because they are afraid that maintenance tasks will not be performed properly and the maintenance department will be forced to "fix" the problems that production employee create (Lawrence 1999). However, in the TPM implementation process, employees must be provided not only with skills and technical skills related to work but also with the quality improvement and behavioral training, therefore, training objectives should include the systematic knowledge development, skill, and attitude required, creating the "I produce, inspect, and maintain" mentality (Ahuja and Khamba 2008a). In such a way that each person assumes the responsibilities in the activities that he/she has to perform in the framework of the TPM implementation.

- *Leadership attitudes in the TPM performance by senior managers*: The biggest challenge for the organization is to make a radical transformation in the organization culture to ensure the general employees' participation in the maintenance and improvement of manufacturing performance through TPM initiatives. Also, concentrated efforts must be made by senior management to motivate the organization culture by making employees aware of the true TPM potential as well as communicating its contributions (Ahuja and Khamba 2008a, b). In addition, the senior management participation must be demonstrated through actions that are necessary and not only by words or statements. Thus, senior management must encourage leadership to influence anyone else's behavior (Talib and Rahman 2010). In a way that a supportive cultural environment and intelligent leadership from managers are fundamental requirement for achieving high-quality maintenance (Eti et al. 2006a).
- *Meetings between maintenance and production*: In order to ensure a trouble-free production, maintenance is inevitably necessary, therefore, maintenance can improve the production performance, however, excessive maintenance can interrupt processing and decrease the production rate, then, production control and maintenance management are closely related (Kang and Subramaniam 2018). As a result, it is important that there is an appropriate communication between these departments, in order that they know which of them is working on and thus avoid any unwanted activity that interferes with the development of their activities. Consequently, it is advisable to hold informative meetings between the maintenance and production departments to share information, discuss problems, and exchange ideas, in that case, both departments are favored.
- *Promotion of workers participation in maintenance activities by managers*: The involvement of people in each maintenance activity is necessary to ensure that everyone understands about the designated activities. Also, the combination of employee participation and senior management is able to develop a maintenance culture (Attri et al. 2013a). In addition, this participation encourages employees to increase the level of commitment that is reflected in each activity or work performed, and even in the desire to lead a larger staff to be involved in the decision-making as well as take the risk of any improvement in the maintenance quality (Sani et al. 2012).
- *Management focuses on quality and maintenance*: The senior management needs to have a strong commitment to the TPM implementation program, and it

must strive to develop multilevel communication mechanisms for all employees, explaining the importance and benefits from the entire program, and spreading the TPM benefits to the organization, to the employees while linking TPM to the strategy and the general organization objectives (Ahuja and Khamba 2008a).

– *Management involvement in maintenance projects*: Senior management has the primary responsibility to prepare an adequate and supportive environment before the official TPM launch within their organization, only with the top management full commitment and enthusiasm, changes in the corporate strategy can be designed and executed properly, therefore, the successful TPM implementation requires the senior management support, commitment, and participation (Attri et al. 2013b).

3.2.4 Clients

In today's competitive markets, customer participation has played a crucial role in the TPM adopting process in the industries, because the industries because industries are more focus on improve their services in order to fulfill customers' needs and demands (Ahuja and Khamba 2008a). Because the search for quality maintenance is directed towards the client achievement and satisfaction in order to deliver the highest quality products (Singh et al. 2013), since a deficient strategy will cause team stoppages that would provide as a consequence difficulties in the established contracts fulfillment with clients (Gosavi 2006). Also, due to this situation, customers play a significant role in maintenance in general as well, which is why the following aspects related to customer satisfaction and fulfillment must be relevant:

– *Consider the customer's opinion on maintenance issues*: The feedback that the customer can provide is significant, not only in delivery and quality terms (McKone et al. 2001) because it allows to find aspects that can be improved to accomplish some demands. In addition, the involvement and communication that exists between the client and the company are fundamental elements for the establishment of relationships that will allow the company to take specific measures linked to maintenance and avoid problems such as defects, delays, stoppages, customer's loss, and, therefore, lost in performance and profits.
– *Avoid noncompliance with customers due to lack of maintenance*: One of the aspects that should be considered besides the product quality and service, it is the attention towards customers, as a result, the organization should strive to create products with "zero defects" and "zero customer complains" by supporting and maintaining team conditions through strategic initiatives to maintain quality (Ahuja 2009).
– *Delaying maintenance activities to be satisfied with production*: Current production planning and control systems from manufacturing companies do not include future-oriented maintenance strategies that allow accurate maintenance prediction tasks, showing inefficient production processes due to unpredictable

machine downtimes, fluctuating waiting times, and a large number of urgent orders (Glawar et al. 2018).

– *Costs for noncompliance of orders because of poor maintenance*: Incorrectly selected maintenance dates significantly influence the productivity of a production system, therefore, an ideal solution cannot be found between the costs related to maintenance and production (Glawar et al. 2018). Therefore, it is significant that the arranged dates are contemplated and respected to avoid problems, but it is also necessary that each person acknowledges the importance that designated activities have.

3.3 Activities Associated with Operational Factor

3.3.1 Preventive Maintenance Implementation

In the past, the preventive maintenance execution was considered as an activity that did not add value to the industrial process, however, nowadays, it is a special requirement to increase the machinery and equipment life cycle used in the industry (Singh et al. 2013).

In addition, PM activities are scheduled activities that are carried out before the product fails. Also, PM may be perfect, which restores the product to the "almost a brand new" state or may be imperfect, restoring the product to a state that is between "as good as new" and "as bad as old".

In general, PM can help reduce the frequent unexpected repairs even when the failure rate is by increasing nature (Gosavi 2006). Some elements to consider in the PM execution include the following aspects:

– *Exclusive days to perform maintenance*: Organizational performance improvement not only relies on its processes and approaches but also people participation, engagement, and ownership in the TPM implementation initiatives (Nakamura 2007). TPM initiative starts from autonomous maintenance, which involves operators from the lower level in the manufacturing hierarchy to perform minor maintenance tasks through day-to-day routine (Lai Wan and Tat Yuen 2017).
– *Maintenance as a strategy to achieve the quality and programming of activities*: PM actions are executed to reduce the probability of failure or extend the product useful life (Anand et al. 2018). In addition, performing PM produces a reduction in inventory and costs related to quality (Anis et al. 2008). Although PM plays an important role in the compliance with guarantee periods (Chen and Chien 2007), However, the guarantees lead to the loss in income due to the repair cost in the items, this loss can be minimized by providing a PM, but providing preventive maintenance, the manufacturer also has to provide some maintenance cost, therefore, it is better for the manufacturer to perform PM only

if the cost reduction in the warranty service is greater than the additional cost incurred with the preventive maintenance (Chen and Chien 2007).

- *Days reserved for maintenance activities*: Choosing an appropriate maintenance policy is a critical activity in manufacturing systems as it affects the performance of companies' equipment (Hemmati et al. 2018), since a good maintenance program has a great impact on machine reliability and availability (Savsar 1997).
- *Maintenance department helps operators to perform preventive maintenance in their machines*: In order to the maintenance activities to be performed in an appropriate way, there must be a responsibility, discipline, and standardization from the employees, this to achieve the autonomous maintenance (Morales Méndez and Rodriguez 2017). In a way that the key autonomous maintenance element is the machine operator mentality, that is, the operator will take care of the machine if he/she feels that the machine belongs to him/her (Singh et al. 2013). As a result, while the operators clean their machines they will know them better and will develop skills to see or detect their deterioration, such as oil leaks, vibrations or unusual noises. As time goes on, they will be able to perform essential frontline care tasks as well as some minor maintenance tasks within the limits of their own abilities, this process will be completed along with the maintenance personnel, who will be free to apply their technical skills when necessary (Willmott and McCarthy 2001a). Since in TPM the machine operator is able to perform many of the daily tasks with simple maintenance and fault location, teams are created that include a technical expert (often an engineer or a maintenance technician) and operators (Kiran 2017).
- *Operators are informed about the maintenance performed*: The operator can understand the machinery and identify potential problems, correct them before they can affect production, therefore, it will decrease downtime and reduce production costs, any defects are presented to the supervisor to guarantee their immediate attention (Kiran 2017). Additionally, TPM defines the operators and the maintenance personnel responsibilities and that each one has all the necessary skills to carryout their functions, likewise, TPM emphasizes the adequate and continuous training, and the maintenance department has the responsibility to train the operators in minor maintenance routines (Kiran 2017). Also, the maintenance staff works with the operator to improve some aspects, in the same way, they support the operator regarding the standards for inspections, lubrication, minor adjustments or care (Moore 2007).
- *Machine maintenance statistics publication*: Visual management (or visual communication) must be implemented in a comprehensive manner to boost the company productivity by increasing the employee's effectiveness through the effective exchange of information and encouraging the workers to participate in the development of this information (Ahuja 2009). In a way that it is possible to communicate and easily observe status and levels in where the equipment and machinery are located, and thus know which important measures must be taken.
- *Simple access to historical maintenance and productivity information*: Due to the maintenance success, it is important to establish an information system that

tracks the past and current equipment performance (McKone et al. 2001). It is usually a plan that is based mainly on historical data and recommendations from the manufacturers (Mosaddar and Shojaie 2013). Since, the predictive maintenance aims to operate the equipment for the maximum possible period of time before performing the maintenance, planning, and executing maintenance, activities related to condition monitoring data and development estimations based on historical data and experience, as well as statistical information (Ravnestad et al. 2012).

– *Quality control in the products generated by each machine*: The TPM implementation not only contributes to the initial increasement in equipment availability but also to efficiency, in addition, a team backed up by TPM guarantees a higher product quality (Morales Méndez and Rodriguez 2017). Also, in the TPM development, the maintenance personnel is responsible for the long-term maintenance planning as well as the equipment preparation status, this planning approach generally dedicates time to scheduled maintenance activities, assigns tasks to specific people and inspects the labor and machinery quality (McKone et al. 2001).
– *Machinery malfunction identification and recording*: TPM requires a rigorous record about each event occurred in the corresponding machine, which facilitates the continuous information monitoring, therefore that immediate and effective measures can be taken to guarantee 100% availability (Morales Méndez and Rodriguez 2017). In addition, databases are used to maintain records that involve a continuous and disciplined registration about the machine's performance, and sometimes there are dedicated teams committed to perform these activities (Morales Méndez and Rodriguez 2017).

3.3.2 Total Productive Maintenance Implementation (Implementation Style)

Performance maintenance is a widely used activity as a support function that is not directly related to production, but it is required to allow machines and equipment to function properly as well as allow the products development to fulfill the defined standards and requirements. Normally, to develop maintenance in an appropriate manner, the appropriate maintenance strategy must be established; an activity that is as relevant as it is complicated since it involves the consideration of several topics such as safety, cost, added values, and viability.

In addition, among the most important strategies, it can be listed the corrective maintenance, preventive maintenance based on time (time-based preventive maintenance), maintenance based on conditions, and preventive maintenance (Kirubakaran and Ilangkumaran 2016). Also, the selection of such strategy involves some maintenance-related activities execution, which must be carried out in an effective manner to ensure that the equipment failures decrease, as well as the associated costs. Among them, the following are mentioned:

- *Training and education for the maintenance personnel*: Through individual training and education, the maintenance culture development is influenced (Sani et al. 2012). In fact, with the promotion of an adequate training and an education program, it is possible to update and improve both, the skills and the technical capabilities for the production and maintenance personnel (Abhishek et al. 2014). Because it has a direct impact on the employee's performance, since they will have a greater awareness and thus will be more careful when executing the maintenance tasks in assets and facilities, they will be more motivated to perform their work correctly, while improving their knowledge about maintenance work (Sani et al. 2012).
- *Monitoring and control on the maintenance program*: The TPM objective is to keep the plant and equipment in good condition without interfering with the daily process, for this, a program that includes mainly prevention and predictive maintenance is required (Kiran 2017). Also, an appropriate maintenance program provides the equipment with an environment where its design can perform efficiently and reliably (Forsthoffer 2017). An effective plan includes only those necessary tasks to achieve the established objectives, therefore, it does not program additional tasks that increase maintenance costs without a corresponding increasement in the inherent reliability level of protection (Smith and Mobley 2011). Finally, predictive maintenance is based on diagnostics and health management technology, which means that the equipment remaining life can be predicted (Raza and Ulansky 2017).
- *The objectives of TPM not obtained are not explained to the operators*: TPM is a process that changes corporate culture and permanently improves and maintains the overall effectiveness of equipment through active involvement of operators and all other members of the organization (Smith and Hawkins 2004a), therefore, one of the most important aspects of the TPM program is the communication of its results and initiatives (Willmott and McCarthy 2001b).
- *Lack of knowledge of the operators in the handling of equipment and machinery under their responsibility*: Maintenance technology is evolving continuously, TPM operators need continuing training (Tsai 1996). Initially, they must acquire and maintain broad, generic skills for routine maintenance of their machines as well as skills in teambuilding and leadership (Tsai 1996). Since the modern technocratic society and manufacturing practice determine radically new requirements to a professional qualification level of employees engaged in manufacture (Petukhov and Steshina 2015).
- *Senior and maintenance staff commitment with the machine functionality*: The organization, procedures, and practices instituted to regulate the activities and maintenance demands in an industrial company are not their responsibility to guarantee satisfactory results, the senior executive and staff must influence all functional activities (Haroun and Duffuaa 2009). In addition, the commitment to a program by members from the upper level management team is necessary, as

well as empowering employees to initiate corrective actions for noncompliance aspects with the system or process under their administration, it means that TPM is initiated as a "top-down" exercise but only it is implemented successfully if there is a "bottom-up" way of participation (Eti et al. 2004).

- *Management leadership when executing TPM programs*: Maintenance performance can never exceed the quality of its leadership and direction, good leadership is derived from teamwork, which is the essence of success in any company (Haroun and Duffuaa 2009), within the maintenance programs, the managers are responsible for carrying out these tasks. In addition, the TPM administrator will work under the TPM office and will be in charge of the activities effective management in the organization, in the first instance, they must acquire this information and knowledge in order to make the decision to introduce TPM in the company (Mwanza and Mbohwa 2015). Also, management understanding, commitment, support, and participation is vital within TPM (Smith and Hawkins 2004b) which is reflected in the basic policies and objectives establishment, and they are in charge of creating a master plan for the program development (Mwanza and Mbohwa 2015).
- *Leadership by production and engineering when executing TPM programs*: TPM has been widely recognized as a strategic weapon to improve manufacturing performance by improving the production facilities effectiveness, likewise, TPM describes a synergistic relationship between all the organization functions, but particularly between production and maintenance, for the product quality continuous improvement, operational efficiency, productivity, and safety (Ahuja 2009). Also, TPM changes the organization structure to break the traditional barriers between maintenance and production, encourage improvement by looking at multiple perspectives between production and maintenance areas.
- *Leadership by maintenance managers when executing TPM programs*: Leadership is based on the senior management commitment that acts as an internal resource to speed up the attitude that a person has to perform and understand maintenance tasks very well (Sani et al. 2012). Maintenance tasks will not be considered a burden, but a good practice that should be implemented for the next chance, this is where the need for leadership qualities in a leader demonstrates a serious commitment to the work done by providing a work plan that everyone can easily understand and follow (Sani et al. 2012). In addition, the synergy and will from the people involved with the maintenance area is necessary, that is, the senior management, the maintenance and production managers, the operating personnel, and all the personnel in charge of maintenance (Rodrigues and Hatakeyama 2006).
- *Communication between production and maintenance on equipment availability and maintenance performance*: Communication plays an important role that involves providing information on maintenance work practices for all members in the organization, so that all members in the organization understand the importance of the maintenance facilities and assets (Sani et al. 2012). Within the TPM environment, a high level of communication between operators,

maintenance personnel, and engineers are very vital, since along with the teamwork, they are the basis for the start of the autonomous maintenance (Abhishek et al. 2014).

- *Operators know the schedule for equipment maintenance operation*: Objective of autonomous maintenance is operation of equipment without breakdown, versatile and flexible operators to operate, and maintain other equipment and eliminating the defects at source through active employee participation (Singh et al. 2013). To do this, it is important that checklist is prepared and following this checklist is mandatory for all the operators (Singh et al. 2013). It is the responsibility of a shop's floor in-charge to see that every operator fills up this checklist (Singh et al. 2013).
- *Knowledge about the critical systems of failure in the machines*: The production performance is affected by machine failures, which interrupt the production and may cause losses (Kang and Subramaniam 2018). Therefore, machine operators and manufacturers must be able to track the condition of the machines based on the characteristic values and evaluate the remaining monitored machines useful life (Neugebauer et al. 2011).
- *Maintenance programs focused on the machinery systems and components useful life*: TPM is designed to maximize the equipment effectiveness (improving efficiency) by establishing a comprehensive productive maintenance system that covers the equipment entire life, covering all the fields related to the equipment (planning, usage, maintenance, etc.) and with the employees participation from senior management to workers in the plant, in order to promote the productive maintenance through the management of the motivation or voluntary activities in small groups (Tsuchiya 1992). In addition, the maintenance programs should focus on carrying out activities that allow maintaining the machinery and equipment in optimal conditions to avoid unexpected breakdowns, speed losses, and quality defects that arise from the activities during the process (Abhishek et al. 2014).
- *Differences between the periods of useful life of the parts and components provided by the suppliers and the products by the company in the daily performance of the equipment*: Spare parts are common inventory stock items, which are needed to maintain equipment and the cost takes a large share of product life cycle cost (Hu et al. 2018). Spare parts can be characterized by their own life cycles which are associated with the life cycle of the final products that utilize them (Fortuin and Martin 1999). The nonavailability of spare parts, as and when required for repairs, will result in a great financial loss, therefore, spare parts management plays an important role in achieving the desired equipment availability at a minimum cost (Hu et al. 2018).
- *Informative meetings by those responsible for maintenance*: During the TPM preliminary stages and activities execution, those responsible must hold information sessions. For example, once the execution is finished, the manager reports the work plans developed and the activities carried out by the TPM structure, the policies, the basic objectives as well as the master plan, finally, the

activities that were performed by the operators are reported in order that its commitment reach the established objectives (Chan et al. 2005).

– *Investment in updated tools*: One of the main problems affecting effective initiatives to improve maintenance in manufacturing organizations is the industry low performance in the maintenance enrollment and behavior patterns, poor management of spare parts, and deployment inefficient maintenance improvements in future production systems. Additionally, it can be attributed to the inadequate use of computerized maintenance management systems and the minimal role of information technology in solving problems related to maintenance (Ahuja 2009). Therefore, the implementation of an effective computerized maintenance management system is strongly recommended to achieve a significant improvement in maintenance performance in the manufacturing industry (Ahuja 2009).
– *Contemplate maintenance as purchase criteria*: Large investments in operations or maintenance functions can improve the manufacturing system performance as well as the competitive position in the organization's market (Patrik and Magnus 1999). Since the operators are the ones who are in constant interaction with the machines, maintenance personnel must also be considered to consider the necessary functional elements where maintenance is required according to its capabilities and functionality, therefore, their opinion should be considered before the new equipment acquisition.

3.3.3 Technological Status

As a matter of fact, in this globalization era, the manufacturing strategy is changing rapidly, where the innovative strategies that became vital for each manufacturing organization are interested in adopting advanced manufacturing technologies in the organization to produce high-quality products at low levels as well as costs at the shortest delivery time (Singh and Kumar 2013). In this sense, to survive and face the competition, more advanced manufacturing technology is used and automated, having more maintenance personnel and investing a larger amount of budget on it. In addition, they have more proactive maintenance policies, better planning and control systems, and more decentralized maintenance organization structures compared to others (Pinjala et al. 2006).

Therefore, increasing the technology level in the company increases the TPM application power and at the same time, the manufacturing performance level (McKone et al. 2001). As a result, in terms of technology, it is relevant to consider the following aspects:

– *Advanced technology*: The modern technology effective application can only be achieved through people, starting with the operators and those in charge of that technology, and not only through the systems, hence, the TPM emergence as the enabling tool to maximize the effectiveness of the established team as well as maintaining the optimal relationship between workers and their machines (Bon and Lim 2015).

- *Update and focus on maintaining new generations of technology*: Uncertainties in the current economic environment and changing consumer preferences have significantly increased the importance of implementing advanced manufacturing technology (Thakur and Jain 2006), which has been recognized as a strategic tool that gives companies a competitive advantage (Ragavan and Punniyamoorthy 2003), consequently, to maintain a competitive place in the market, it is essential that an advanced manufacturing company has a solid maintenance management system that can control its maintenance costs at the lowest level and maintain its overall efficiency of the equipment at the most high level (Tu et al. 2001).
- *Efficient use of new technologies and modern maintenance systems*: The importance of the maintenance function has increased because it plays an important role in the retention and improvement of system availability and safety, and in the product quality (Tsang 2002), the maintenance practice has changed significantly due to the developments in equipment design, information and communication technology, cost pressures, risks acceptance, and customer failures (Hodkiewicz and Pascual 2006), therefore, an integrated high-level maintenance system that have multiple subsystems requires the collaboration of multiple stakeholders, such as departments or units, to improve resources, information exchange, and maintenance practices (Syafar and Gao 2013) to stay updated and perform the indicated activities for each team.
- *Learning and updating equipment after installation*: The equipment is becoming more sophisticated and automated, this increases the need for guaranteed output quality, low power consumption, and operational safety, therefore, it becomes essential to design an appropriate training system in order that the established objectives are completed, it is desirable that each company divides its own training system, in particular, to adapt it to its equipment (Duffuaa and Raouf 2015). Also, it is necessary to update such training systems as fast as technology changes (Chan et al. 2005).
- *Improvement of existing machinery and equipment*: TPM can improve a company's technology base by improving equipment technology and improving employee skills (McKone et al. 2001). Therefore, the technology improvement rate increases exponentially as dynamic innovation becomes significant to create competitive advantages (Ercan and Kayakutlu 2014).
- *Equipment development and patents registration*: Manufacturing companies are executing research projects throughout the company or in collaboration with allies, industrial companies are supporting individual ideas looking for innovation, in this way, each researcher has the right to protect each new idea in the market (Ercan and Kayakutlu 2014). In addition, patents and licenses are not only to protect inventions, but they provide global protection for any intellectual asset because global patent applications are fundamental for investment in research and development (Ercan and Kayakutlu 2014).

– *Generating competitive advantages for machinery and equipment*: In today's manufacturing world, advanced manufacturing technologies play a critical role for the industry and organization growth, the implementation of advanced manufacturing technologies offers the advantage of mass production on customer's requests in a very short time with the latest advanced technologies help where efficiency can be maintained (Nath and Sarkar 2017). In addition, working together with other initiatives such as TQM, JIT, or cellular manufacturing help to improve performance and competitiveness (Ahuja and Khamba 2008a). Finally, the implementation of advanced manufacturing technologies results in positive outcomes such as improved quality, flexibility, improved competitive advantages, and increased productivity (Singh and Kumar 2013).
– *Specialized software usage for maintenance processes*: In praxis, most information technology systems to support maintenance services focus on organizational issues such as planning, programming, and coordinating the necessary resources. Nowadays, they are mainly used to provide information online such as maintenance instructions (Uhlmann et al. 2013). When adopting a computerized system, the calculation of performance indicators is easier and allows analysis and comparisons with the objectives (Oliveira et al. 2016).

3.3.4 Layout

The facilities layout design is a crucial task in the manufacturing systems, since in this activity the facilities location, workstations, inspection, and cleaning areas are determined to achieve several objectives such as maximizing the design performance, dimensional errors minimization, and the products shape according to the device design, the raw material transportation costs, parts, tools, work in process, and finished products between the facilities, the total routes, employee morale increasement, the injury and property damage risk, decreasement in the average time delay for each machine as well as total time in the system (Azadeh et al. 2013).

Furthermore, an effective layout design is achieved through collective effort of several skills and abilities, where among them, the following can be listed:

– *Proximity between processes and machinery*: Layout drastically affects production performance, a well-designed layout can help to reduce transportation, facilitate control, and have more comfort (Nguyen and Do 2016; Huang et al. 2002). Understanding the system productivity generally involves the complex layout analysis and understanding where the equipment interconnection of many pieces exists (Huang et al. 2002). In addition, the layout design from some machines types that perform different functions must be done in a tight sequence (usually U-shaped) to allow the flow of a single piece and the flexible human effort deployment (Anvari et al. 2014).
– *Machinery grouping according to the family of produced products*: With the group technology implementation, which is a manufacturing concept that

provides an alternative approach for batch-oriented production systems, it leads to cell manufacturing, where the machine shop is divided into several manufacturing cells (Saeed and Ming 2004). Each cell is composed of machines that are functionally different and are dedicated to the parts processing or families with similar parts. Finally, the success of a cell manufacturing system depends on how the machines and parts are arranged (Saeed and Ming 2004).

- *Plant distribution to facilitate maintenance*: The maintenance capacity is a significant system feature that can make maintenance convenient, fast, and economical, because the design is the stage where the maintenance capacity of the future system is almost determined, therefore, it is important that designers take maintenance into account during their work (Luo et al. 2014). The components layout design is a significant step in the whole process of designing the multiple components system, which has an important impact on the component maintenance capacity (Luo et al. 2014). In addition, Layout facility planning focuses on the disposal plans design for different facilities to achieve operational efficiency; a well-designed layout can greatly reduce the total operation cost (Jiang et al. 2014), from 10 to 30% (Raman et al. 2009), minimizing the material handling cost, which is the ultimate goal of a facility distribution layout (Wang et al. 2011). Also, this importance is reinforced by the long-term decisions consequences and the cost of having to redesign the plant.
- *Equipment and heavy machinery movement within the facilities for maintenance*: The layout design discipline is the part from the plant design that determines how the equipment and the support structures that comprise it should be presented, that is why designers must consider several key criteria in their designs, among which are efficient, reliable and safe plant operation, effective, economical and ergonomic safe space use, and convenient access for the equipment maintenance process by the total elimination or partial or on-site repair, among others (Moran 2017).
- *Adequate signs in the plant related to maintenance*: Visual controls are simple signs that provide an immediate understanding of a situation or condition, as they are very efficient, self-regulating, and managed by workers (Ahuja 2009). In addition, a visual workplace should include hundreds of visual control devices, where a visual device is a mechanism or device designed intentionally to influence, direct or limit behavior by making information vital to the task easily available at a glance without request it, as a consequence, the main visual controls objective is to organize a work area in order that people (even external ones) can determine if things are going well or not, without the help of an expert (Ahuja 2009).

3.3.5 Warehouse Management

The TPM application has not only proven to provide benefits directly to the production or maintenance area; one of them is the inventories reduction (Abhishek et al. 2014), because it facilitates the obtaining of the products in the desired quantities at the first intention (Eti et al. 2006b). In this sense, the department store and inventory management performance and participation play an essential role as they help to increase the advantages. Therefore, it is important to consider the following activities for a satisfactory performance:

- *Warehouse areas cleaning*: An effective warehouse is one that is clean and tidy, storage areas must be kept rubble free, good maintenance prevents damage, and facilitates the item storage and order picking processes (Ross 2015b).
- *Organized and appropriate space for parts and components within the warehouse*: The warehouse design must address several problems, in addition to the items allocation in storage locations with the purpose that the warehouse is being able to support its operations (Mohsen 2002). In this way, there should be a concerned about the functional warehouse areas arrangement, determine the number of dock sites, the entry and exit points, determine the number of corridors, their dimensions, their orientation, estimate the space requirements, design the flow pattern, and the picking zones shape (Mohsen 2002).
- *Daily inspections to avoid missing*: The TPM implementation promotes the detailed records maintenance for the machine, which in turn allows an efficient maintenance orientation of the resources both in time and in the spare parts availability (Morales Méndez and Rodriguez 2017).
- *Appropriate material and parts coding*: Automatic identification systems minimize or eliminate the human operator participation in the information summary through the optical and radio technologies use that input information directly into the storage systems. Also, the importance of automatic identification systems in modern storage has two ways: (1) it minimizes or eliminates the need for human activity to collect data and (2) it significantly increases the data collection accuracy and speed (Ross 2015b).
- *Arrangement and adaptation of spaces for material or parts with special needs (heating, cooling, humidity*, etc.): The inventory storage in the warehouse locations can be arranged together with physically related items since they are stored according to their physical characteristics, which normally require a storage and handling equipment. For instance, the steel barrels usage to store temperature-controlled wheels or freezers to store food products (Ross 2015b).
- *Registration of changes in the components and tools location*: In order to respond to changes in the business environment, it usually develops a decision support system with the warehouse data use, which includes product, information, location, city, supplier, time, and sales (Huang et al. 2014). For example, goods can be moved several times to and from the internal check-in areas, storage, reservation, and storage, movement can happen between temporary, fixed, dynamic, and reservation locations to select them according to (Ross 2015b). Finally, the movement is often achieved through the usage of several types of handling equipment material (Ross 2015b).

References

Abhishek J, Rajbir B, Harwinder S (2014) Total productive maintenance (TPM) implementation practice: a literature review and directions. Int J Lean Six Sigma 5(3):293–323. https://doi.org/10.1108/IJLSS-06-2013-0032

Ahuja IPS (2009) Total productive maintenance. In: Ben-Daya M, Duffuaa SO, Raouf A, Knezevic J, Ait-Kadi D (eds) Handbook of maintenance management and engineering. Springer, London, pp 417–459. https://doi.org/10.1007/978-1-84882-472-0_17

Ahuja IPS, Khamba JS (2008a) Strategies and success factors for overcoming challenges in TPM implementation in Indian manufacturing industry. J Qual Maint Eng 14(2):123–147. https://doi.org/10.1108/13552510810877647

Ahuja IPS, Khamba JS (2008b) Total productive maintenance: literature review and directions. Int J Qual Reliab Manag 25(7):709–756. https://doi.org/10.1108/02656710810890890

Anand A, Singhal S, Panwar S, Singh O (2018) Optimal price and warranty length for profit determination: an evaluation based on preventive maintenance. In: Kapur PK, Kumar U, Verma AK (eds) Quality, IT and business operations: modeling and optimization. Springer, Singapore, pp 265–277. https://doi.org/10.1007/978-981-10-5577-5_21

Anis C, Nidhal R, Mehdi R (2008) Simultaneous determination of production lot size and preventive maintenance schedule for unreliable production system. J Qual Maint Eng 14 (2):161–176. https://doi.org/10.1108/13552510810877665

Anvari A, Zulkifli N, Sorooshian S, Boyerhassani O (2014) An integrated design methodology based on the use of group AHP-DEA approach for measuring lean tools efficiency with undesirable output. Int J Adv Manuf Technol 70(9):2169–2186. https://doi.org/10.1007/s00170-013-5369-z

Attri R, Grover S, Dev N, Kumar D (2013a) Analysis of barriers of total productive maintenance (TPM). Int J Syst Assur Eng Manag 4(4):365–377. https://doi.org/10.1007/s13198-012-0122-9

Attri R, Grover S, Dev N, Kumar D (2013b) An ISM approach for modelling the enablers in the implementation of total productive maintenance (TPM). Int J Syst Assur Eng Manag 4(4):313–326. https://doi.org/10.1007/s13198-012-0088-7

Azadeh A, Motevali Haghighi S, Asadzadeh SM, Saedi H (2013) A new approach for layout optimization in maintenance workshops with safety factors: the case of a gas transmission unit. J Loss Prev Process Ind 26(6):1457–1465. https://doi.org/10.1016/j.jlp.2013.09.014

Bamber CJ, Sharp JM, Hides MT (1999) Factors affecting successful implementation of total productive maintenance: a UK manufacturing case study perspective. J Qual Maint Eng 5 (3):162–181. https://doi.org/10.1108/13552519910282601

Bon AT, Lim M (2015) Total productive maintenance in automotive industry: issues and effectiveness. In: 2015 international conference on industrial engineering and operations management (IEOM), pp 1–6, 3–5 Mar 2015. https://doi.org/10.1109/ieom.2015.7093837

Bruch J, Bellgran M (2013) Critical factors for successful user-supplier integration in the production system design process. In: Emmanouilidis C, Taisch M, Kiritsis D (eds) Advances in production management systems. Competitive manufacturing for innovative products and services. Springer, Berlin, Heidelberg, pp 421–428

BS (1992) Technical manuals. British Standards Institution, London

Chan FTS, Lau HCW, Ip RWL, Chan HK, Kong S (2005) Implementation of total productive maintenance: a case study. Int J Prod Econ 95(1):71–94. https://doi.org/10.1016/j.ijpe.2003.10.021

Chand G, Shirvani B (2000) Implementation of TPM in cellular manufacture. J Mater Process Technol 103(1):149–154. https://doi.org/10.1016/S0924-0136(00)00407-6

Chen JA, Chien YH (2007) Renewing warranty and preventive maintenance for products with failure penalty post-warranty. Qual Reliab Eng Int 23(1):107–121

Chidambaranathan S, Muralidharan C, Deshmukh SG (2009) Analyzing the interaction of critical factors of supplier development using interpretive structural modeling—an empirical study. Int J Adv Manuf Technol 43(11):1081–1093. https://doi.org/10.1007/s00170-008-1788-7

Duffuaa SO, Raouf A (2015) Maintenance training. In: Duffuaa SO, Raouf A (eds) Planning and control of maintenance systems: modelling and analysis. Springer International Publishing, Cham, pp 213–221. https://doi.org/10.1007/978-3-319-19803-3_9

Ercan S, Kayakutlu G (2014) Patent value analysis using support vector machines. Soft Comput 18 (2):313–328. https://doi.org/10.1007/s00500-013-1059-x

Eti MC, Ogaji SOT, Probert SD (2004) Implementing total productive maintenance in Nigerian manufacturing industries. Appl Energy 79(4):385–401. https://doi.org/10.1016/j.apenergy.2004.01.007

Eti MC, Ogaji SOT, Probert SD (2006a) Impact of corporate culture on plant maintenance in the Nigerian electric-power industry. Appl Energy 83(4):299–310. https://doi.org/10.1016/j.apenergy.2005.03.002

Eti MC, Ogaji SOT, Probert SD (2006b) Reducing the cost of preventive maintenance (PM) through adopting a proactive reliability-focused culture. Appl Energy 83(11):1235–1248. https://doi.org/10.1016/j.apenergy.2006.01.002

Forsthoffer MS (2017) Predictive and preventive maintenance (Chapter 11). In: Forsthoffer MS (ed) Forsthoffer's more best practices for rotating equipment. Butterworth-Heinemann, Burlington, pp 501–546. https://doi.org/10.1016/B978-0-12-809277-4.00011-5

Fortuin L, Martin H (1999) Control of service parts. Int J Oper Prod Manag 19(9):950–971

Glawar R, Karner M, Nemeth T, Matyas K, Sihn W (2018) An Approach for the integration of anticipative maintenance strategies within a production planning and control model. Procedia CIRP 67:46–51. https://doi.org/10.1016/j.procir.2017.12.174

Gosavi A (2006) A risk-sensitive approach to total productive maintenance. Automatica 42 (8):1321–1330. https://doi.org/10.1016/j.automatica.2006.02.006

Haroun AE, Duffuaa SO (2009) Maintenance organization. In: Ben-Daya M, Duffuaa SO, Raouf A, Knezevic J, Ait-Kadi D (eds) Handbook of maintenance management and engineering. Springer, London, pp 3–15. https://doi.org/10.1007/978-1-84882-472-0_1

Helander A, Möller K (2008) How to become solution provider: system supplier's strategic tools. J Bus-Bus Mark 15(3):247–289. https://doi.org/10.1080/15470620802059265

Hemmati N, Rahiminezhad Galankashi M, Imani DM, Farughi H (2018) Maintenance policy selection: a fuzzy-ANP approach. J Manuf Technol Manag. https://doi.org/10.1108/jmtm-06-2017-0109

Hnaien F, Yalaoui F, Mhadhbi A, Nourelfath M (2016) A mixed-integer programming model for integrated production and maintenance. IFAC-PapersOnLine 49(12):556–561. https://doi.org/10.1016/j.ifacol.2016.07.694

Hodkiewicz M, Pascual R (2006) Education in engineering asset management—current trends and challenges. In: International physical asset management conference, pp 28–31

Hu Q, Boylan JE, Chen H, Labib A (2018) OR in spare parts management: a review. Eur J Oper Res 266(2):395–414. https://doi.org/10.1016/j.ejor.2017.07.058

Huang SH, Dismukes JP, Shi J, Su Q, Wang G, Razzak MA, Robinson DE (2002) Manufacturing system modeling for productivity improvement. J Manuf Syst 21(4):249–259. https://doi.org/10.1016/S0278-6125(02)80165-0

Huang Y-S, Duy D, Fang C-C (2014) Efficient maintenance of basic statistical functions in data warehouses. Decis Support Syst 57:94–104. https://doi.org/10.1016/j.dss.2013.08.003

Jamali G, Ebrahimi M, Abbaszadeh MA (2010) TQM implementation: an investigation of critical success factors. In: 2010 IEEE, pp 112–116

Jiang S, Ong SK, Nee AYC (2014) An AR-based hybrid approach for facility layout planning and evaluation for existing shop floors. Int J Adv Manuf Technol 72(1):457–473. https://doi.org/10.1007/s00170-014-5653-6

Joshi SP, Verma R, Bhasin HV, Kharat MG, Kharat MG (2016) Structural equation modelling of determinants of buyer-supplier relationship improvement strategies: case of Indian manufacturing firms. Asia Pac J Manag Res Innov 12(2):95–108

Jugraj Singh R, Inderpreet Singh A (2017) 5S—a quality improvement tool for sustainable performance: literature review and directions. Int J Qual Reliab Manag 34(3):334–361. https://doi.org/10.1108/IJQRM-03-2015-0045

Kadir KA, Tam OK, Ali H (2011) Patterns of supplier learning: case studies in the Malaysian automotive industry. Asian Acad Manag J 16(1)

Kang K, Subramaniam V (2018) Integrated control policy of production and preventive maintenance for a deteriorating manufacturing system. Comput Ind Eng. https://doi.org/10.1016/j.cie.2018.02.026

Khanlari A, Mohammadi K, Sohrabi B (2008) Prioritizing equipments for preventive maintenance (PM) activities using fuzzy rules. Comput Ind Eng 54(2):169–184. https://doi.org/10.1016/j.cie.2007.07.002

Kim CS, Djamaludin I, Murthy DNP (2004) Warranty and discrete preventive maintenance. Reliab Eng Syst Saf 84(3):301–309. https://doi.org/10.1016/j.ress.2003.12.001

Kiran DR (2017) Total productive maintenance (Chapter 13). In: Kiran DR (ed) Total quality management. Butterworth-Heinemann, Burlington, pp 177–192. doi:https://doi.org/10.1016/B978-0-12-811035-5.00013-1

Kirubakaran B, Ilangkumaran M (2016) Selection of optimum maintenance strategy based on FAHP integrated with GRA–TOPSIS. Ann Oper Res 245(1):285–313. https://doi.org/10.1007/s10479-014-1775-3

Krause DR, Ellram LM (1997) Critical elements of supplier development: the buying-firm perspective. Europ J Purch Suppl Manag 3(1):21–31. https://doi.org/10.1016/S0969-7012(96)00003-2

Lai Wan H, Tat Yuen L (2017) Total productive maintenance and manufacturing performance improvement. J Qual Maint Eng 23(1):2–21. https://doi.org/10.1108/JQME-07-2015-0033

Lawrence JJ (1999) Use mathematical modeling to give your TPM implementation effort an extra boost. J Qual Maint Eng 5(1):62–69. https://doi.org/10.1108/13552519910257078

Luo X, Yang Y-M, Ge Z-X, Wen X-S, Guan F-J (2014) Layout problem of multi-component systems arising for improving maintainability. J Cent South Univ 21(5):1833–1841. https://doi.org/10.1007/s11771-014-2129-7

Manzini R, Regattieri A, Pham H, Ferrari E (2010) Introduction to Maintenance in Production Systems. In: Manzini R, Regattieri A, Pham H, Ferrari E (eds) Maintenance for Industrial Systems. Springer, London, pp 65–85. https://doi.org/10.1007/978-1-84882-575-8_4

McKone KE, Schroeder RG, Cua KO (2001) The impact of total productive maintenance practices on manufacturing performance. J Oper Manag 19(1):39–58. https://doi.org/10.1016/S0272-6963(00)00030-9

Michał M (2016) The autonomous maintenance implementation directory as a step toward the intelligent quality management system. Manag Syst Prod Eng 4(24):6

Mobley RK (2004) Scheduled preventive maintenance (Chapter 5). In: Maintenance fundamentals, 2nd edn. Butterworth-Heinemann, Burlington, pp 45–54. https://doi.org/10.1016/B978-075067798-1/50026-1

Mohamed Sabry S, Ali HA (2014) Critical success factors for total productive manufacturing (TPM) deployment at Egyptian FMCG companies. J Manuf Technol Manag 25(3):393–414. https://doi.org/10.1108/JMTM-09-2012-0088

Mohsen Hassan (2002) A framework for the design of warehouse layout. Facilities 20(13/14):432–440. https://doi.org/10.1108/02632770210454377

Moore R (2007) Total productive maintenance (Chapter 9). In: Moore R (ed) Selecting the right manufacturing improvement tools. Butterworth-Heinemann, Burlington, pp 173–191. https://doi.org/10.1016/B978-075067916-9/50010-2

Morales Méndez JD, Rodriguez RS (2017) Total productive maintenance (TPM) as a tool for improving productivity: a case study of application in the bottleneck of an auto-parts machining line. Int J Adv Manuf Technol 92(1):1013–1026. https://doi.org/10.1007/s00170-017-0052-4

Moran S (2017) The discipline of layout in context (Chapter 2). In: Process plant layout, 2nd edn. Butterworth-Heinemann, Oxford, pp 9–35. https://doi.org/10.1016/B978-0-12-803355-5.00002-0

Mosaddar D, Shojaie AA (2013) A data mining model to identify inefficient maintenance activities. Int J Syst Assur Eng Manag 4(2):182–192. https://doi.org/10.1007/s13198-013-0148-7

Mwanza BG, Mbohwa C (2015) Design of a total productive maintenance model for effective implementation: case study of a chemical manufacturing company. Procedia Manuf 4:461–470. https://doi.org/10.1016/j.promfg.2015.11.063

Nakamura S (2007) TPM in semiconductor plants. JIPM-Solutions

Nath S, Sarkar B (2017) Performance evaluation of advanced manufacturing technologies: a De novo approach. Comput Ind Eng 110:364–378. https://doi.org/10.1016/j.cie.2017.06.018

Neugebauer R, Fischer J, Praedicow M (2011) Condition-based preventive maintenance of main spindles. Prod Eng Res Devel 5(1):95–102. https://doi.org/10.1007/s11740-010-0272-z

Nguyen M-N, Do N-H (2016) Re-engineering assembly line with lean techniques. Procedia CIRP 40:590–595. https://doi.org/10.1016/j.procir.2016.01.139

Ni J, Gu X, Jin X (2015) Preventive maintenance opportunities for large production systems. CIRP Ann 64(1):447–450. https://doi.org/10.1016/j.cirp.2015.04.127

Oliveira M, Lopes I, Rodrigues C (2016) Use of maintenance performance indicators by companies of the industrial hub of Manaus. Procedia CIRP 52:157–160. https://doi.org/10.1016/j.procir.2016.07.071

Patrik J, Magnus L (1999) Evaluation and improvement of manufacturing performance measurement systems—the role of OEE. Int J Oper Prod Manag 19(1):55–78. https://doi.org/10.1108/01443579910244223

Peng K (2012) Equipment management in the post-maintenance era: a new alternative to total productive maintenance (TPM). Taylor & Francis

Petukhov I, Steshina L (2015) Training personalization for operators of complex equipment. Procedia Soc Behav Sci 186:1240–1247. https://doi.org/10.1016/j.sbspro.2015.04.067

Pinjala SK, Pintelon L, Vereecke A (2006) An empirical investigation on the relationship between business and maintenance strategies. Int J Prod Econ 104(1):214–229. https://doi.org/10.1016/j.ijpe.2004.12.024

Ragavan P, Punniyamoorthy M (2003) A strategic decision model for the justification of technology selection. Int J Adv Manuf Technol 21(1):72–78. https://doi.org/10.1007/s001700300008

Raman D, Nagalingam SV, Lin GCI (2009) Towards measuring the effectiveness of a facilities layout. Robot Comput Integr Manuf 25(1):191–203. https://doi.org/10.1016/j.rcim.2007.06.003

Ravnestad G, Panesar SS, Kayrbekova D, Markeset T (2012) Improving periodic preventive maintenance strategies using condition monitoring data. In: Frick J, Laugen BT (eds) Advances in production management systems. Value networks: innovation, technologies, and management, Berlin, Heidelberg, 2012. Springer, Berlin, Heidelberg, pp 260–267

Raza A, Ulansky V (2017) Modelling of predictive maintenance for a periodically inspected system. Procedia CIRP 59:95–101. https://doi.org/10.1016/j.procir.2016.09.032

Rodrigues M, Hatakeyama K (2006) Analysis of the fall of TPM in companies. J Mater Process Technol 179(1):276–279. https://doi.org/10.1016/j.jmatprotec.2006.03.102

Rosimah S, Sudirman I, Siswanto J, Sunaryo I (2015) An autonomous maintenance team in ICT network system of indonesia telecom company. Procedia Manuf 2:505–511. https://doi.org/10.1016/j.promfg.2015.07.087

Ross DF (2015a) Procurement and supplier management. In: Ross DF (ed) Distribution planning and control: managing in the era of supply chain management. Springer US, Boston, MA, pp 531–604. https://doi.org/10.1007/978-1-4899-7578-2_11

Ross DF (2015b) Warehouse management. In: Ross DF (ed) Distribution planning and control: managing in the era of supply chain management. Springer US, Boston, MA, pp 605–685. https://doi.org/10.1007/978-1-4899-7578-2_12

Saeed Z, Ming L (2004) Comprehensive machine cell/part family formation using genetic algorithms. J Manuf Technol Manag 15(6):433–444. https://doi.org/10.1108/17410380410547843

Sani SIA, Mohammed AH, Misnan MS, Awang M (2012) Determinant factors in development of maintenance culture in managing public asset and facilities. Procedia Soc Behav Sci 65:827–832. https://doi.org/10.1016/j.sbspro.2012.11.206

Savsar M (1997) Simulation analysis of maintenance policies in just-in-time production systems. Int J Oper Prod Manag 17(3):256–266. https://doi.org/10.1108/01443579710159897

Setchi R, White D (2003) The development of a hypermedia maintenance manual for an advanced manufacturing company. Int J Adv Manuf Technol 22(5):456–464. https://doi.org/10.1007/s00170-002-1513-x

Singh H, Kumar R (2013) Measuring the utilization index of advanced manufacturing technologies: a case study. IFAC Proc Vol 46(9):899–904. https://doi.org/10.3182/20130619-3-RU-3018.00395

Singh R, Gohil AM, Shah DB, Desai S (2013) Total productive maintenance (TPM) implementation in a machine shop: a case study. Procedia Eng 51:592–599. https://doi.org/10.1016/j.proeng.2013.01.084

Smith R, Hawkins B (2004a) Total productive maintenance (TPM) (Chapter 3). In: Smith R, Hawkins B (eds) Lean maintenance. Butterworth-Heinemann, Burlington, pp 55–104. https://doi.org/10.1016/B978-075067779-0/50003-0

Smith R, Hawkins B (2004b) Lean maintenance: reduce costs, improve quality, and increase market share. Life cycle engineering series. Elsevier Science

Smith R, Mobley RK (2011) Rules of thumb for maintenance and reliability engineers. Elsevier Science, United States

Syafar F, Gao J (2013) Mobile collaboration technology in engineering asset maintenance—what technology, organisation and people approaches are required? In: Mustofa K, Neuhold EJ, Tjoa AM, Weippl E, You I (eds) Information and communication technology, Berlin, Heidelberg, 2013. Springer, Heidelberg, pp 173–182

Talib F, Rahman Z (2010) Critical success factors of TQM in service organizations: a proposed model. Serv Mark Q 31(3):363–380. https://doi.org/10.1080/15332969.2010.486700

Thakur LS, Jain VK (2006) Advanced manufacturing techniques and information technology adoption in India: a current perspective and some comparisons. Int J Adv Manuf Technol 36 (5):618. https://doi.org/10.1007/s00170-006-0852-4

Tsai JW (1996) Operator empowerment for total productive maintenance. IFAC Proc Vol 29 (1):795–798. https://doi.org/10.1016/S1474-6670(17)57758-7

Tsang AH (2002) Strategic dimensions of maintenance management. J Qual Maint Eng 8(1):7–39

Tsuchiya S (1992) Quality maintenance: zero defects through equipment management. Productivity Press

Tu PYL, Yam R, Tse P, Sun AO (2001) An integrated maintenance management system for an advanced manufacturing company. Int J Adv Manuf Technol 17(9):692–703. https://doi.org/10.1007/s001700170135

Uhlmann E, Otto F, Geisert C (2013) Improving maintenance services for machine tools by integrating specific software functions. In: Shimomura Y, Kimita K (eds) The philosopher's stone for sustainability, Berlin, Heidelberg, 2013. Springer, Heidelberg, pp 459–464

Wan S, Li D, Gao J, Roy R, He F (2018) A collaborative machine tool maintenance planning system based on content management technologies. Int J Adv Manuf Technol 94(5):1639–1653. https://doi.org/10.1007/s00170-016-9829-0

Wang L, Keshavarzmanesh S, Feng H-Y (2011) Reconfigurable facility layout design for job-shop assembly operations. In: Wang L, Ng AHC, Deb K (eds) Multi-objective evolutionary optimisation for product design and manufacturing. Springer, London, pp 365–384. https://doi.org/10.1007/978-0-85729-652-8_13

Willmott P, McCarthy D (2001a) Techniques to deliver the TPM principles (Chapter 4). In: Total productivity maintenance, 2nd edn. Butterworth-Heinemann, Oxford, pp 62–77. https://doi.org/10.1016/B978-075064447-1/50007-2

Willmott P, McCarthy D (2001b) Case studies (Chapter 11). In: Willmott P, McCarthy D (eds) Total productivity maintenance, 2nd edn. Butterworth-Heinemann, Oxford, pp 212–246. https://doi.org/10.1016/B978-075064447-1/50014-X

Yan W, Huang Y, Wang Y, He J (2015) Supplier selection for equipment manufacturing under the background of free trade zone. In: Zhang Z, Shen ZM, Zhang J, Zhang R (eds) LISS 2014, Berlin, Heidelberg, 2015. Springer, Heidelberg, pp 1391–1397

Zhang Y, Andrews J, Reed S, Karlberg M (2017) Maintenance processes modelling and optimisation. Reliability Engineering & System Safety (168):150–160. https://doi.org/10.1016/j.ress.2017.02.011

Chapter 4
Benefits Associated with the TPM Implementation in the Industry

Abstract Throughout this chapter, the main benefits that can be obtained with the effective TPM implementation are exposed. In addition, these benefits have been identified from the literature review previously written, including a total of 22 benefits (observed variables) divided into three categories: Benefits for the company, productivity benefits, and safety benefits. Also, a brief description of each element is presented.

4.1 Introduction

TPM has been shown to be a tool that provides several benefits to companies if it is implemented in an appropriate manner. Obviously, these benefits may vary depending on a range of factors, such as the type of company, the implementation time, and even the company commitment itself. In general, the benefits that arise from the implementation have been cataloged according to the area where they directly impact, as productivity (*P*), quality (*Q*), cost (*C*), delivery (*D*), security (*S*), and morale (*M*) (Ahuja 2009). In this research, the benefits have been classified into the following categories:

- *Benefits for the company*: They are those that help the company in general; in this sense, all the activities that provide a better working environment for both employees and company partners are included; the relevant aspects are
  - Working environment,
  - Employee morale,
  - Work culture,
  - Learning culture, and
  - Communication networks.
- *Productivity Benefits*: Those activities that allow to increase the performance are considered as well as the machinery and equipment reliability; therefore, the defects reduction translates into a better products quality. Some of the activities in this section include the following topics:

J. R. Díaz-Reza et al., *Impact Analysis of Total Productive Maintenance*,
https://doi.org/10.1007/978-3-030-01725-5_4

  - Elimination of losses,
  - Reliability and equipment availability,
  - Maintenance costs,
  - Product quality,
  - Technology, and
  - Competitiveness.

- *Security Benefits*: Security is a crucial element within the work areas because in a safe place, workers will feel more comfortable and perform activities in a better way. However, TPM not only provides benefits to workers and the company but also seeks to connect them with concern and commitment toward the environment. In addition, TPM benefits the global security by focusing on the following critical elements:
  - Environmental conditions,
  - Accident prevention culture, and
  - Problem identification and solving.

4.2 Benefits for the Company

Throughout the literature, it has been demonstrated that TPM is a tool that provides attractive results for a company when it decides to incorporate it into its strategy. However, adapting TPM may provide multiple benefits to the business, staff, and team (Levitt 2010), but it is not a simple activity that offers the results from night to morning; it is a process that requires a holistic intervention, since diverse parties participation and integration is required, such as operators, engineers, managers, and senior managers, where each of them has different responsibilities, obligations, and interests, which at least requires a period of time between 3 to 5 years to show significant (Ahuja 2009). In order to allow a better development and implementation, the following critical elements must be considered:

- *Improves the work environment quality*: In TPM, the simple cleaning work is transformed into high-quality standards of a company; consequently, the high-quality standards give pride to the working environment (Kiran 2017). Since TPM is aimed to improve companies' competitiveness and it encompasses a powerful structured approach to changing the employees' mindset, it makes a visible change in the organizations work culture.
- *Better operations control*: TPM helps the administration to develop new policies and operating strategies to improve production performance in order to take advantage of the company's potential in the highly competitive manufacturing environment nowadays (Ng et al. 2013).
- *Increases employee morale*: The TPM objective is to increase production and, at the same time, increase employee morale and job satisfaction (Agustiady and Cudney 2016). Management's commitment throughout the TPM implementation

is very crucial, as this will maintain the team members' morale, to encourage them to continue their hard work, since they realize that senior management would take seriously the TPM results (Sun et al. 2003). Likewise, the TPM implementation depends on a large extent of competencies and employees' motivation in order to significantly improve the productive systems through continuous improvement and Kaizen events (Ahuja 2009).

- *Creates a responsibility culture, discipline, and respect toward standards*: The effective TPM implementation in the first instance requires a rational change in the culture, as well as in the employees' behavior, in other words, operators, engineers, maintenance technicians, and, of course, senior managers (Attri et al. 2013). The TPM initiative is aimed at improving the companies competitiveness as well as encompasses a powerful structured approach to change employees mentality, which creates a visible change in the work culture in organizations (Ahuja 2009).
- *Permanent learning*: Empowering the workforce causes the development of a bright, cheerful, and relaxed workplace for production people; the growth work habits, the technical skills development, and the multifunctional team promotion create an enthusiastic workforce to improve the company both competitive power and image (Chan et al. 2005).
- *Creates an environment where participation, collaboration, and creativity are a reality*: With the TPM implementation, the members from the autonomous maintenance become aware of the TPM benefits in the operational availability increasement and understand that they are giving their key role in the company productivity; similarly, they accept the conditions from the team in charge and act as contributing taxpayers in the company (Chan et al. 2005).
- *Appropriate staff schedule sizing*: Operators, maintenance personnel, shift supervisors, planners, and senior management should be included in a team; each person becomes a "stakeholder" in the process and it is encouraged to do their best to contribute to the team effort and success (Sun et al. 2003).
- *Efficient communication networks*: From the benefits mentioned, also a significant improvement is acquired in the equipment availability and performance within the plant, as well as from a better communication between the employees (Elaine and Maged 2013).

4.3 Productivity Benefits

- *Eliminate losses that affect plants productivity*: TPM is a maintenance process developed to improve productivity by making processes more reliable with less wastes (Kiran 2017).
- *Improve equipment reliability and availability*: One of the main TPM objectives is to improve the equipment reliability and performance; since it plays a key role in the organization competitiveness, it has a direct impact on quality and cost (Agustiady and Cudney 2016). In fact, the equipment effectiveness is an

indicator about the amount of waste that exists as a result of issues related to the equipment (Agustiady and Cudney 2016).

- *Maintenance costs reduction*: The costs reduction as TPM objective can be explained in a very simple way: if more products of the same equipment are produced, in conventional countable terms, the cost of each production unit decreases (Szwejczewski and Jones 2013).
- *Improve the final product quality*: Avoiding equipment breakdowns and standardizing them produces less variability and increases the products quality. TPM is, and still is, one of the bases for the quality movement (Kiran 2017).
- *Low financial costs for spare parts*: The TPM strategic plans have had a huge impact on the final results, along with the significant capacity improvement, while reducing not only the maintenance costs but also the general operating costs (Elaine and Maged 2013).
- *Improve company's technology*: TPM improves the industry, technological base, and by improving the plant technology, therefore, it helps to improve manufacturing performance (McKone et al. 2001).
- *Increase responsiveness to market movements*: The TPM implementation not only leads to efficiency and effectiveness increasement, but also prepares the plant to face the challenges by competitive worldwide economies as well as to achieve a world-class manufacturing environment, since those plants may face competitive manufacturing challenges in the twenty-first century by adopting and practicing TPM as a maintenance strategy (Kumar et al. 2006).
- *Create competitive capabilities from the factory*: The operator is responsible for the primary care of the plant because a sense of ownership is cultivated in it, through the introduction of an autonomous maintenance and performing tasks that include cleaning, routine inspection, lubrication, adjustments, and minor repairs, as well as cleaning the local workspace (Eti et al. 2004).

4.4 Safety Benefits

- *Improve environmental conditions*: The TPM strategic plans have had a great impact on the final results, along with a significant capacity improvement while reducing not only the maintenance cost but also the general operating costs, which has resulted in the creation of secure and environmentally safe workplaces (I.P.S and Pankaj 2009).
- *Preventive culture of negative events for health*: Security promotion activities such as security slogan competition; security poster, essay, speech, poetry competition; security month, celebration of the week; and reporting on the best initiatives from near misses may go along toward improving safety in the workplace (Ahuja 2009).

– *Increase in the capacity to identify potential problems and search for corrective actions*: Through series of TPM implementation steps, team members must manage the maintenance and simple equipment, repairs, which develops skills and a sense of problem-solving (Chan et al. 2005).
– *Acknowledge why there are certain rules, rather than how to perform a task*: Adequate safety training and awareness among all employees must be provided against neglected work attitudes, and employees must be motivated to follow safety standards (Ahuja 2009).
– *Prevention and elimination of potential accidents causes*: Through the safety base implementation, health, and environment, a safe work environment is guaranteed, because an appropriate work environment is provided and incidents, injuries, and accidents are eliminated (Wickramasinghe and Asanka 2016).
– *Radically eliminate pollution sources*: Autonomous maintenance is a characteristic that defines TPM, since it starts from the premise that approximately 70% of breakdowns and loss performance are due to deterioration or equipment contamination (Szwejczewski and Jones 2013). Particularly, there is an accelerated deterioration: deterioration that can be avoided through proper maintenance practices (Szwejczewski and Jones 2013).

References

Agustiady TK, Cudney EA (2016) Total productive maintenance: strategies and implementation guide. CRC Press

Ahuja IPS (2009) Total productive maintenance. In: Ben-Daya M, Duffuaa SO, Raouf A, Knezevic J, Ait-Kadi D (eds) Handbook of maintenance management and engineering. Springer, London, pp 417–459. https://doi.org/10.1007/978-1-84882-472-0_17

Ahuja IPS, Pankaj K (2009) A case study of total productive maintenance implementation at precision tube mills. J Qual Maint Eng 15(3):241–258. https://doi.org/10.1108/13552510910983198

Attri R, Grover S, Dev N, Kumar D (2013) An ISM approach for modelling the enablers in the implementation of total productive maintenance (TPM). Int J Syst Assur Eng Manage 4 (4):313–326. https://doi.org/10.1007/s13198-012-0088-7

Chan FTS, Lau HCW, Ip RWL, Chan HK, Kong S (2005) Implementation of total productive maintenance: a case study. Int J Prod Econ 95(1):71–94. https://doi.org/10.1016/j.ijpe.2003.10.021

Elaine A, Maged E (2013) TPM implementation in large and medium size organisations. J Manufact Technol Manage 24(5):688–710. https://doi.org/10.1108/17410381311327972

Eti MC, Ogaji SOT, Probert SD (2004) Implementing total productive maintenance in Nigerian manufacturing industries. Appl Energy 79(4):385–401. https://doi.org/10.1016/j.apenergy.2004.01.007

Kiran DR (2017) Chapter 13—Total productive maintenance. In: Kiran DR (ed) Total quality management. Butterworth-Heinemann, pp 177–192. https://doi.org/10.1016/B978-0-12-811035-5.00013-1

Kumar SR, Dinesh K, Pradeep K (2006) Manufacturing excellence through TPM implementation: a practical analysis. Ind Manage Data Syst 106(2):256–280. https://doi.org/10.1108/02635570610649899

Levitt J (2010) TPM reloaded: total productive maintenance. Industrial Press

McKone KE, Schroeder RG, Cua KO (2001) The impact of total productive maintenance practices on manufacturing performance. J Oper Manage 19(1):39–58. https://doi.org/10.1016/S0272-6963(00)00030-9

Ng KC, Chong KE, Goh GGG (2013) Total productive maintenance strategy in a semiconductor manufacturer: a case study. In: IEEE international conference on industrial engineering and engineering management, 10–13 Dec 2013, pp 1184–1188. https://doi.org/10.1109/ieem.2013.6962598

Sun H, Yam R, Wai-Keung N (2003) The implementation and evaluation of total productive maintenance (TPM)—an action case study in a Hong Kong manufacturing company. Int J Adv Manufact Technol 22(3):224–228. https://doi.org/10.1007/s00170-002-1463-3

Szwejczewski M, Jones M (2013) Cost Reduction through total productive maintenance. In: Szwejczewski M, Jones M (eds) Learning from world-class manufacturers. Palgrave Macmillan UK, London, pp 66–83. https://doi.org/10.1057/9781137292308_4

Wickramasinghe GLD, Asanka P (2016) Effect of total productive maintenance practices on manufacturing performance: investigation of textile and apparel manufacturing firms. J Manufact Technol Manage 27(5):713–729. https://doi.org/10.1108/JMTM-09-2015-0074

Part III
Research Problem, Objectives and Methodology

Chapter 5
Definition of the Problem and Objective of the Research

Abstract In this chapter, the research problem that is addressed in the book is described and based on that the objectives of the book are established, where the limitations and scopes are defined according to the accessed data.

5.1 Critical Success Factors of TPM

Nowadays, there are some studies that have reported the critical success factors of TPM, as well as its benefits. For instance, Shen (2015) has done a compilation of studies in the last 10 years, where critical factors and their authors can be observed.

- Wu and Seddon (1994) report nine factors, where the following can be found: long-term commitment to TPM from senior managers, continuous and obvious concentration and support to TPM from senior managers, Union participation, have a complete TPM aide and organization on line, carry out effective feasibility studies, develop effective lead-in plan, develop a suitable strategy for the company's environment, equipment and products, acquire the support within the company, reach cognition on TPM, and strengthen it continuously.
- Patterson et al. (1996) report 10 critical success factors of TPM, where among them the following can be found: the support and commitment from senior managers, promotion and establishment of some kind of team culture, full empowerment toward the employees, overall employee involvement, high willingness for the involvement from the operators to the maintenance work, acquire the consensus from all employees within the company, technical trainings for the operators, cognition and support for TPM activities by maintainers, take advantage of the equipment maintenance records, and seize the improvement opportunities and high value on long-term benefits by the senior management.
- Windle (1993) reports five factors, the following are found: measurable policies, targets and effectiveness, clear management plans and factory management implementation, carry out high-quality, high-effective educational trainings, a

J. R. Díaz-Reza et al., *Impact Analysis of Total Productive Maintenance*,
https://doi.org/10.1007/978-3-030-01725-5_5

PPM director who can solve problems, overlook plans implementation, take accountability, and make all employees understand the meaning of TPM promptly.

- Williamson (1997) describes seven factors, among them the following can be found: if TPM implementation steps were carried out thoroughly, the support and resolution from senior managers, all employees consensus within the company, establish a specific department to carry out TPM tasks, the operating department must take out more inspections to related industries as well as attend workshops, establish all employees consensus within the company, all level leaders and directors should attend more seminars or educational trainings, personnel and equipment integration, and reasonable improvement.
- Chand and Shirvani (2000) report a total of five factors, where the following can be found: establish a project team along with a combination of the current system, establish overall and long-term targets, set awards for challenges, carry out educational training, and guidance from professionals.
- Rodrigues and Hatakeyama (2006) mention four factors, among them the following can be described: launch the 5S movement and carry out a complete implementation, organizational operation of small groups during self-maintenance, continuous educational training, and cooperate with the TPM implementation, support and participation from all directors.
- Katila (2000) reports eight factors: education and training about TPM, establishment of a maintenance system, real supervision from senior directors, lead-in education about TPM, plan the promotional TPM organization properly, establish thoughtful preventive maintenance policies, good maintenance data record or maintenance status, and an upgrade in maintenance management technologies.
- Khanlari et al. (2008) have mentioned six factors, which are the following: enough understanding and support for TPM from senior managers, proper candidate for promoting TPM, start with cadre demonstration (career demonstration), activate enthusiastic employees participation, get full understanding on the basic conditions that equipment should have, a specialized equipment department should have enough professional knowledge, as well as effective supportive live activities, and there should be personnel promoting teachers within factories.

Considering that the critical success factors of TPM have been identified, one of the problems addressed in this book is how to measure them through observable variables; that is, the FCE of TPM that have been listed above are dimensions that must be measured through other variables in the industrial field. For example, it is known that the level of commitment from the senior management is important, but the problem relies on determining how to measure it to know the level of application that it has, also it is pointed out that education and training are vital when implementing TPM, but how to notice if that is executing. Thus, one of the problems that exist is the lack of knowledge about how to measure the level of performance of these FCE.

5.2 TPM Benefits

As a matter of fact, there is no company that would implement TPM in its production lines if a series of benefits are not guaranteed since its implementation requires many resources, which are often not enough. Therefore, there should always be a cost/benefit analysis before implementing any TPM tool.

Fortunately, there are also reports that indicate the benefits that can be obtained from TPM. For example, Willmott and McCarthy (2001) describe a list of benefits grouped into three categories: organizational, safety, and productivity. In addition, organizational benefits are those that are totally associated with the company and its internal structure, safety benefits refer to those that help to reduce the risk of accidents and incidents for the operators, finally, productivity benefits are associated with the efficiency indexes in the company.

On the one hand, Cua et al. (2001) declared that the following TPM benefits may be obtained: reduction of spaces in the production lines, better performance of the product quality when manufactured with calibrated machines, production lower cost, reduction of the production time, it facilitates the delivery of small orders or flexibility in the volume, deliveries are executed on time to the client, and a preservation culture of the production means is achieved.

On the other hand, McKone et al. (2001) claimed that the manufacturing processes' performance depends on the compliance with the maintenance plans and programs, the transportation of materials within the company with just-in-time systems (JIT), and the quality that was achieved, which generates a relational model. Similarly, Swanson (2001) declares that maintenance must be part of the company's strategy in order to generate the benefits it offers, which implies a great commitment from the company's senior management. Also, Seth and Tripathi (2006) state that the TQM is required to achieve quality in the processed products, which later it translates into benefits for the companies and Konecny and Thun (2011) indicate that those two terms are closely related and that quality is the only aspect that the client can observe. Finally, Kamath and Rodrigues (2016) perform an analysis where it is shown that these two factors (quality and TQM) must be applied simultaneously in the company.

5.3 Research Problem

According to the previous information, several aspects that are worthy of being researched can be observed. In addition, it is mentioned that there are many critical success factors of TQM, but there is not indicated which are the variables that must be measured or valued in the production lines in order to know the level where these factors are achieved. In the same way, it is noticed that several benefits associated with TQM are mentioned, unfortunately, these studies mostly refer to work isolated performed, in other words, there are studies that report FCE and

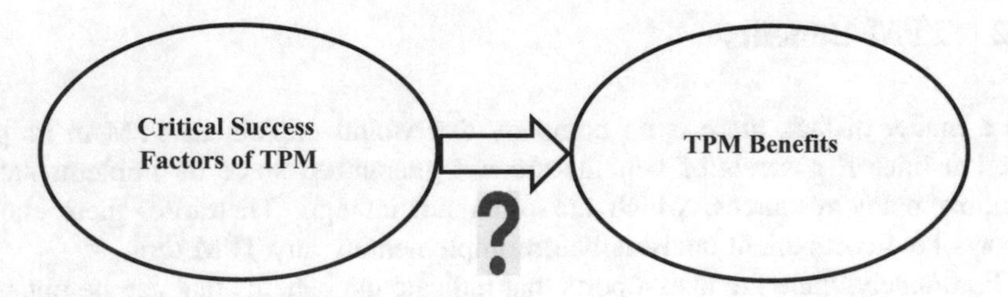

Fig. 5.1 Research problem

studies that report the benefits, but each on their own. Based on the previous statements, the problem is that there are not enough investigations linking these FCE of TPM with the benefits obtained that allow determining what the effect is between them.

Moreover, Fig. 5.1 helps to identify the research problem, where the FCE of TPM are the independent variables, since their execution is a function from human resources and their programs, also, the dependent variables are the benefits obtained. Logically, the problem is that the effect between these variables is unknown, therefore it is represented by a question mark.

5.4 Research Objective

As it was mentioned above, there are FCE of TPM and benefits obtained after its implementation that have been clearly identified in the literature review, consequently, the general objective of this research is the following.

5.4.1 General Objective

Quantify the effect between the critical success factors of TPM and the benefits that are obtained after its implementation in a productive system.

5.4.2 Specific Objectives

In order to fulfill the previous objective, a series of specific objectives must be achieved, such as those indicated below:

- Identify the observed variables in the production systems that indicate that the success factors of the TPM are achieved.

- Identify the benefits obtained when implementing TPM in a production system.
- Generate and apply a survey in an industrial productive sector.
- Generate hypotheses that relate the FCE of TPM with each other, as well as the FCE of TPM with the benefits obtained and between the benefits.
- Statistically validate the hypotheses with the data obtained.
- Conclude regarding the results obtained from the analysis of the relationship between the FCE of TPM and its benefits.
- Create recommendations for managers and people responsible for the decision-making process associated with TPM.

References

Cua K, McKone K, Schroeder R (2001) Relationships between implementation of TQM, JIT, and TPM and manufacturing performance. J Oper Manage 19(6):675–694. https://doi.org/10.1016/S0272-6963(01)00066-3

Chand G, Shirvani B (2000) Implementation of TPM in cellular manufacture. J Mater Process Technol 103(1):149–154. https://doi.org/10.1016/S0924-0136(00)00407-6

Kamath NH, Rodrigues LLR (2016) Simultaneous consideration of TQM and TPM influence on production performance: a case study on multicolor offset machine using SD model. Perspect Sci 8:16–18. https://doi.org/10.1016/j.pisc.2016.01.005

Katila P (2000) Applying total productive maintenance-TPM principles in the flexible manufacturing systems. lulea tekniska university, Luleå, Sweden

Khanlari A, Mohammadi K, Sohrabi B (2008) Prioritizing equipments for preventive maintenance (PM) activities using fuzzy rules. Comput Ind Eng 54(2):169–184. https://doi.org/10.1016/j.cie.2007.07.002

Konecny PA, Thun J-H (2011) Do it separately or simultaneously—an empirical analysis of a conjoint implementation of TQM and TPM on plant performance. Int J Prod Econ 133(2):496–507. https://doi.org/10.1016/j.ijpe.2010.12.009

McKone KE, Schroeder RG, Cua KO (2001) The impact of total productive maintenance practices on manufacturing performance. J Oper Manage 19(1):39–58. https://doi.org/10.1016/S0272-6963(00)00030-9

Patterson J, Fredendall L, Kennedy W, McGee A (1996) Adapting total productive maintenance to Asten. Inc. Prod Invent Manage J 37(4):32–36

Rodrigues M, Hatakeyama K (2006) Analysis of the fall of TPM in companies. J Mater Process Technol 179(1–3):276–279. https://doi.org/10.1016/j.jmatprotec.2006.03.102

Seth D, Tripathi D (2006) A critical study of TQM and TPM approaches on business performance of Indian manufacturing industry. Total Qual Manage Bus Excellence 17(7):811–824. https://doi.org/10.1080/14783360600595203

Shen CC (2015) Discussion on key successful factors of TPM in enterprises. J Appl Res Technol 13(3):425–427. https://doi.org/10.1016/j.jart.2015.05.002

Swanson L (2001) Linking maintenance strategies to performance. Int J Prod Econ 70(3):237–244. https://doi.org/10.1016/S0925-5273(00)00067-0

Williamson T (1997) Improve organization performance with total productive maintenance. Plant Eng 46:110–114

Willmott P, McCarthy D (2001) Assessing the true costs and benefits of TPM (Chapter 2). In: Total productivity maintenance, 2nd Edn. Butterworth-Heinemann, Oxford, pp 17-22. doi: http://dx.doi.org/10.1016/B978-075064447-1/50005-9

Windle W (1993) TPM: more alphabet soup or a useful plant improvement concept? Plant Eng-Chicago 47:62–63

Wu B, Seddon JJM (1994) An anthropocentric approach to knowledge-based preventive maintenance. J Intell Manuf 5(6):389–397. https://doi.org/10.1007/BF00123658

Chapter 6
Methodology

Abstract This chapter presents the methodology that will be used to solve the issue previously mentioned, which intends to connect the critical success factors of total productive maintenance and their benefits where hypotheses are developed and statistically tested. Also, it explains how an instrument or questionnaire is designed to gather information from the industry field, the process validation to analyze obtained data, and the statistical tests to find relationships among variables.

6.1 Introduction

Figure 6.1 is a flowchart that shows each activity that was used to reach the stated objective. In addition, it starts with a literature review for questionnaire design, and finishes with a structural equation modeling series validation where the analyzed variables are associated. Moreover, the previous activities are explained and justified in the following paragraphs.

6.2 Literature Review

The first activity is linked to the literature review in order to identify two relevant aspects: The critical success factors of total productive maintenance (TPM) and its benefits when it is integrated on assembly lines production. Additionally, to carry out this task some research databases were used, such as ScienceDirect, Scopus, Ingenta, Springer, and EBSCOhost, among others.

Furthermore, the research database keywords are TPM benefits, critical success factors of TPM, TPM implementation, and preventive maintenance, among others. Also, with the present literature review it is possible to develop a list of activities that are required for the TPM implementation and another list of benefits.

J. R. Díaz-Reza et al., *Impact Analysis of Total Productive Maintenance*,
https://doi.org/10.1007/978-3-030-01725-5_6

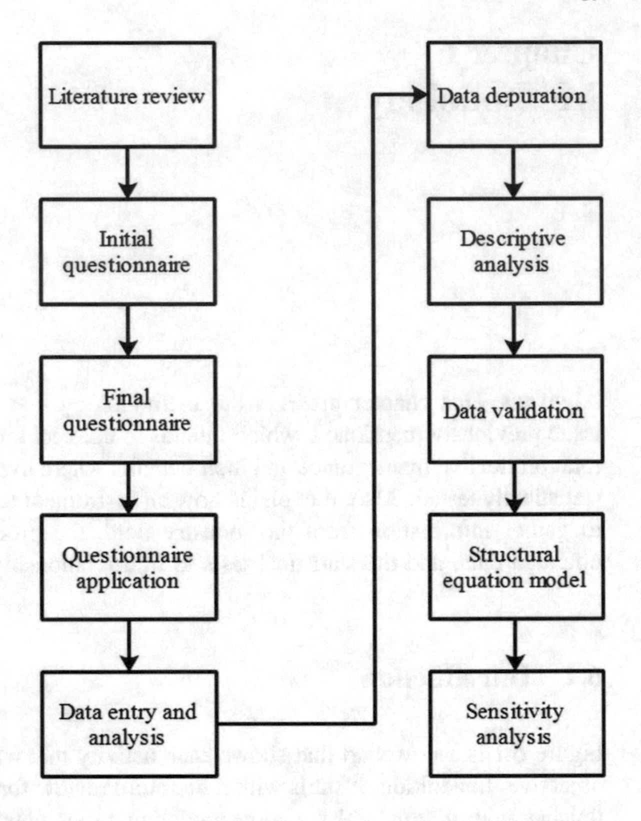

Fig. 6.1 Proposed methodology

The current literature review process has represented a rationale about the research that is being carried out (Alcaraz et al. 2014; Avelar-Sosa et al. 2014), since it is based on previously reported tasks that are looking forward to get adapted to the regional context.

6.3 Initial Questionnaire: Experts Validation

In order to consider TPM and its benefits as successful, a questionnaire is created applying required activities that have been mentioned in the literature review section to gather data from the industry sector in study.

The required activities to implement TPM are grouped into categories, as well as the obtained benefits to develop an initial version of the questionnaire. However, these activities and benefits have been retrieved from researches in other countries with different contexts, and frequently with different industrial sectors. Therefore, a validation process is carried out by professional experts to be adapted to the regional context.

Additionally, academics, business maintenance managers in the region, and two experts in the Spanish and English languages are invited to assess an initial version of the questionnaire, and are asked to evaluate the following aspects:

- Pragmatic equivalence in translation;
- Clear and concise research questions;
- Imposition degree in each research question;
- Appropriate research questions according to the participants' background;
- Appropriate vocabulary according to the participants' environment; and
- Experts predilections.

After this validation process by professional experts, considering all their recommendations as well, a final version of the questionnaire is established, which is described in the following section.

6.4 Final Questionnaire

The final questionnaire is developed after the rationale through the literature review and the validation by judges previously explained. This questionnaire is separated into sections linked to the data that was intended to obtain, which are described in the following section.

6.4.1 Section 1. Demographic Data

The present section is about the participants information; it allows researchers to have a better identification and classification from the obtained data.

This section is focused on the following aspects:

- The period that the participant has been working on the current job, which allows knowing the level of experience when giving answers.
- The participant industrial sector, since the industry that has the questionnaire addressed is a set of companies from the automotive, medical, electrical, and electronic industries, among others.
- The participant's gender, since gender equity is widely attempted in these companies.
- Hierarchical position that the participant has in the company, since they can be department managers, a member of the workforce, engineers, and supervisors, each one of them from the maintenance department.

6.4.2 Section 2. Critical Success Factors of Total Productive Maintenance

In this section, critical success factors or activities that have been discussed in the literature review through the rationale, and that have been modified and adapted after the validation process by professional judges are incorporated. Moreover, the activities have been divided into categories, which had a series of items to be assessed. The categories are the following:

- Aspects related to the labor culture, integrated by 13 items;
- Aspects related to equipment suppliers, integrated by 8 items;
- Relationship among the implementation of preventive maintenance in the company, with 9 items;
- Aspects regarding how to apply TPM in the company, with 16 items;
- The relationship of the company status toward technology, with 8 items;
- How the industry administration is managed in relation to TPM, with 6 items;
- The relationship between the influence and participation of clients in TPM, with 4 items;
- Aspects related to the Layout for maintenance, with 5 items; and
- Respecting materials in stock and spare parts in storage, 13 items.

As it has been shown, this section deals with technical, administrative, and procedural aspects that companies have followed to implement TPM in their production lines. The total questions or success factors associated with the success of TPM are 82.

6.4.3 Section 3. TPM Benefits

The TPM benefits that are presented in the questionnaire are those that have been validated by the literature review, and by professional judges for an appropriate adaptation to the environment. Also, they have been classified into three categories, which are listed below:

- Benefits for the organization, with 8 items;
- Security benefits, with 6 items; and
- Productivity benefits, 8 items.

As it has been illustrated, a total of 22 benefits are analyzed to identify if the organization and its information flow, the safety of workers, and personnel performing maintenance tasks, and equally important, those who are associated with the company performance and its productivity, have improved by implementing TPM.

Table 6.1 Rating scale

Score range	1	2	3	4	5
Rating category	Never	Rarely	Sometimes	Often	Always

6.4.4 Rating Scale

In order to answer each item response from each category about critical success factors of TPM and its benefits, the five-point Likert scale is used, which includes a score range from one to five; their rating category is represented in Table 6.1.

For instance, the score 1 points out that this activity or critical success factor is never performed during the implementation process of TPM or this benefit is not obtained because of it. Similarly, score 5 points out that the activity is always performed during the implementation process of TPM or the benefit is always obtained because of this tool.

It is important to mention that the Likert scale was chosen due to its ease of interpretation (Cua et al. 2001) as well as the high levels of familiarity from the respondents, and because there are reports of its reliability (Yaşlıoğlu et al. 2014; Swafford et al. 2006). Furthermore, it has been applied on similar researches by Avelar-Sosa et al. (2015) who used it to rate aspects from the supply chain in the maquila industry, also by Alsyouf (2009) to know the status of implementation maintenance systems in the Swedish automotive industry, and by Attri et al. (2013) it was needed to determine the barriers that can be encountered while implementing TPM.

6.5 Questionnaire Application

The application of the questionnaire to maquiladora companies requires a series of tasks, which are defined in the section below.

6.5.1 Sample

Through different media, managers and personnel who are working and in charge of a maintenance department clearly defined in the Mexican maquila industry are contacted. In addition, to consider a company as a candidate to participate with its staff in the study, the next inclusion rules are followed:

- Companies must have within their organizational structure a maintenance department that allows access to their logs.

- Companies must be recognized as maquila companies by IMMEX (Mexican Maquiladora Export Industry).
- Companies must have at least two production lines where product changes are made periodically.
- Surveyed participants should be familiar with maintenance programs; therefore, the sample is limited to identifying the engineering managers, technicians, and maintenance operators.

Also, it is relevant to mention that in this first stage a stratified sampling was carried out, since the managers who were candidates to participate were already identified (Arnab 2017a; Singh et al. 2016).

6.5.2 *Gathering Data*

The managers from companies are contacted by email or telephone to be invited to participate in the present study; for those who have accepted, dates and times were settled for an interview related to the critical success factors of TPM and its benefits.

The appointment was on the date and time set; however, not all of them were able to answer the survey at that time, so a new appointment is established. In addition, the interview layoff is mainly due to unexpected events in the production lines, which had to be attended by them in a personal way. Also, if throughout the second date for the interview was canceled, then the manager was asked to provide the name of someone else working in the same maintenance department to answer the questionnaire.

Finally, in case that after three interview attempts to apply the questionnaire, it was not possible to obtain the information, then this particular case was meant to be canceled due to the amount of time it required to be invested on. Also, it is necessary to mention that once the interview was finished, some participants shared information about some people in the same job position working in other companies who might support the study. Additionally, the managers always mentioned who were the maintenance engineers, technicians, and operators working in the company, and that they could have the knowledge to participate in the study as well. Consequently, in this stage, the snowball sampling method was implemented (Ompad et al. 2008; Serrano Sanguilinda et al. 2017).

6.6 Data Entry and Analysis

The data collected from the questionnaires is entered into the software SPSS v.24 for analysis, and then a database is designed in it, where each row represents a case and each line to an observed variable or item. This software has been chosen because of its non-complexity of use and the amount of analysis it allows to perform as well (Preacher and Hayes 2004). However, it is essential to say that analysis had been done with spreadsheets, where a series of graphics were also created, because they are friendlier than other software.

The usage of SPSS is widely reported in other studies. For instance, on analysis of regressions by main components (Liu et al. 2003), on analysis of obtained data from the maquiladora industry and the application of the just-in-time compilation (JIT) (Alcaraz et al. 2014), and on the analysis of JIT key success factors (García et al. 2014).

6.7 Data Depuration

The data from the questionnaires may have a series of errors that must be checked before proceeding to their analysis, even during the input process it is possible that other errors will be made. For this reason, a data purification is carried out, focusing mainly on three aspects: identification of missing values, identification of outliers, and a standard deviation analysis, which are explained in the section below.

6.7.1 Identification of Missing Values

As it was previously mentioned, the data in the questionnaires has been entered into a database in the SPSS v.24 software; their routines are used to identify missing values. Research seems to consider that an observed item or variable has a lost value if an answer has not been given, which may be by default at the time of inputting data or because the participant did not know the answer and left it blank.

In addition, the missing values in each of the questionnaires were counted and a certain percentage of unanswered questions were over 10%, consequently, that case was eradicated (Dray and Josse 2015). However, if the proportion of missing values was under this amount, researchers would proceed to estimate the average of each of the observed variables; therefore, this value was substituted for that missed value, since the scale that was used for the valuation is ordinal (Toutenburg and Srivastava 1998; Bagnaschi et al. 2015).

6.7.2 Identification of Outliers

Another important task in the data purification process is the identification of extreme values, commonly known as outliers, which can be defined as values that are outside the range of the scale that was used. It is significant to remember that in the present research, a Likert scale is used that uses values that are between one and five, and as a result, values outside that range are considered outliers. In addition, the existence of these values is mainly due to errors in the inputting process, since the questionnaire is designed graphically, and it is not possible to make an error.

In order to identify these values, each of the items or variables that are observed is standardized, where the average is zero and the standard deviation is one. Also, the standardized values that are superior in absolute value to four, are considered as outliers (Schubert et al. 2014; Wu et al. 2013). Similarly, box-and-whisker diagrams were made for each of the items, where the boxes represent the first and third quartiles and the endings are illustrated in the lower or upper part, which allows a graphic and quick visualization of the data distribution (Hekimoglu et al. 2015).

Furthermore, these outliers were replaced by the median to avoid preconceptions while carrying out the data analysis, since the scale that was used for the valuation was ordinal (Jin et al. 2006). However, it is relevant to mention that there are many other methods to perform the replacement of outliers and missing values, but still there was the limitation from the scale that was used.

6.7.3 Standard Deviation Analysis

The third assessment aim is to identify those noncommitted participants who always respond with the same judgment on each item in the questionnaire. Additionally, this analysis is performed on a spreadsheet, which is exported directly from the SPSS V.24 software, because each of the cases collects the sample standard deviation, where the characteristics associated with this statistician are employed. If a participant always answers with the same value to all the questions, then the standard deviation for that questionnaire will be zero, or if the answers have very similar values, then that standard deviation will be very close to zero.

In the present research, questionnaires that had a standard deviation of less than 0.5 on the ordinal scale were eradicated from the analysis, since they indicated that the participant had answered with similar characters in all the items (Arnab 2017b; Adhikary 2016).

6.7.4 Normal Distribution

In order to relate the latent variables or key success factors of TPM categories with its benefits, it is assumed that there is dependency between them, and the quantify measurement for it is pursued, which is done by regression techniques that require the principles of normality to meet in each observed variable (Jönsson 2011).

However, these principles are not always achieved and in this case, the Jarque–Bera test is implemented as a normality test, since it takes into account the sample size, the degrees of freedom, the index of asymmetry, and the kurtosis from the sample (Raïssi 2017). Also, histograms are made for each of the analyzed items, where a curve of normal distribution for a visual analysis is adjusted, which in an immediate examination gives an idea about the type of distribution that is available.

The implementation of these tests have been reported by Kim (2016) who displays a new improved index, Bear and Knobe (2017) report a descriptive analysis from the competency of several indexes, also Shalit (2012) describes an analysis for settings where normality should be portrayed in the data when using ordinary least squares techniques.

6.7.5 Homoscedastic Analysis

When a regression technique is applied, models are generated to fit data and make predictions. Thus, real values may be acquired with the predicted models and values previously generated on the model. Moreover, the difference between these two values produces the model residuals, which must be homogeneously distributed and without patterns so that homoscedasticity is able to exist.

In the current study, residual graphs are used to identify waste patterns and homoscedasticity. In addition, this method has been applied in recent studies reported by Pastor et al. (2018).

6.7.6 Collinearity Analysis

Because the relationships between the critical success factors and its benefits are from a multivariable point of view, it is convenient to first analyze the collinearity of the observed variables that are part of each latent variable, and then the collinearity that they have when integrated into a model.

Although there are many methods to predict collinearity, variance inflation factors (VIF) are used in this research (Vu et al. 2015), which are based on the eigenvalues associated with the matrix of correlations, where it is expected that the VIF are less than 5, although there are reports in which values over 10 are accepted. Nowadays, there are many studies where VIF are used to detect multicollinearity, for instance, Hsieh and Lavori (2000) apply it in a proportional hazard analysis study to determine the efficiency of sample sizes, Wang (2017) implements it to identify collinearity in associated factors with residential electric power consumption.

6.8 Descriptive Analysis

The descriptive analysis from the obtained data was carried out in two different stages; the first phase refers to a descriptive analysis of the sample based on the demographic data from the first section of the questionnaire. On the other hand, the second one refers to the descriptive analysis of the assessments in sections 2 and 3 of the questionnaire, which refer to the critical success factors of the TPM and its benefits. Moreover, these analyzes are briefly discussed in the section below.

6.8.1 *Sample Descriptive Analysis*

The descriptive analysis of the sample is carried out to identify preconceptions in the participants in order to achieve a data classification. In addition, in this stage, tables related to the participant gender were obtained, frequency graphs of the hierarchical positions that the respondents have within the company, as well as years of experience (Cooper et al. 2006; Biresselioglu et al. 2017). Similarly, crossed graphs were done where two variables or more intervened, to review the grouping of data and tendencies.

Also, this type of analysis has been reported in other researches, where it has been necessary to create categories and classifications of the collected data, such as in Kaur et al. (2016).

6.8.2 *Data Descriptive Analysis*

The descriptive analysis of the data intends covering two important aspects of each one of the items in the questionnaire, which refer to central tendency and dispersion measurements, which are discussed in the following section.

6.8.2.1 Measure of Central Tendency: The Median

Although it has already been mentioned that the median has been used as a measure of central tendency to replace the missing values and outliers that could be found in the database, there has not been a broad justification of the reason for its usage, which is mainly because the Likert scale in which the items were valued is of the ordinal type.

In addition, depending on the median of each item, an interpretation of it may be done (Adamson and Prion 2013a):

- If the median value is high, then it may be said that this task is important for the company, and thus it achieves an adequate implementation of TPM; in the same way, if it is a benefit, it indicates that this is always or almost always obtained after implementing TPM (Clark-Carter 2010).

- On the other hand, if the median has a low value, then it may be said that this task or critical success factor is not important in the TPM implementation process, or that this benefit is almost never obtained because of its implementation.
- Values around three indicate that this task is regularly important or that this benefit is regularly obtained.

Applications in which the use of the median is reported as a measure of central tendency can be consulted in Díaz-Reza et al. (2017) for studies associated with SMED techniques, in Alcaraz et al. (2016) to identify the main benefits that are obtained from the implementation of systems just in time, and recently in Realyvásquez et al. (2016) to analyze the environmental and psychosocial factors that may affect the performance of a company.

6.8.2.2 Statistical Dispersion

As a measure of dispersion in this research, the interquartile range is used, which requires that all quarters are estimated, and it represents the difference between the third and the first quartile. In fact, in the box-and-whisker diagrams that have been used for the identification of outliers, the box from this graph represents the interquartile range.

Additionally, it has been decided to use this measure, since as mentioned before, and the obtained data is on an ordinal scale. Also, from the analysis of the values in the interquartile range, the following can be concluded (Tastle and Wierman 2007):

- If the interquartile range values are nearly to zero, it will indicate that all participants have avoided a similar value for the same critical success factor of TPM or its benefit, indicating prominent levels of agreement or consensus regarding the values that belong to them. Furthermore, it is attempted that participants make judgments with very similar values, so it is expected that the interquartile ranges have values close to zero.
- On the contrary, if the values of the interquartile range are high, then this demonstrates that participants provided very different values for the same item, which can be translated as a lack of consensus or agreement in relation to the expected true value, and therefore these types of values are not desired.

Currently, there are many studies where the implementation of the interquartile range is addressed or where it is used as a measure of dispersion of the values; for instance, Midiala et al. (2016) use it in a study conducted on the critical success factors of associated tasks with the continuous improvement of production processes, García-Alcaraz et al. (2016) implement it in the flexibility study in a supply chain and its impact on the quality, and performance of it, whereas Kang and Lee (2005) apply it to analyze convergences of values and statistical parameters.

6.9 Data Validation

Once the data has been depurated and outliers free and missing values, then the next step is to validate it from a statistical point of view. In addition, it is relevant to recall that two types of validations have been completed; the first is done through the literature review and the second by professional judges. Also, in this specific case where the data is already collected, the validation is done from a statistical point of view.

Furthermore, in this type of validation, several indexes have been used, which are listed below and have been applied to each of the variables categories previously mentioned, and they will be named latent variables:

- *R*-squared,
- Adjusted *R*-squared,
- *Q*-squared,
- Integrated reliability,
- Cronbach's alpha index,
- Average variance extracted (AVE), and
- Full collinearity variance inflation factor.

6.9.1 R-*Square and Adjusted* R-*Squared*

These two indexes are used to measure the parametric predictive validity in a latent dependent variable, and in other words, it is explained by other variables. Also, it is intended that the values in these indexes are always more than or up to 0.02, which indicates that the independent latent variables must explain at least 2% of the variability in a latent dependent variable.

Equally important, in this index high values are target, preferably close to the unit, since this would reveal that the latent dependent variables are over 100% by independent latent variables. However, it is also worth mentioning that high values of *R*-square in this type of studies bring a lot of attention, since they are indicating that five are having collinearity issues between the variables (Ratzmann et al. 2016).

Additionally, the difference between *R*-square and adjusted *R*-square is that the first one does not consider the size of the sample for its estimation, while the second one does. Also, it is recommended that the difference between these two parameters will not be more than 5%; otherwise, it would imply that there are problems associated with the size of the sample that is being analyzed (Biancolillo et al. 2017).

6.9.2 Q-Square

Likewise, the *R*-square and the adjusted *R*-square, this index helps to measure the predictive validity that a latent dependent variable may have, however, this index is nonparametric. Also, the main objective is that the obtained values in this index are over zero and similar in magnitude to those from *R*-square. Although it seems irrational to indicate that the values of *Q*-square must be positive, it is convenient to mention that it is possible that these values are negative, which illustrates radical problems in the model structure and that the relationship between the variables is meaningless. In Mihail and Kloutsiniotis (2016) a study using *Q*-square as a nonparametric index predictive reliability is reported.

6.9.3 Integrated Reliability

The present index is associated with a latent variable, which is also known as the Dillon–Goldstein rho coefficient and is often referred to as a modified or standardized Cronbach's alpha index. Due to its quantification formula, it frequently releases values over those from Cronbach's alpha, and also one of its main characteristics is that it considers the factorial analysis of the items in its estimation.

Moreover, it is employed to measure the internal reliability of latent variables and values over 0.7 are expected. Also, values under 0.7 in this index are indicating that there are items that do not belong to the latent variable in evaluation; therefore, an iterative process must be executed to determine which of those should be removed or added to another latent variable, and it means that it increases its value. In addition, studies where the usage of this index is mentioned are in Lee and Hallak (2017) to determine the role of education in restaurants performances and in Fakih et al. (2016) to analyze the customers attitudes.

6.9.4 Cronbach's Alpha Index

As well as the integrated reliability index, it is helpful to assess the latent variables internal reliability that is settled by items. In addition, values over 0.7 are expected in order to consider that a construct or variable has internal validity also (Adamson and Prion 2013b), if the values are lower, an iterative analysis to look for improvement is pursued, where items are deleted, and the index is reestimated. Furthermore, in this research, there is a routine in the SPSS v.24 software where those sensitivity analysis and improvement from the Cronbach's alpha index are removed; therefore, this method is performed in an iterative manner.

Moreover, this present index is a modification of a coefficient determination used in regression analysis, where an extreme value decreases the efficiency of it, offering a low value that may decrease the explanatory power of the model. Additionally, there is a comparable situation occurring in the estimation of Cronbach's alpha index, because if an item does not belong to the latent variable that is being analyzed, then it will have a low value and must be removed from that variable or it will be added to another one to improve it. Also, the only difference in the coefficient determination is that the Cronbach's alpha index considers the estimated number of questions in the latent variable.

In addition, this index has been widely used in scientific reports, for instance, Cazan and Indreica (2014) it is used in their study that attempts to determine the need of cognition process and approaches in learning among university students, Abdel-Maksoud et al. (2010) is implemented to identify the employee morale, non-financial performance measures, the deployment of innovative management practices, and workshops participation in Italian manufacturing companies and recently, de Guimarães et al. (2017) it is performed in an empirical study where the impact of clean production projects are measured in the project management area.

6.9.5 Average Variance Extracted (AVE)

The AVE index is associated with a latent variable composed of several items, and it provides the evaluation of the discriminant validity, although sometimes it is also used to assess the convergent validity. Additionally, values up to 0.5 are preferred when it is evaluated as a metric for convergent validation, since higher values show that this latent variable is properly represented by the items that comprise it. Also, it is used to determine the discriminant validity, Kock (2015) suggests the following criteria: for each latent variable, the square root of the average variance extracted must be over any of the correlations involving that latent variable.

Moreover, the application of this index is reported in several researches to identify the convergent and discriminant validity. For instance, Chowdhury (2016) uses it in a study associated with spiritual well-being, Farooq et al. (2018) implement it to validate the service quality according to customer satisfaction in Malaysian airlines, and Blanco-Oliver et al. (2016) apply it to microfinance analysis.

6.9.6 Full Collinearity Variance Inflation Factor

When a multivariate analysis is performed, as in this case, where there are many observed variables integrated into latent variables linked to each other, it is convenient to discard collinearity issues. Also, in this investigation, the variance inflation indexes are used to detect this type of problem and five is set as the maximum cutting-stock value (Reguera-Alvarado et al. 2016). In addition, this current index is analyzed in two kinds of analysis; when the observed variables in the latent variables are evaluated, and when these are integrated into a model.

Furthermore, this index has been reported in other investigations, for example, Nitzl (2016) apply it in the study of financial administration to assess dependency between observed variables, Lee and Hallak (2017) implement it in a study related to the staff educational level and restaurants efficiency while Roni et al. (2015) use it on behavior analysis.

6.10 Structural Equation Modeling

Once the latent variables depuration and validating process has been performed, they are integrated into different structural equation models in the WarpPLS 5® software, which has algorithms based on partial least squares (Kock 2015). Also, WarpPLS 5® is being used, since it is widely recommended if the data is on an ordinal scale, it does not require data normality, and it offers reliable results when working with small samples (Rasoolimanesh et al. 2015; Ekrot et al. 2016).

Moreover, one advantage of using the structural equation modeling is that it is a third-generation technique within the family of linear regression, and it allows to relate latent variables that are integrated by other variables, as can be seen in the group of critical success factors of TPM and its benefits. In other words, it allows relating variables that are integrated by other observed variables, which cannot be done in multiple regression, where only one variable observed as a dependent variable can be related to several independent variables (Tenenhaus et al. 2005). In addition, another advantage offered using the structural equation modeling is that it accepts the same latent variable to portray independent or dependent roles, since a network of interrelated variables is constructed.

In the structural equation modeling, a specific symbology is used (Schubring et al. 2016):

- An ellipse represents a framework where the latent variables are integrated by observed variables.
- An arrow represents the relationship between two latent variables.
- Rectangles represent observed o-valued variables; in this case, they represent the critical success factors of TPM and its benefits.

Figure 6.2 illustrates that two latent variables are related; one is independent and the other is dependent, and also two of the similar elements are shown, such as the latent variable and the arrows that link them, which represents a hypothesis that is explained in the section below.

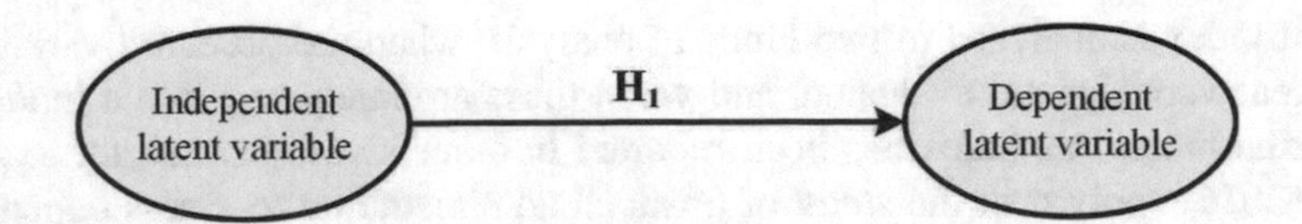

Fig. 6.2 A simple model

6.10.1 The Modeling Hypothesis

Figure 6.2 exemplifies a relationship between two variables, where a hypothesis is established and, in this research each of the established relationships between the latent variables are assigned, which are validated with 95% of reliability. Additionally, the indicated hypothesis in Fig. 6.2 could be disclosed in the next form:

H_1: the independent latent variable has a direct and positive effect on the independent latent variable.

However, this model is very simple and in the present research, it is aimed to generate latent variables that integrate more than two latent variables as a result, they are more complex. In addition, Fig. 6.3 illustrates a model that integrates three latent variables with three relationships among them, and in other words, three hypotheses.

Furthermore, the definition of these hypotheses is based on a series of theoretical foundations that allow researchers to define an inference about their relationship, although it is very important to take into account the following aspects (Schlittgen et al. 2016):

- The events temporality, first an act is planned and then a result is obtained. In this case, first an activity associated with total productive maintenance is performed, and therefore, the benefits are acquired.

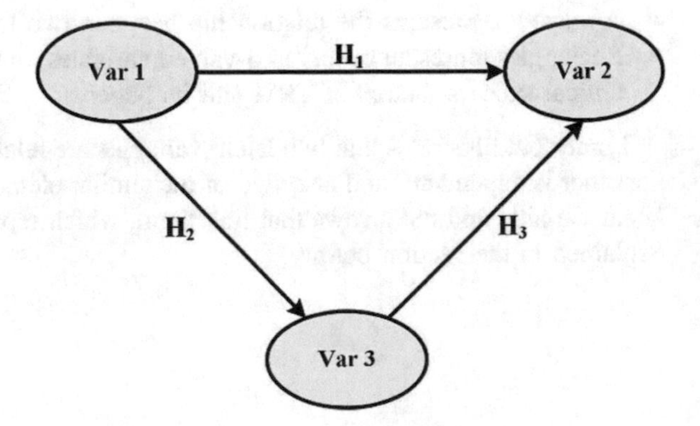

Fig. 6.3 A complex model

- The theoretical foundations that are able to portray the relationship, and in other words, that each one of the hypotheses must be based on previous investigations, where the items that integrate them interfere among each other.

In Fig. 6.3 it is represented that there are three relationships that present hypotheses in the model, which are called direct effects, but there are also other effects that are analyzed in this investigation; the list of the analyzed effects in this investigation is as follows:

- Direct effects,
- Indirect effects, and
- Total effects.

Figure 6.4 illustrates an evaluated model that will be mentioned later to explain and interpret diverse types of effects found in the structural equation modeling, which integrates four latent variables and has a higher level of complexity.

6.10.2 Direct Effects

Figure 6.3 shows that latent variables are represented by an arrow that links them directly, which describes a hypothesis, if there are no other intermediate variables in that relationship, it is said that the effect is direct. Logically, there is a variable where the arrow starts, which is called independent latent variable, and it has an

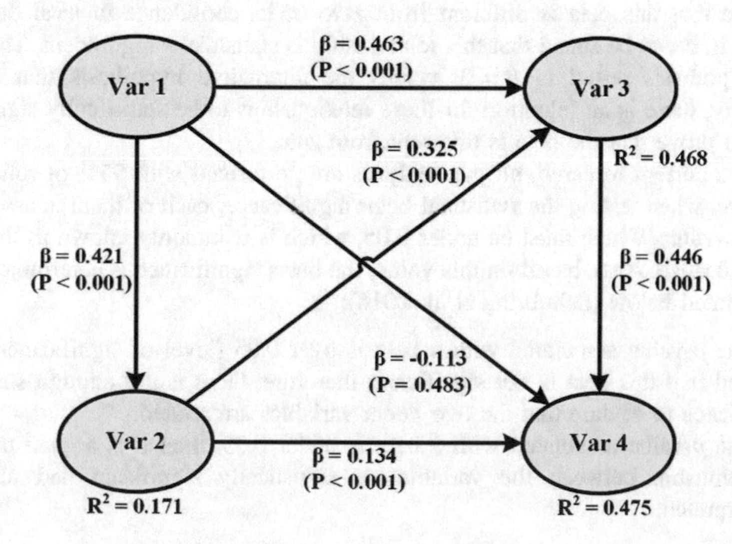

Fig. 6.4 An evaluated complex model

effect on the variable where the arrow finishes and it is called the dependent latent variable (Kazár 2014).

The effects from one latent variable in another one are measured by estimations through partial least squares medians, and each of the relations or hypotheses is associated with a value of beta (β). Therefore, there will be as many betas as hypotheses or relationships between variables proposed in the model. For instance, in Fig. 6.4 each relation (hypothesis) has a beta and an associated *p*-value. In addition, that beta represents an intensity of change that is related to the independent latent variant with the dependent latent variable, which is expressed in standard deviations (Reguera-Alvarado et al. 2016). For example, if the beta is found connected in Fig. 6.2 out of 0.530, then it could be said that there is enough statistical evidence to state that the independent latent variable has a direct and positive effect on the latent dependent variable, since when the first variable increases its standard deviation by one unit, the second variable goes up to 0.530 units.

Similarly, according to the data from Fig. 6.4 it can be seen that the direct effect that exists between Var 1 and Var 2 is 0.421, which indicates that when Var 1 increases its standard deviation in one unit, the second one as well in 0.421 units, and a similar interpretation may occur for all other relationships that are between the latent variables and are part of this model.

However, it is important to know that if the intensity of change or beta value that has been determined is statistically significant, and thus a hypothesis test is always carried out to determine if the beta value is equal to zero or different (Sarstedt et al. 2014). Moreover, if the beta value would have been equal to zero or its confidence interval included that value, then it would be claimed that this relationship is not statistically significant. On the other hand, if after the hypothesis analysis and test it is found that this beta is different from zero or its confidence interval does not include it, it can be stated that this relationship is statistically significant. Thus, the null hypothesis tested is $\beta = 0$, versus the alternative hypothesis that $\beta \neq 0$. Logically, there is an intention for these relationships to be statistically significant to try to prove that the beta is different from zero.

In the current research, all the analyzes are performed with 95% of reliability; therefore, when testing the statistical betas significance, each of them is associated with a *p*-value, which must be under 0.05, which is commonly known as the level of significance. Also, based on this value, the betas significance is determined, as it is illustrated below (Schubring et al. 2016):

- If the *p*-value associated with a beta is over 0.05 (level of significance), it is stated that this beta is not significant; therefore, there is not enough statistical evidence to declare that the two latent variables are related.
- If the *p*-value associated with a beta is under 0.05, then it is argued that this relationship between the variables is statistically significant and must be interpreted.

6.10.3 Indirect Effects

Figure 6.3 illustrates three direct effects represented by the three hypotheses, and it is possible that variable one influences variable two through variable three, which occurs through the two segments, that is, in the relationship between variable one and variable three and then continue from variable three to variable two. In this case, it is said that variable one has an indirect effect on variable two using two segments, where variable three is a mediating variable (Richter et al. 2016). Another example can be showed using Fig. 6.4, where Var 1 has a direct effect on Var 2, and also it influences Var 3. Thus, there is a conclusion showing that there is an indirect effect between Var 1 and Var. 3, where the average variable is Var 2.

Moreover, as it can be seen, in this model where there is only three latent variables interjected, which is relatively simple; there is only one indirect effect, but there will be times when there are many indirect effects and they are seen through more of a mediating variable; in other words, where many segments are used to link a latent independent variable with a latent dependent variable. Additionally, in Fig. 6.4 it can be observed that the variable Var 1 has several indirect effects on the variable Var 4, which can be 2:03 segments, where Var 2 and Var 3 are used as mediating variables.

Furthermore, in this research, only the total amount from the indirect effects or from all the segments that have an independent variable on a dependent variable will be reported, which will be expressed through a value of beta, and it will be associated with a *p*-value to determine its statistical significance. Also, the value interpretation is the same that has been performed for direct effects, where it is always intended to have *p*-values under 0.05.

In addition, this type of indirect effects analysis is very relevant in praxis, because sometimes the relationships between the variables are not direct, since the relationship is not statistically significant based on the *p*-value, but the third mediating variable presence is required in order to have a statistically significant effect (Nitzl 2016). In addition, this entails that the direct effect that is analyzed between an independent latent variable and a latent dependent variable is not statistically significant, but it may be in the presence of other variables, which indicates that the indirect effect is statistically significant.

6.10.4 Total Effects

Total effects represent the total amount from the indirect and direct effects (Tenenhaus et al. 2005; Kock 2015). Each of these total effects is associated with a value of the beta and to discover its statistical significance, a *p*-value is also integrated. Also, the relevance about this type of analysis is that it allows researchers to know the total magnitude of an independent latent variant on a latent dependent variable, which is often not observed directly, because it requires the presence of other variables. To illustrate an example of total effect, Fig. 6.4 is appropriate where the variable Var 1 has a direct effect on Var 3, but an indirect effect using the mediating variable Var 2; the total amount of effects will be the direct effect plus the indirect effect.

6.10.5 Effects Size

Figure 6.4 shows that each of the latent dependent variables is associated with an *R*-square value as a measure of the variance explained by the independent latent variables in that variable. In the relationship between Var 1 and Var 2, it is observed that there is an *R*-square equal to 0.171, where Var 1 is the only variable that explains Var 2, so that the variable is responsible for its *R*-square value. However, frequently a dependent latent variant is explained by more than one independent latent variable, such as with Var 3 that is portrayed by two variables and Var 4 that is explained by three variables.

On the previous case, Var 3 has an *R*-square value of 0.468 and it is convenient to unscramble that value to determine what the value of Var 1 and Var 2 is. Also, the same situation occurs with Var 4, which is affected by Var 1, Var 2 and Var 3; the setting of the *R*-square value is known as the size of the effect and it helps to determine which of all the independent variables that affect the dependent variable is the most important. However, it is relevant to mention that the idea associated with the importance of independent latent variables can also be judged based on the size of the betas that affect it in each of the relationships.

Furthermore, in this research, a value of the beta is always reported for each of the relationships, the *p*-value is associated to the statistical test of its significance and the *R*-square value from the latent dependent variables. However, before proceeding to interpret a model, it is necessary to identify if the integration of each one of the latent variables in the model has a series of efficiency indexes, which are illustrated below.

6.10.6 Efficiency Modeling Indexes

Before to begin with the structural equation modeling interpretation, it is essential to determine if it fulfills all the required efficiency indexes. Also, it is relevant to recall that until now, the validity indexes required for each of the latent variables have been indicated, but the model validity indexes have not been presented when all those variables have been integrated. Additionally, in this investigation, these indexes are considered and explained in the following section.

6.10.6.1 Average Path Coefficient (APC)

The regression coefficients shown in Fig. 6.4 are commonly called beta, and it helps to validate a hypothesis in a direct effect. However, this assessment is specifically for two latent variables that are related between them. In addition, the APC is the average of all the coefficients or betas that are part of a structural equation modeling, that is, in the total amount of values from all the coefficients divided by the number of coefficients existing in the model (Reguera-Alvarado et al. 2016).

Moreover, to perform the significance statistical test of this globalized index, it is associated with a *p*-value, which must be under 0.05, since all the estimations made in this research are done with 95% of reliability. Also, it is important to mention that frequently direct effects can be negative, which decreases the PAC value because it is an average tool.

6.10.6.2 Average *R*-Squared (ARS)

It has been previously referenced that each of the dependent latent variables, which are explained by others, is associated with an *R*-square value, which is specific for the variable that is being explained. However, to understand the predictive validity level of the model in general, the middle *R*-square is used, which is recommended to be over 0.02 (Kock 2015); to clarify, the model has to be at 2% of predictive validity.

Furthermore, ARS is associated with a *p*-value to define its significance level, and as in the previous statistical tests, it is intended that this is under 0.05, since the reliability level where estimations are made is 95%. Also, it is important to say that frequently when two models are poorly planned and if they lack common sense, negative *R*-square values are found, which is unacceptable, but it indicates that some adjustments must be made, since it is inadmissible.

6.10.6.3 Average Adjusted *R*-Squared (AARS)

This index has a similar interpretation to the ARS, since it helps in the predictive validity of the total model. As mentioned above, the difference between ARS and AARS is that the first one considers the sample size, because the number of freedom degrees is contemplated for its calculation. Also, values up to 0.02 are required in AARS and its decision is based on the associated *p*-value (Kock 2015).

6.10.6.4 Average Block Variance Inflation Factor (AVIF) and Average Full Collinearity (AFVIF)

It has been previously mentioned that inflation variance indexes can be used to measure the linearity within a latent variable, that is, with the linearity that may exist between the different items or variables observed. However, this analysis is for each of the latent variables, and now the AVIF and AFVIF represent a general average of the inflation indexes from the variance among the latent variables that integrate the model (Kock 2015).

Moreover, it is recommended that this index is under or equal to 3.3, although it is also commented that values over 10 can be accepted. Also, in the specific case of this investigation, it has been decided to accept inflation rates of variance that are under five.

6.10.6.5 Tenenhaus Goodness of Fit (GoF)

One of the main concerns regarding a regression model is knowing the adjustment level that the data may have in a model. Technically, in the regression model there is an index called coefficient of determination to notice the adjustment level that the data has in the model, and in the case of structural equation modeling there is a multivariate model aspect, the Tenenhaus goodness-of-fit index is used (Schubring et al. 2016; Kock 2015).

Moreover, it is recommended that a model has values over 0.36 to state that there is an adequate adjustment of the data obtained in the model. Also, in the case that this index is not achieved, interactive tests should be performed to identify errors within the latent variables, as well as the meaning of the relationships between them.

6.10.6.6 Other Indexes

In this research, four other indexes have been studied to help to determine if the direction of the relationships is adequate, since it could be that the direction proposed in the model is not appropriate. (Tenenhaus et al. 2005; Kock 2015). In other words, the dependent latent variable is a total dependent latent variable, which is quickly corrected by changing the direction of the relationship, and hence the importance of its analysis. The indexes that have been analyzed are the following:

- Simpson's paradox ratio (SPR),
- *R*-squared contribution ratio (RSCR),
- Statistical suppression ratio (SSR), and
- Nonlinear bivariate causality direction ratio (NLBCDR).

It is recommended that these indexes have values over 0.7, although the ideal values are equal to one, which indicates that 100% of the established relationships between the hypotheses are correct.

6.11 Sensitivity Analysis

The objective of a structural equation model is to determine the dependence between the variables that integrate it. In this book, the β values are used to determine that dependence expressed in standard deviations, and thus, if a β value of 0.36 is found in a model, it indicates that when the independent variable increases its standard deviation in one unit, the second variable increases in 0.36 units. However, this data is often not enough and requires more details to know the level of dependence between the variables.

In this case, the partial least squares technique is implemented to evaluate the proposed models, which use standardized values, and hence the β value is expressed in standard deviations. However, this data conversion facilitates the sensitivity analysis to acknowledge the relationships between the variables, since it is assumed that the data has a normal distribution and probabilities of occurrence where certain events can be estimated.

Moreover, a sensitivity analysis is performed for each of the relationships between the variables in the models, which are represented by the hypotheses. In addition, since the variables are standardized, it is assumed that a value of $Z > 1$ indicates that this variable has a high level of occurrence, while a value of $Z < -1$ indicates that this variable has a low level of occurrence. Therefore, in this research estimations are described for each of the following scenarios:

- The probability that variables are at their high and low scenarios simultaneously ($Z > 1$ and $Z < -1$).
- The probability that both variables (independent and dependent) are at their low level ($Z < -1$) simultaneously.
- The probability that both variables (independent and dependent) are at their high level ($Z > 1$) simultaneously.
- The probability that the independent variable is at its low level ($Z < -1$) while the dependent variable is at its high level ($Z > 1$) simultaneously.
- The probability that the independent variable is at its high level ($Z > 1$) while the dependent variable is at its low level ($Z < -1$).
- The probability that the dependent variable is presented at its low level ($Z < -1$) because the independent variable is at its high level ($Z > 1$).
- The probability that the dependent variable is presented at its high level ($Z > 1$) because the independent variable is at its low level ($Z < -1$).

Analyzing the probabilities of occurrence from the variables' scenarios, the optimistic scenarios where both variables have positive levels were determined, as well as the risks of being at a pessimistic level where both variables are at their low levels were analyzed. In addition, those probabilities of occurrence from the scenarios allow the maintenance manager to take decisions to promote or avoid them.

References

Abdel-Maksoud A, Cerbioni F, Ricceri F, Velayutham S (2010) Employee morale, non-financial performance measures, deployment of innovative managerial practices and shop-floor involvement in italian manufacturing firms. Br Account Rev 42(1):36–55. https://doi.org/10.1016/j.bar.2010.01.002

Adamson KA, Prion S (2013a) Making sense of methods and measurement: measures of central tendency. Clin Simul Nurs 9(12):e617–e618. https://doi.org/10.1016/j.ecns.2013.04.003

Adamson KA, Prion S (2013b) Reliability: measuring internal consistency using cronbach's α. Clin Simul Nurs 9(5):e179–e180. https://doi.org/10.1016/j.ecns.2012.12.001

Adhikary AK (2016) Variance estimation in randomized response surveys (Chap. 12). In: Chaudhuri A, Christofides TC, Rao CR (eds) Handbook of statistics, vol 34. Elsevier, pp 191–208. https://doi.org/10.1016/bs.host.2016.01.010

Alcaraz JLG, Macías AAM, Luevano DJP, Fernández JB, López AdJG, Macías EJ (2016) Main benefits obtained from a successful JIT implementation. Int J Adv Manuf Technol 86(9):2711–2722. https://doi.org/10.1007/s00170-016-8399-5

Alcaraz JLG, Maldonado AA, Iniesta AA, Robles GC, Hernández GA (2014) A systematic review/survey for JIT implementation: Mexican maquiladoras as case study. Comput Ind 65 (4):761–773. https://doi.org/10.1016/j.compind.2014.02.013

Alsyouf I (2009) Maintenance practices in Swedish industries: survey results. Int J Prod Econ 121 (1):212–223. https://doi.org/10.1016/j.ijpe.2009.05.005

Arnab R (2017a) Stratified sampling (Chap. 7). In: Arnab R (ed) Survey sampling theory and applications. Academic Press, pp 213–256. https://doi.org/10.1016/b978-0-12-811848-1.00007-8

Arnab R (2017b) Variance estimation: complex survey designs (Chap. 18). In: Arnab R (ed) Survey sampling theory and applications. Academic Press, pp 587–643. https://doi.org/10.1016/b978-0-12-811848-1.00018-2

Attri R, Grover S, Dev N, Kumar D (2013) Analysis of barriers of total productive maintenance (TPM). Int J System Assur Eng Manag 4(4):365–377. https://doi.org/10.1007/s13198-012-0122-9

Avelar-Sosa L, García-Alcaraz JL, Castrellón-Torres JP (2014) The effects of some risk factors in the supply chains performance: a case of study. J Appl Res Technol 12(5):958–968. https://doi.org/10.1016/S1665-6423(14)70602-9

Avelar-Sosa L, García-Alcaraz JL, Vergara-Villegas OO, Maldonado-Macías AA, Alor-Hernández G (2015) Impact of traditional and international logistic policies in supply chain performance. Int J Adv Manuf Technol 76(5):913–925. https://doi.org/10.1007/s00170-014-6308-3

Bagnaschi E, Cacciari M, Guffanti A, Jenniches L (2015) An extensive survey of the estimation of uncertainties from missing higher orders in perturbative calculations. J High Energy Phys 2015 (2):133. https://doi.org/10.1007/jhep02(2015)133

Bear A, Knobe J (2017) Normality: part descriptive, part prescriptive. Cogn 167:25–37. https://doi.org/10.1016/j.cognition.2016.10.024

Biancolillo A, Næs T, Bro R, Måge I (2017) Extension of SO-PLS to multi-way arrays: SO-N-PLS. Chemometr Intell Lab Syst 164:113–126. https://doi.org/10.1016/j.chemolab.2017.03.002

Biresselioglu ME, Yelkenci T, Ozyorulmaz E, Yumurtaci IÖ (2017) Interpreting Turkish industry's perception on energy security: a national survey. Renew Sustain Energy Rev 67:1208–1224. https://doi.org/10.1016/j.rser.2016.09.093

Blanco-Oliver A, Irimia-Dieguez A, Reguera-Alvarado N (2016) Prediction-oriented PLS path modeling in microfinance research. J Bus Res 69(10):4643–4649. https://doi.org/10.1016/j.jbusres.2016.03.054

Cazan A-M, Indreica SE (2014) Need for cognition and approaches to learning among university students. Procedia Soc Behav Sci 127:134–138. https://doi.org/10.1016/j.sbspro.2014.03.227

Chowdhury RMMI (2016) Religiosity and voluntary simplicity: the mediating role of spiritual well-being. J Bus Ethics. https://doi.org/10.1007/s10551-016-3305-5
Clark-Carter D (2010) Measures of central tendency. In: Peterson P, Baker E, McGaw B (eds) International encyclopedia of education, 3rd edn. Elsevier, Oxford, pp 264–266. https://doi.org/10.1016/b978-0-08-044894-7.01343-9
Cooper CJ, Cooper SP, del Junco DJ, Shipp EM, Whitworth R, Cooper SR (2006) Web-based data collection: detailed methods of a questionnaire and data gathering tool. Epidemiol Perspect Innovations 3(1):1. https://doi.org/10.1186/1742-5573-3-1
Cua KO, McKone KE, Schroeder RG (2001) Relationships between implementation of TQM, JIT, and TPM and manufacturing performance. J Oper Manag 19(6):675–694. https://doi.org/10.1016/S0272-6963(01)00066-3
de Guimarães JCF, Severo EA, Vieira PS (2017) Cleaner production, project management and strategic drivers: an empirical study. J Clean Prod 141:881–890. https://doi.org/10.1016/j.jclepro.2016.09.166
Díaz-Reza J, García-Alcaraz J, Mendoza-Fong J, Martínez-Loya V, Macíaz-Jiménez E, Blanco-Fernández J (2017) Interrelations among SMED stages: a causal model. Complexity 2017:10. https://doi.org/10.1155/2017/5912940
Dray S, Josse J (2015) Principal component analysis with missing values: a comparative survey of methods. Plant Ecol 216(5):657–667. https://doi.org/10.1007/s11258-014-0406-z
Ekrot B, Kock A, Gemünden HG (2016) Retaining project management competence—antecedents and consequences. Int J Proj Manag 34(2):145–157. https://doi.org/10.1016/j.ijproman.2015.10.010
Fakih K, Assaker G, Assaf AG, Hallak R (2016) Does restaurant menu information affect customer attitudes and behavioral intentions? A cross-segment empirical analysis using PLS-SEM. Int J Hosp Manag 57:71–83. https://doi.org/10.1016/j.ijhm.2016.06.002
Farooq MS, Salam M, Fayolle A, Jaafar N, Ayupp K (2018) Impact of service quality on customer satisfaction in Malaysia airlines: a PLS-SEM approach. J Air Transp Manag 67:169–180. https://doi.org/10.1016/j.jairtraman.2017.12.008
García-Alcaraz JL, Adarme-Jaimes W, Blanco-Fernández J (2016) Impact of human resources on wine supply chain flexibility, quality, and economic performance. Ing E Investig 36(3):8. https://doi.org/10.15446/ing.investig.v36n3.56091
García JL, Rivera L, Blanco J, Jiménez E, Martínez E (2014) Structural equations modelling for relational analysis of JIT performance in maquiladora sector. Int J Prod Res 52(17):4931–4949. https://doi.org/10.1080/00207543.2014.885143
Hekimoglu S, Erdogan B, Erenoglu RC (2015) A new outlier detection method considering outliers as model errors. Exp Tech 39(1):57–68. https://doi.org/10.1111/j.1747-1567.2012.00876.x
Hsieh FY, Lavori PW (2000) Sample-size calculations for the cox proportional hazards regression model with nonbinary covariates. Control Clin Trials 21(6):552–560. https://doi.org/10.1016/S0197-2456(00)00104-5
Jin W, Tung AK, Han J, Wang W (2006) Ranking outliers using symmetric neighborhood relationship. In: Advances in knowledge discovery and data mining. Springer, Berlin, pp 577–593
Jönsson K (2011) A robust test for multivariate normality. Econ Lett 113(2):199–201. https://doi.org/10.1016/j.econlet.2011.06.018
Kang SJ, Lee M (2005) Q-convergence with interquartile ranges. J Econ Dyn Control 29(10):1785–1806. https://doi.org/10.1016/j.jedc.2004.10.004
Kaur H, Chaudhary S, Choudhary N, Manuja N, Chaitra TR, Amit SA (2016) Child abuse: cross-sectional survey of general dentists. J Oral Biol Craniofac Res 6(2):118–123. https://doi.org/10.1016/j.jobcr.2015.08.002
Kazár K (2014) PLS path analysis and its application for the examination of the psychological sense of a brand community. Procedia Econ Finance 17:183–191. https://doi.org/10.1016/S2212-5671(14)00893-4

Kim N (2016) A robustified Jarque-Bera test for multivariate normality. Econ Lett 140:48–52. https://doi.org/10.1016/j.econlet.2016.01.007

Kock N (2015) WarpPLS 5.0 User Manual, ScriptWarp Systems. Laredo, TX, USA

Lee C, Hallak R (2017) Investigating the moderating role of education on a structural model of restaurant performance using multi-group PLS-SEM analysis. J Bus Res. https://doi.org/10.1016/j.jbusres.2017.12.004

Liu RX, Kuang J, Gong Q, Hou XL (2003) Principal component regression analysis with spss. Comput Methods Programs Biomed 71(2):141–147. https://doi.org/10.1016/S0169-2607(02)00058-5

Midiala OV, Luis GAJ, Aracely MMA, Valeria ML (2016) The impact of managerial commitment and kaizen benefits on companies. J Manuf Technol Manag 27(5):692–712. https://doi.org/10.1108/JMTM-02-2016-0021

Mihail DM, Kloutsiniotis PV (2016) The effects of high-performance work systems on hospital employee's work-related well-being: evidence from greece. Eur Manag J 34(4):424–438. https://doi.org/10.1016/j.emj.2016.01.005

Nitzl C (2016) The use of partial least squares structural equation modelling (PLS-SEM) in management accounting research: directions for future theory development. J Account Lit 37:19–35. https://doi.org/10.1016/j.acclit.2016.09.003

Ompad DC, Galea S, Marshall G, Fuller CM, Weiss L, Beard JR, Chan C, Edwards V, Vlahov D (2008) Sampling and recruitment in multilevel studies among marginalized urban populations: the IMPACT studies. J Urban Health 85(2):268. https://doi.org/10.1007/s11524-008-9256-0

Pastor E, Soliveres S, Vilagrosa A, Bonet A (2018) Intraspecific leaf shape at local scale determines offspring characteristics. J Arid Environ. https://doi.org/10.1016/j.jaridenv.2017.12.013

Preacher KJ, Hayes AF (2004) SPSS and SAS procedures for estimating indirect effects in simple mediation models. Behav Res Methods Instrum Comput 36(4):717–731. https://doi.org/10.3758/bf03206553

Raïssi H (2017) Testing normality for unconditionally heteroscedastic macroeconomic variables. Econ Model. https://doi.org/10.1016/j.econmod.2017.10.015

Rasoolimanesh SM, Jaafar M, Kock N, Ramayah T (2015) A revised framework of social exchange theory to investigate the factors influencing residents' perceptions. Tour Manag Perspect 16:335–345. https://doi.org/10.1016/j.tmp.2015.10.001

Ratzmann M, Gudergan SP, Bouncken R (2016) Capturing heterogeneity and PLS-SEM prediction ability: alliance governance and innovation. J Bus Res 69(10):4593–4603. https://doi.org/10.1016/j.jbusres.2016.03.051

Realyvásquez A, Maldonado-Macías AA, García-Alcaraz J, Cortés-Robles G, Blanco-Fernández J (2016) Structural model for the effects of environmental elements on the psychological characteristics and performance of the employees of manufacturing systems. Int J Environ Res Public Health 13(1):104. https://doi.org/10.3390/ijerph13010104

Reguera-Alvarado N, Blanco-Oliver A, Martín-Ruiz D (2016) Testing the predictive power of PLS through cross-validation in banking. J Bus Res 69(10):4685–4693. https://doi.org/10.1016/j.jbusres.2016.04.016

Richter NF, Cepeda G, Roldán JL, Ringle CM (2016) European management research using partial least squares structural equation modeling (PLS-SEM). Eur Manag J 34(6):589–597. https://doi.org/10.1016/j.emj.2016.08.001

Roni SM, Djajadikerta H, Ahmad MAN (2015) PLS-SEM approach to second-order factor of deviant behaviour: constructing perceived behavioural control. Procedia Econ Finance 28:249–253. https://doi.org/10.1016/S2212-5671(15)01107-7

Sarstedt M, Ringle CM, Smith D, Reams R, Hair JF (2014) Partial least squares structural equation modeling (PLS-SEM): a useful tool for family business researchers. J Fam Bus Strat 5(1):105–115. https://doi.org/10.1016/j.jfbs.2014.01.002

Schlittgen R, Ringle CM, Sarstedt M, Becker J-M (2016) Segmentation of PLS path models by iterative reweighted regressions. J Bus Res 69(10):4583–4592. https://doi.org/10.1016/j.jbusres.2016.04.009

Schubert E, Zimek A, Kriegel H-P (2014) Local outlier detection reconsidered: a generalized view on locality with applications to spatial, video, and network outlier detection. Data Min Knowl Disc 28(1):190–237. https://doi.org/10.1007/s10618-012-0300-z

Schubring S, Lorscheid I, Meyer M, Ringle CM (2016) The PLS agent: predictive modeling with PLS-SEM and agent-based simulation. J Bus Res 69(10):4604–4612. https://doi.org/10.1016/j.jbusres.2016.03.052

Serrano Sanguilinda I, Barbiano di Belgiojoso E, González Ferrer A, Rimoldi SML, Blangiardo GC (2017) Surveying immigrants in southern europe: Spanish and Italian strategies in comparative perspective. Comp Migr Stud 5(1):17. https://doi.org/10.1186/s40878-017-0060-4

Shalit H (2012) Using OLS to test for normality. Stat Probab Lett 82(11):2050–2058. https://doi.org/10.1016/j.spl.2012.07.004

Singh S, Sedory SA, del Mar Rueda M, Arcos A, Arnab R (2016) Tuning in stratified sampling (Chap. 8). In: Singh S, Sedory SA, del Mar Rueda M, Arcos A, Arnab R (eds) A new concept for tuning design weights in survey sampling. Academic Press, pp 219–256. https://doi.org/10.1016/b978-0-08-100594-1.00008-5

Swafford PM, Ghosh S, Murthy N (2006) The antecedents of supply chain agility of a firm: scale development and model testing. J Oper Manag 24(2):170–188. https://doi.org/10.1016/j.jom.2005.05.002

Tastle WJ, Wierman MJ (2007) Consensus and dissention: a measure of ordinal dispersion. Int J Approx Reason 45(3):531–545. https://doi.org/10.1016/j.ijar.2006.06.024

Tenenhaus M, Vinzi VE, Chatelin Y-M, Lauro C (2005) PLS path modeling. Comput Stat Data Anal 48(1):159–205. https://doi.org/10.1016/j.csda.2004.03.005

Toutenburg H, Srivastava VK (1998) Estimation of ratio of population means in survey sampling when some observations are missing. Metrika 48(3):177–187. https://doi.org/10.1007/pl00003973

Vu DH, Muttaqi KM, Agalgaonkar AP (2015) A variance inflation factor and backward elimination based robust regression model for forecasting monthly electricity demand using climatic variables. Appl Energy 140:385–394. https://doi.org/10.1016/j.apenergy.2014.12.011

Wang E (2017) Decomposing core energy factor structure of U.S. residential buildings through principal component analysis with variable clustering on high-dimensional mixed data. Appl Energy 203:858–873. https://doi.org/10.1016/j.apenergy.2017.06.105

Wu G, Pawlikowska I, Gruber T, Downing J, Zhang J, Pounds S (2013) Subgroup and outlier detection analysis. BMC Bioinform 14(17):A2. https://doi.org/10.1186/1471-2105-14-s17-a2

Yaşlıoğlu MM, Şap Ö, Toplu D (2014) An investigation of the characteristics of learning organizations in turkish companies: scale validation. Procedia - Soc Behav Sci 150:726–734. https://doi.org/10.1016/j.sbspro.2014.09.037

Part IV
Validation and Analysis of Data

Chapter 7
Validation of Variables

Abstract This chapter deals with the validation process from the latent variables that integrate the structural equation modeling. A total of eight complex models are generated, involving nine variables associated with critical factors of total productive maintenance (TPM) and three latent variables related to obtained benefits. Also, the validity indexes for each final model are reported, where the observed variables that are part of each latent variable are presented, since many were not considered in order to increase the reliability.

7.1 Introduction

In order to start describing and analyzing the gathered data from the questionnaires, it is convenient to carry out their validation process. For example, Annex 1 illustrates the questionnaire used to obtain information from companies, which shows nine latent variables associated with the critical factors of total productive maintenance (TPM) and three latent variables associated with obtained benefits from its proper implementation. However, it is possible that in the first latent variable associated with Labor Culture, which initially has 13 items or observed variables in the questionnaire, during the validation process, they could be reduced to only 6 because there are discrepancies in some of them and they are eliminated.

Thus, the objective of this chapter is to explain the validation that is carried out with the nine latent variables to identify which of the items are those that remain for further analysis. Additionally, in case that some variable is reduced to a smaller number of items, it will be indicated in each of them and those items are those implemented in the generation of eight complex models that are distributed in other chapters. The latent variables have been classified into three categories to integrate them in the models, as illustrated below:

- Human factor

– *Work culture,*
– *Suppliers,*

J. R. Díaz-Reza et al., *Impact Analysis of Total Productive Maintenance*,
https://doi.org/10.1007/978-3-030-01725-5_7

- *Management commitment, and*
- *Customers.*

- Operational factor

- *Implementation of PM,*
- *Implementation of TPM,*
- *Technological status,*
- *Layout, and*
- *Warehouse management.*

- Benefits

- *Fort he organization,*
- *Safety, and*
- *Productive.*

It can be said that this chapter is divided into two large sections; first the validation indexes from the variables in each of the models are indicated; they are discussed and then, for each of the latent variables, are analyzed, and the items or observed variables that compose them and that have passed the validation process are listed.

Also, it is important to mention that the variations are integrated by each of the structural equation modelings and it is possible that a latent variable occurs in more than one model; therefore, the associated data with its convergent validity will often be repeated, the linearity and internal validity indexes; however, the predictive validation indexes are different in each model, since the latent dependent variables are explained by different independent latent variables.

7.2 Validation of Variables in Models

As it was previously mentioned, there are a total of eight complex models involving nine latent variables associated with the critical success factors of total productive maintenance and three latent variables associated with the obtained benefits from its implementation; for that reason, each model is described and the validity indexes of the variables that intervene in them are illustrated as well.

In addition, for readers who are interested in identifying the validation indexes and their meaning, it is recommended to read Chap. 8 again, where they are described. In this section, only the results are interpreted without going deeply into the concepts. Also, the indexes that are used to validate each of the latent variables are those indicated in Table 7.1, where the type of validity that allows to demonstrate is expressed.

Table 7.1 Validation indexes and type of validity

Index	Type of validity
R-squared coefficients	Parametric predictive validity
Adjusted *R*-squared coefficients	
Q-squared coefficients	Predictive nonparametric validity
Composite reliability coefficients	Internal validity
Cronbach's alpha coefficients	
Average extracted variances	Convergent validity
Full collinearity VIFs	Collinearity

7.2.1 Model One

In this model, there are four latent variables that refer to the following:

- *Work culture,*
- *Suppliers,*
- *Management commitment, and*
- *Customers.*

Table 7.2 presents the validation indexes obtained for each of the latent variables; according to this, the following can be concluded:

- The dependent latent variables have enough parametric and nonparametric predictive validity according to the values of square *R*, adjusted square *R*, and square *Q*.
- There is enough internal validity according to Cronbach's alpha indexes and compound reliability.
- There are no collinearity problems from the inside of the latent variables.
- There is an adequate variance extraction in the variables, although *Work culture* has only 0.479 and it has been decided that it will remain in the data.

Table 7.2 Validity indexes from model one

Coefficient	*Work culture*	*Suppliers*	*Management commitment*	*Customers*
R-squared coefficients	0.475	0.171	0.468	
Adjusted *R*-squared coefficients	0.471	0.168	0.465	
Composite reliability coefficients	0.865	0.918	0.942	0.816
Cronbach's alpha coefficients	0.817	0.897	0.926	0.699
Average variances extracted	0.479	0.583	0.729	0.527
Full collinearity VIFs	1.886	1.884	2.058	1.44
Q-squared coefficients	0.479	0.173	0.469	

7.2.2 Model Two

This model is also integrated into four latent variables, which are illustrated below.

- *Work culture,*
- *Management commitment,*
- *Implementation of TPM, and*
- *Technological status.*

Table 7.3 illustrates the reliability indexes for the variables on this model and as a result, the following can be stated:

- There is a parametric and nonparametric predictive validity in latent dependent variables.
- There is validity content according to the Cronbach's alpha index and compound reliability index.
- There are no collinearity issues and there is enough convergent validity according to the average variance extracted, although, *Work culture* is related in this model and is barely less than the recommended value.

7.2.3 Model Three

This model is integrated into four latent variables, which is the first to include benefits. The latent variables that incorporated it are the following:

- *Benefits for the company,*
- *Suppliers,*

Table 7.3 Validity indexes from model two

Coefficient	*Work culture*	*Management commitment*	*Implementation of TPM*	*Technological status*
R-squared coefficients	0.367		0.698	0.565
Adjusted *R*-squared coefficients	0.366		0.696	0.562
Composite reliability coefficients	0.865	0.942	0.944	0.946
Cronbach's alpha coefficients	0.817	0.926	0.935	0.935
Average variances extracted	0.479	0.729	0.583	0.688
Full collinearity VIFs	1.81	2.847	3.245	2.554
Q-squared coefficients	0.368		0.699	0.566

Table 7.4 Validity indexes from model three

Coefficient	*Benefits for the company*	*Suppliers*	*Management commitment*	*Productive benefits*
R-squared coefficients	0.394	0.350		0.660
Adjusted *R*-squared coefficients	0.391	0.348		0.657
Composite reliability coefficients	0.950	0.918	0.942	0.950
Cronbach's alpha coefficients	0.939	0.897	0.926	0.939
Average variances extracted	0.702	0.583	0.729	0.732
Full collinearity VIFs	3.029	1.658	1.819	2.892
Q-squared coefficients	0.395	0.350		0.659

- *Management commitment, and*
- *Productive benefits.*

Table 7.4 illustrates the validation indexes from model three; based on it, it is concluded that the required parametric and nonparametric predictive validity values are achieved according to square *R*, adjusted square *R*, and squared *Q*. Likewise, it is observed that there is an adequate convergent validity according to the extracted variance, there is content validity, and there are no collinearity problems, although it should be mentioned that on this occasion the inflation variance indexes in the *benefits for the company* are over 3.

7.2.4 Model Four

This model is integrated into four latent variables, where two benefits are already included. The variables are the following:

- *Benefits for the company,*
- *Suppliers,*
- *Layout, and*
- *Security benefits.*

Table 7.5 shows the validity indexes from the latent variables in this model; it can be said that there is a parametric and nonparametric predictive validity in those that are dependent; all variables have content and convergent validity, and there are no collinearity problems. However, it is worth mentioning that the *safety benefits and benefits for the company* have values higher than 3, and 3.3 was the maximum value established.

Table 7.5 Validity indexes from model four

Coefficient	*Benefits for the company*	*Suppliers*	*Layout benefits*	*Safety*
R-squared coefficients	0.347		0.312	0.677
Adjusted *R*-squared coefficients	0.344		0.31	0.675
Composite reliability coefficients	0.950	0.918	0.907	0.945
Cronbach's alpha coefficients	0.939	0.897	0.872	0.93
Average variances extracted	0.702	0.583	0.662	0.74
Full collinearity VIFs	3.201	1.626	1.614	3.051
Q-squared coefficients	0.349		0.313	0.677

7.2.5 Model Five

This model is also integrated into four latent variables, where those associated with productive processes are included. The variables are the following:

- Implemetation of PM,
- Implementation of TPM,
- Technological Status, and
- Layout.

Table 7.6 illustrates the validity indexes of the variables that are into the model, where it can be observed that parametric and nonparametric predictive validity is accomplished, with the content and convergent. In the same way, according to the inflation rates of the variance, there are no collinearity problems and the model can be properly analyzed.

Table 7.6 Validity indexes from model five

Coefficient	*Implementation of PM*	*Implementation of TPM*	*Technological status*	*Layout*
R-squared coefficients	0.423	0.659		0.395
Adjusted R-squared coefficients	0.42	0.657		0.393
Composite reliability coefficients	0.902	0.944	0.946	0.907
Cronbach's alpha coefficients	0.873	0.935	0.935	0.872
Average variances extracted	0.569	0.583	0.688	0.662
Full collinearity VIFs	2.062	2.933	2.538	1.837
Q-squared coefficients	0.425	0.661		0.396

7.2.6 *Model Six*

Similarly, this model is integrated into four latent variables, where aspects related to warehouse administration, implementation of preventive maintenance, and two benefits are included. The variables are the following:

- *Benefits for the company,*
- *Warehouse management,*
- *Implementation of PM, and*
- *Productive benefits.*

Table 7.7 presents the validity indexes obtained from the latent variables in this model. It is seen that the requirements of parametric and nonparametric predictive validity are met according to the values of square *R*, adjusted square *R*, and square *Q*. Similarly, it is observed that there are no content validity problems, since the compound reliability index and the Cronbach's alpha index are over 0.7. Also, it is observed that there are no convergent validity or collinearity problems within the latent variables.

7.2.7 *Model Seven*

This model is integrated into four latent variables, where two obtained benefits from the implementation of TPM are presented as well as two that are associated with technical and operational aspects. The latent variables are the following:

Table 7.7 Validity indexes from model six

Coefficient	*Benefits for the company*	*Warehouse management*	*Implementation of PM*	*Productive benefits*
R-squared coefficients	0.358	0.336		0.631
Adjusted *R*-squared coefficients	0.355	0.334		0.628
Composite reliability coefficients	0.950	0.919	0.902	0.956
Cronbach's alpha coefficients	0.939	0.894	0.873	0.948
Average variances extracted	0.702	0.653	0.569	0.732
Full collinearity VIFs	3.148	1.689	1.588	2.921
Q-squared coefficients	0.359	0.336		0.665

Table 7.8 Validity indexes from model seven

Coefficient	*Benefits for the company*	*Implementation of TPM*	*Technological status*	*Safety benefits*
R-squared coefficients	0.363	0.548		0.681
Adjusted R-squared coefficients	0.359	0.547		0.678
Composite reliability coefficients	0.950	0.944	0.946	0.945
Cronbach's alpha coefficients	0.939	0.935	0.935	0.93
Average variances extracted	0.702	0.583	0.688	0.74
Full collinearity VIFs	3.151	2.437	2.318	3.094
Q-squared coefficients	0.364	0.549		0.68

- *Benefits for the company,*
- *Implementation of TPM,*
- *Technological status, and*
- *Safety benefits.*

The outcomes from the validation of the variables in this model are described in Table 7.8, where it can be verified that the predictive validity requirements about convergent validity and content validity are achieved. Likewise, there are no collinearity problems in the latent variables; therefore, they can be analyzed in the model.

7.2.8 Model Eight

The current model is special, because it is a second-order model, where the critical success factors of TPM are integrated into two latent variables and the three benefits are included into a single variable. Thus, the latent variables of this model are the following:

- Human factor

– *Work culture,*
– *Suppliers,*
– *Management commitment, and*
– *Customers.*

Table 7.9 Validity indexes from model eight

Coefficient	*Human factor*	*Operational factor*	*Benefits*
R-squared coefficients		0.756	0.467
Adjusted R-squared coefficients		0.755	0.464
Composite reliability coefficients	0.877	0.924	0.956
Cronbach's alpha coefficients	0.812	0.897	0.931
Average variances extracted	0.643	0.709	0.878
Full collinearity VIFs	4.335	4.217	1.815
Q-squared coefficients		0.756	0.469

- Operational factor

– *Implantation of PM,*
– *Implementation of TPM,*
– *Technological status,*
– *Layout, and*
– *Warehouse management.*

- Benefits

– *Organizational,*
– *Safety, and*
– *Productive.*

Table 7.9 illustrates the efficiency indexes from the second-order model, where it can be seen that there is enough predictive validity in the dependent latent variables, because the values of square R, adjusted squared R, and square Q are greater than 0.02. Similarly, it is concluded that there is sufficient content validity; since the Cronbach's alpha indexes and compound reliability are over 0.7, there are no collinearity problems due to the fact that the VIFs are under 5, and finally, there is enough validity convergent according to the average variance extracted. Therefore, it is concluded that the variables particularly have the required validity and it can be proceeded with the analysis of the model.

7.3 Observed Variables in Latent Variables

As mentioned above, during the validation process of latent variables, it is often necessary to eliminate observed variables that are redundant or cause some type of conflict linked to the collinearity or any of the validity indexes. Also, Table 7.10 shows a summary of the items that initially had the latent variable in the questionnaire before the validation, as well as the items that remain in the variable after the validation process and when the number variables were eliminated.

Table 7.10 Summary of items in the latent variables

Latent variable	Questionnaire items	Validated items	Eliminated items
Work culture	13	7	6
Suppliers	8	8	0
Management commitment	6	6	0
Customers	4	4	0
Implementation of PM	9	7	2
Implementation of TPM	16	12	4
Layout	5	5	0
Warehouse management	13	6	7
Organizational benefits	8	8	0
Safety benefits	6	6	0
Productive benefits	8	8	0

In addition, it is observed that in four latent variables, items or variables observed have been eliminated in order to improve their validity indexes. To sum up, a total of 19 items were removed.

The following list or observed variables are those that integrate each of the latent variables after the validation process and they are those that are included into the structural equation modeling.

Work Culture

- Tools and accessories organization and placement in designated places,
- Clean and organized work áreas,
- Departments and company cleaning,
- Training for employees to carry out multiple activities,
- Registration in work logs and maintenance for each machine,
- Coordination of the maintenance schedule with the production department, and
- Order and cleanliness promotion by senior management.

Suppliers

- Providing maintenance manuals for sold machinery and equipment,
- Provide technical advice on the sold,
- Sign contracts with the buyer about the equipment maintenance,
- Establishment of the conditions in the guarantees for the sold equipment,
- Installation and start-up from the acquired equipment,
- Training programs between the supplier and the buyer,
- Provide training at the supplier's facilities, and
- Provide training in the buyer's facilities.

Implementation of PM

- Maintenance as a strategy to achieve the quality and programming of activities,
- Maintenance department helps operators to perform preventive maintenance in their machines,
- Operators are informed about the maintenance performed,
- Machine maintenance statistics publication,
- Simple access to historical maintenance and productivity information,
- Quality control in the products generated by each machine, and
- Machinery malfunction identification and recording.

Implementation of TPM

- Training and education for the maintenance personnel,
- Monitoring and control on the maintenance program,
- Senior and maintenance staff commitment with the machine functionality,
- Management leadership when executing TPM programs,
- Leadership by production and engineering when executing TPM programs,
- Leadership by maintenance managers when executing TPM programs,
- Communication between production and maintenance on equipment availability and maintenance performance,
- Knowledge about the critical systems of failure in the machines,
- Maintenance programs focused on the machinery systems and components useful life,
- Informative meetings by those responsible for maintenance,
- Investment in updated tools, and
- Contemplate maintenance as purchase criteria.

Technological Status

- Advanced technology,
- Update and focus on maintaining new generations of technology,
- Efficient use of new technologies and modern maintenance systems,
- Learning and updating equipment after installation,
- Improvement of existing machinery and equipment,
- Equipment development and patents registration,
- Generating competitive advantages for machinery and equipment, and
- Specialized software usage for maintenance processes.

Layout

- Proximity between processes and machinery,
- Machinery grouping according to the family of produced products,
- Plant distribution to facilitate maintenance,
- Equipment and heavy machinery movement within the facilities for maintenance,
- Adequate signs in the plant related to maintenance?

- Appropriate signs.

Warehouse Management

- Warehouse areas cleaning,
- Organized and appropriate space for parts and components within the warehouse,
- Daily inspections to avoid missing,
- Appropriate material and parts coding,
- Arrangement and adaptation of spaces for material or parts with special needs (heating, cooling, humidity, etc.), and
- Registration of changes in the components and tools location.

Management Commitment

- The TPM responsibility from heads department,
- Leadership attitudes in the TPM performance by senior managers,
- Meetings between maintenance and production,
- Promotion of workers participation in maintenance activities by managers,
- Management focuses on quality and maintenance, and
- Management involvement in maintenance projects.

Customers

- Consider the customer's opinion on maintenance issues,
- Avoid noncompliance with customers due to lack of maintenance,
- Delaying maintenance activities to be satisfied with production, and
- Costs for noncompliance of orders because of poor maintenance.

Benefits for the Company

- Management involvement in maintenance projects;
- Better operations control;
- Increases employee morale;
- Creates a responsibility culture, discipline, and respect toward standards;
- Permanent learning;
- Creates an environment where participation, collaboration, and creativity are a reality;
- Appropriate staff schedule sizing; and
- Efficient communication networks.

Productive Benefits

- Eliminate losses that affect plants productivity,
- Improve equipment reliability and availability,
- Maintenance costs reduction,
- Improve the final product quality,

- Low financial costs for spare parts,
- Improve company's technology,
- Increase responsiveness to market movements, and
- Create competitive capabilities from the factory.

Safety Benefits

- Improve environmental conditions,
- Preventive culture of negative events for health,
- Increase in the capacity to identify potential problems and search for corrective actions,
- Acknowledge why there are certain rules, rather than how to perform a task,
- Prevention and elimination of potential accidents causes, and
- Radically eliminate pollution sources.

7.4 Conclusions

In this chapter, a total of twelve latent variables have been validated, nine associated with the critical success factors of TPM and three linked to the obtained benefits from an adequate implementation. In general, the following can be concluded:

- A total of 19 observed variables in four items were eliminated with the aim of improving the validity indexes in the latent variables.
- All latent variables have required validity indexes, except for the *Work culture* that does not get a convergent validity, although it was decided to integrate it into the analysis due to its importance in the implementation process of TPM.
- Only two latent variables have values of variance inflation rates over 3.3, recommended value; however, it can be used to a maximum of 5, also they are related with the second-order model and it refers to the Human and Operational Factor variables.
- In general, it is concluded that latent variables can be integrated into the model.

Chapter 8
Descriptive Analysis

Abstract This chapter presents the descriptive analysis of the data obtained from the applied questionnaires in this research, which is developed in two sections: the first part refers to the descriptive analysis of the sample with demographic data. On the other hand, the second section shows the items in each of the latent variables, where the median is used as a measure of central tendency and the interquartile lag as a measure of dispersion. In addition, managers are able to identify the most remarkably observed variables from a univariate point of view using this type of data.

8.1 Descriptive Analysis of the Sample

From January to June 2017, the survey was applied to the manufacturing industry in Mexico, and a total of 397 cases were collected, where only 368 were valid, since 31 were canceled because they had over 10% of missing data. In addition, analyzes that were previously carried out are counted with the obtained data as well.

8.1.1 Gender and Industrial Sector

Table 8.1 illustrates the gender and industrial sector where participants belong. Also, according to the gathered data, it is observed that the most surveyed industrial sector is the automotive industry with 167 participants out of 331 in both demographic aspects (37 participants did not answer any of the items analyzed in the table). Moreover, another relevant sector is electronics with 54 participants and the electric sector with 45. However, it is important to notice that the aeronautical sector, which is new to Mexico, starts to have a representation with five respondents who participated in this study.

Furthermore, it is also observed that 77 of the participants who answered both questionnaires were women and 254 men who are in all the industrial sectors. Also, the previous statement is interesting, because there are more and more women

J. R. Díaz-Reza et al., *Impact Analysis of Total Productive Maintenance*,
https://doi.org/10.1007/978-3-030-01725-5_8

Table 8.1 Descriptive analysis: gender and industrial sector

Industrial sector	Female	Male	Total
Automotive	42	125	167
Electronic	8	46	54
Electric	9	36	45
Other	12	23	35
Medical	5	20	25
Aeronautic	1	4	5
Total	77	254	331

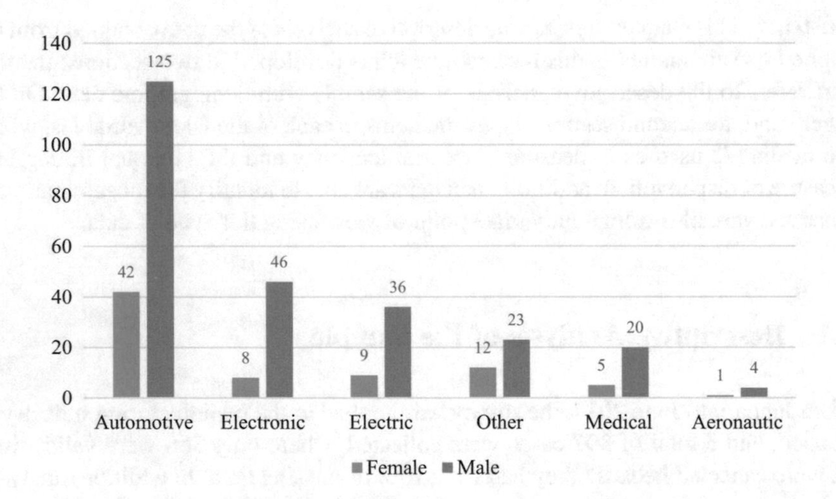

Fig. 8.1 Participants industrial sector and gender

taking high positions in the industry sector, and as it is mentioned by Hadjimarcou et al. (2013), the incorporation of more women in productive processes in Mexico is just beginning, since countries like China have a long tradition in this. However, this phenomenon is recent, because in the past decade it seemed that the maquiladoras were a place for men development only, such as Wilson (2002) states.

Figure 8.1 illustrates a graphic distribution of the participants industrial sectors and gender.

8.1.2 Participants Jerarquical Job Position and Work Experience

In Table 8.2, two new demographic variables of the sample that was surveyed are represented, which are arranged in descending order according to the total row. Also, it is observed that 339 participants answered the 2 questions that were asked,

Table 8.2 Job position and work experience

Position	Year of experience					Total
	>1 and <2	>2 and <3	>3 and <5	>5 and <10	>10	
Technician	45	27	26	35	13	146
Operator	33	18	19	9	3	82
Engineer	8	10	19	15	3	55
Supervisor	6	12	11	13	6	48
Manager	0	1	1	1	2	5
Other	1	1	1	0	0	3
Total	93	69	77	73	27	339

and on this opportunity, technicians are the group who answered most of the survey as well as the operators, which indicates that the data was provided by people who are directly involved with maintenance tasks.

Likewise, it is observed that only five maintenance managers have participated in this investigation, which indicates that a significant percentage of participants had operational positions and that they were the ones who knew the maintenance procedures applied to the company's equipment, therefore, the administrative aspects are a convenient area for further research.

Moreover, regarding participants years of experience, it is noticeable that only 27 people have more than 10 years of experience in their position, while in the other categories a distribution is observed. Also, for readers who are interested in knowing the impact of workers experience and educational level in the maquiladora industry, it is recommended to read the research by Utar and Ruiz (2013). Also, a principle of inclusion in the sample was that the members had at least 1 year of experience in their position, thus, it can be argued that the data is from a reliable source.

8.2 Descriptive Analysis: Items in Success Factors of TPM

In the previous chapter, a validation process from the latent variables has been carried out, which are integrated into eight models that will be described in further chapters, and the items that comprise them as well. Also, in this section, a descriptive analysis is done in the final items that remain in these latent variables after the validation process, where the measures of central tendency represented by the median are reported, besides the interquartile range (IR) is also illustrated as a measure of their dispersion. In addition, for a better analysis, each of the latent variables is separately described.

Table 8.3 Descriptive analysis: work culture

Item	Median	IR
Tools and accessories organization and placement in designated places	4.27	1.59
Registration in work logs and maintenance for each machine	4.09	1.68
Coordination of the maintenance schedule with the production department	4.04	1.59
Order and cleanliness promotion by senior management	4.00	1.54
Clean and organized work areas	3.90	1.52
Training for employees to carry out multiple activities	3.89	1.67
Departments and company cleaning	3.78	1.47

Moreover, it is relevant to mention that the items are descending according to the median in each of the tables, which ease the discussion and analysis from the importance levels that they have within the latent variant.

8.2.1 Work Culture

Table 8.3 portrayed the central tendency measures that items have, which integrate the *Work culture* latent variable with seven items. Additionally, *Work culture* has been investigated as a determining success factor of TPM in industries in Nigeria by Eti et al. (2006a) it is considered as an income of economic savings when it is properly transmitted (Eti et al. 2006b). Also, this *Work culture* must be associated with other philosophies applied to production systems such as Six Sigma in order to improve the company's performance (Kumar Sharma and Gopal Sharma 2014).

Furthermore, in general aspects, it is observed that four out of the items have a median over four and three; they have a lower value. In addition, it can be concluded that what is the most important regarding to habits and *Work culture* is that an emphasis has to be done to decrease the usage of tools and accessories to perform the maintenance, because it is usually achieved with the implementation of other techniques that are applied to the production systems related to 5S, since it helps to keep workplaces clean (Jugraj Singh and Inderpreet Singh 2017; Kobayashi et al. 2008). Similarly, it is relevant to indicate that a work maintenance log must be performed on each machine, which will allow to have a historical archive to reach when malfunction sources are required.

Similarly, it is observed that the least performed task is related to clean departments keeping, which is connected to the item in the first place; in other words, it seems that if there is an order and cleanliness culture in the work area, but it is not always performed, therefore, it is an observation area, which is also reported by Park et al. (2012). In addition, the last section is confirmed by the fifth item indicating that all work areas are clean and in order, which has the minimum of the IR and it indicates a greater consensus from participants. In summary, it is detected that respondents consider that the general order in the departments is less important for the TPM than the specific workplaces where they usually are.

8.2.2 *Suppliers*

The suppliers' latent variable is integrated into eight items, their descriptive analysis is illustrated in Table 8.4. Also, it is observed that only two out of the eight items have a median over four, therefore, they are considered highly important, thus, the other six items are regularly important as critical success factors for TPM according to the scale of assessment used.

Moreover, based on the information in Table 8.4, it is discovered that the most relevant aspect for the participants in order to be successful in TPM programs regarding suppliers, it is that they provide manufacturing guarantees for industrial equipment and machine manufacturers who are in the supply chains. In addition, it is recommended to have service contracts with those suppliers, where they arrange maintenance dates that will be carried out, also the education that will be given to employees in the maintenance department, which is agreed with Willmott and McCarthy (2001), that is the structural architecture basics of preserved programs (Kinney 2006). Similarly, it is considered an important aspect that equipment operating manuals be easily handed, since, as it is indicated in the third item, they are critical for technical advices associated with machine usage and handling, as well as the stop of maintenance. Also, it is essential to recall that in general, equipment and machine manufacturers suppliers proposed this advice in the operation initial stage, however, that task is assigned later to the maintenance department, therefore, these manuals are crucial when there is no longer the suppliers support. In addition, it is recommended to have an electronically archived and back up from manuals and maintenance guides, since printed versions are easy to fade away when they are being used by operators.

Table 8.4 Descriptive analysis: suppliers

Item	Median	IR
Establishment of the conditions in the guarantees for the sold equipment	4.08	1.58
Providing maintenance manuals for sold machinery and equipment	4.00	1.64
Provide technical advice on the sold equipment	3.91	1.69
Sign contracts with the buyer about the equipment maintenance	3.79	1.71
Installation and start-up from the acquired equipment	3.72	1.92
Provide training in the buyer's facilities	3.60	1.84
Training programs between the supplier and the buyer	3.58	1.86
Provide training at the supplier's facilities	3.56	1.90

Additionally, it is significant to mention that those two items in the first place are those that have the lowest interquartile ranges, so it is concluded that they are the ones with the greatest consensus among participants; in other words, it is argued that these two items have the true value that has been assigned to them, and that they are necessary for TPM success.

Furthermore, regarding the items with the lowest values in the median, it is noticeable that they are referring to a lack of mutual training programs linked to maintenance processes between the suppliers and the industrial equipment purchasers. Also, they are provided in the company facilities itself when the machinery is already operating in the supply chain and facing environmental problems, since currently, one of the most frequent issues is the connection between different equipment that are or have different manufacturers. In addition, training and education have been found to be one of the critical success factors for TPM and TQM, because they must be properly linked, according to Seth and Tripathi (2006), also, this lack of training has been identified by Nazeri and Naderikia (2017) as a risk factor for maintenance operators, because they will not know how to perform a task, as a consequence, they could have an accident working.

8.2.3 Preventive Maintenance Implementation

This present latent variable is integrated by seven items and their descriptive analysis is portrayed in Table 8.5, where none of them has a median over four and it is observed that all items have medians under four, which indicates that they are more than significant. First of all, the factor associated with maintenance is found as a quality strategy, because a good machine calibration will provide products with the demanded standards. The Second place refers to the monitoring of products quality after a maintenance process, and this is one of the best ways to discover their efficiency, in other words, through machine outcomes and efficiency parameters. Fortunately, there are some studies related to these two variables: For

Table 8.5 Descriptive analysis: preventive maintenance implementation

Item	Median	IR
Maintenance as a strategy to achieve the quality and programming of activities	3.84	1.80
Quality control in the products generated by each machine	3.77	1.89
Machinery malfunction identification and recording	3.74	1.94
Operators are informed about the maintenance performed	3.40	2.13
Maintenance department helps operators to perform preventive maintenance in their machines	3.35	2.18
Machine maintenance statistics publication	3.29	2.36
Simple access to historical maintenance and productivity information	3.17	2.15

instance, Seth and Tripathi (2006) associate India and Kamath industrial quality and maintenance, and Kamath and Rodrigues (2016) integrates them in a research to analyze the relationship of these variables in a company performance.

Furthermore, while reviewing the items with lower values in the median in this category, it is detected that the opportunity areas are related to the historical data about productivity and information access by those who perform the maintenance in the company. Also, the previous statements are verified with the item that is in the penultimate place; it refers to published information linked with each maintenance machine, which must be in front of each other, although the reliability and veracity have to be assessed before publishing it (Hipkin 1996). In other words, these activities are not performed daily in the maquiladora industry, and operators are not aware when the next maintenance is scheduled, although it is possible that managers know this information, it is not shared with machines and equipment employees.

Moreover, it is advisable to keep a log with dates and tasks performed on the equipment, which must be found in a visible place, in order to operators letting the maintenance department know about inspections dates, in this case, decisions are made together regarding possible deviations. However, many computers can store specific information for their operational conditions, which are obtained through sensors that must be properly calibrated to avoid false alarms. In addition, readers who are interested in reviewing the aspects associated with improving the quality of information for online maintenance, consult Arnaiz et al. (2013).

Additionally, it is relevant to mention that the item with the highest median is also the one with the lowest interquartile range, which points out that there is a lot of consensus among participants saying that maintenance is a strategic activity to guarantee the quality of their products.

8.2.4 Total Productive Maintenance Implementation

Since this current latent variable is the most essential and the main topic of this book: it is integrated into 12 observed items or variables and their descriptive analysis, which is shown in Table 8.6. In general terms, it is portrayed that none of the tasks have ratings over four, however, none of them have ratings under three, which indicates that for participants these activities are more than regularly important.

First, it is detected that communication between departments involved with maintenance must carry out an adequate task planning, because it is essential to know the production commitment plans and life components. In addition, the lack of such communication in planning may be for idle time for machines and equipment operators, therefore, it should be intended that information flows through the appropriate channels (Kock et al. 2009). Additionally, this coordination must be directed in an useful manner by a leader, who is generally the maintenance manager, who must be in direct contact with his counterparts in the production area,

Table 8.6 Descriptive analysis: total productive maintenance implementation

Item	Median	IR
Communication between production and maintenance on equipment availability and maintenance performance	3.89	1.63
Leadership by maintenance managers when executing TPM programs	3.83	1.65
Leadership by production and engineering when executing TPM programs	3.81	1.73
Monitoring and control on the maintenance program	3.78	1.65
Senior and maintenance staff commitment with the machine functionality	3.78	1.72
Maintenance programs focused on the machinery systems and components useful life	3.76	1.76
Knowledge about the critical systems of failure in the machines	3.71	1.79
Management leadership when executing TPM programs	3.70	1.81
Contemplate maintenance as purchase criteria	3.67	1.87
Training and education for the maintenance personnel	3.63	1.61
Informative meetings by those responsible for maintenance	3.50	1.97
Investment in updated tools	3.37	2.10

since the importance of leadership in other sectors has been reported in the industry performance (Cumbler et al. 2016; Tjosvold and Tjosvold 2015).

Also, it is noticed that the opportunity areas to improve the implementation of TPM (because they have lower values in the medians), are regarding to informative meetings by maintenance managers, since equipment is frequently requested without any explanation to other departments that are affected by lockouts, which undoubtedly helps to improve the communication channels. In the same way, it is considered that the purchase of adequate tools should be considered in order to carry out these maintenance activities, consequently in a quick way, and at the same time in a safer way for operators. However, these tools may not be physical tools; they may refer to a software that diagnoses the machine's components status (Adeyeri and Mpofu 2017).

In addition, while reviewing the IRs items, it is observed that the lowest value in this latent variable is associated with the training and qualification that the personnel from the maintenance department has, because it is stated that this is a task that is not frequently performed, and that actually has a low valuation in its median, since there is agreement among participants. Also, another item that has a low value in the IR is referring to communication between the production and maintenance departments, and it is in the first place in the median, which states that this item is highly important, and that there is a consensus among participants. However, the highest value of the IR is related to the task in the last place in the median, which indicates that there is not much consensus like in the other latent variables.

8.2.5 Technological Status

The present variable deals with the machinery and technological equipment or modern level in the supply chains, which are the ones that are providing maintenance. Also, this latent variable is valued by eight items and it was not necessary to remove any of them during the validation process. In addition, Table 8.7 illustrates the descriptive analysis of the items that integrate it, and based on this process it can be concluded that the most important aspect for participants is maintenance, since it is seen as a strategy that helps to generate a competitive advantage for the company, because equipment in good condition operating will be able to achieve the quality requirements demanded by the customer, which is because of the adjustments and calibrations that are made. Furthermore, it is observed that the item with the highest median is also the one with the lowest IT, which concludes that there is a consensus among those who were surveyed regarding the level of importance of TPM as a competitive advantage (Barberá et al. 2012), since it allows them to differentiate themselves from the competition and could acquire the expected benefits from TPM.

Nowadays, there are many studies that relate TPM to business strategies, for instance Jasiulewicz-Kaczmarek (2016) claims that the strengths and weaknesses of maintenance programs must be analyzed to avoid risk problems, and that they have to be aligned with the company's vision, and Nouri Gharahasanlou et al. (2017) presents a model based on company risk reduction, since those that have a good maintenance planning are companies that have fewer risks associated with product quality and delivery dates to customers.

Moreover, in this latent variable, there is an item that is currently very relevant, since modern machines and equipment have many integrated sensors that enable their operative diagnosis, which allows a software to perform quick diagnostics, generate reports, and communicate with suppliers for their interpretation and corrective action. Nowadays, there are many examples where decision support systems (DSS) are used that are focused on maintenance applications, for instance, Mourtzis et al. (2017) reports the augmented reality application case applied in order to keep

Table 8.7 Descriptive analysis: technological status

Item	Median	IR
Generating competitive advantages for machinery and equipment	3.80	1.71
Specialized software usage for maintenance processes	3.73	1.89
Learning and updating equipment after installation	3.66	1.84
Improvement of existing machinery and equipment	3.66	1.81
Advanced technology	3.61	1.84
Equipment development and patents registration	3.51	2.17
Update and focus on maintaining new generations of technology	3.48	1.80
Efficient use of new technologies and modern maintenance systems	3.48	1.88

robots in a supply chain operating, and Adeyeri and Mpofu (2017) use a DSS to support the remote monitoring equipment working conditions.

Furthermore, the last place of the eight items analyzed in this latent variable refers to the integration of state-of-the-art technology and its maintenance, which indicates that the surveyed companies only plan little investments in technology, therefore, they do not properly use it in this sector, which indicates that their modern level is not updated, since it is possible that industrial competitors have better machines and equipment. Also, in this area it is essential to mention that maquiladora companies where this research is carried out, have foreign capital incomes and that type of decisions associated with the technological investment is not made within them, but a matter for parent companies (García-Alcaraz et al. 2016). Consequently, it is recommended that maintenance managers implement a technological track process to identify updated innovations in this sector.

8.2.6 Layout

In order to execute an adequate maintenance process, it is necessary to have an appropriate distribution to promote it in the supply chain, even when new equipment is incorporated (Singh et al. 2013). In addition, the latent variable that is associated with the Layout was valued with five items or tasks, and their descriptive analysis is shown in Table 8.8, where it can be noticed in a holistic manner that only one item has a median over four while others are under this value, but up to three.

Furthermore, due to the information in Table 8.8 it is concluded that the relevant aspect is that machines are organized in families or technological groups according to the operations that they execute in supply chains, which allows to a specialist to take care of their maintenance, since they will be oversighting their performance (Chand and Shirvani 2000). Similarly, machines distribution must be checked, because every safety principle must be respected for operators and employees who may approach machines working (Shen 2015).

Moreover, the last place in this latent variable refers to the possibility of being able to be in the halls with heavy machinery to perform their maintenance, which is

Table 8.8 Descriptive analysis: layout

Items	Median	IR
Machinery grouping according to the family of produced products	4.03	1.66
Plant distribution to facilitate maintenance	3.85	1.72
Adequate signs in the plant related to maintenance	3.81	1.91
Proximity between processes and machinery	3.75	1.62
Equipment and heavy machinery movement within the facilities for maintenance	3.75	1.85

a frequent necessity in the maquiladora industry, since there is huge equipment that demands crane support and other systems to lift heavy parts. Also, when the industrial physical distribution cannot guarantee the maintenance operators and supply chains safeness, it provokes the loss of interest working in areas that are considered dangerous for their physical integrity (Rodrigues and Hatakeyama 2006).

8.2.7 Warehouse Management

Managing the spare parts warehouse is vital in maintenance, because technical machine lockouts are frequently incurred due to the lack of replacements or spare parts, which may cause delays in delivery orders already established with the customers (Antosz and Ratnayake 2016). In addition, Table 8.9 illustrates six items that incorporate the latent variable related to the warehouse management, and based on this information, it can be seen that two out of the six items have values over four, which indicates that they are important, on the contrary, four items get values under it, but up to three, which demonstrates that they are more than regularly relevant according to the respondent's opinion.

Also, regarding the values obtained by the items in the median, it is stated that the most significant aspect in the warehouse is to have identification codes for each of the parties, which facilitates their tracking and administration. Additionally, several part suppliers have their own code and frequently they are used to avoid confusion in the placed orders. In the same way, it is recommended to have components catalogs as those that are in operation, which facilitates the replacements in case the original parts are not achieved quickly. Also, the usage of technology such as RFID is currently reported to support the administration and products traceability in the warehouse (Zhou et al. 2017) or the usage of barcodes (Ringsberg 2016).

Moreover, it is also required that the warehouse has preservation systems for components and parts that demand special care, such as temperature and humidity

Table 8.9 Descriptive analysis: warehouse management

Item	Median	IR
Appropriate material and parts coding	4.05	1.59
Arrangement and adaptation of spaces for material or parts with special needs (heating, cooling, humidity, etc.)	4.02	1.67
Organized and appropriate space for parts and components within the warehouse	3.85	1.58
Warehouse areas cleaning	3.76	1.66
Daily inspections to avoid missing	3.76	1.67
Registration of changes in the components and tools location	3.74	1.74

conditions, among others (Makaci et al. 2017). In addition, this factor is crucial in humid environments, since many products have exposed parts, therefore, they may experience oxidation and degradation in extreme situations, thus, adequate packaging has to be included, even, there must be conditions to help to reduce fire risks (Ju 2016; Atieh et al. 2016).

Lastly, an area that must be properly attended is inspection routine to identify missing materials, since an item associated with this activity is in the penultimate place. However, currently, because the warehouse administration is almost always under the control of a software that sends a signal indicating when a minimum established as a security mattress has been reached, which requires that all entries and delivered parts to be recorded (Caridade et al. 2017). Likewise, it is acknowledged that if the materials location that is more frequently used change, it must be registered, which can be done from a software as well; however, there are software elements that, according to the participants, should be improved.

8.2.8 Management Commitment

As mentioned before, management commitment is one of the most relevant successful factors of TPM, since they are establishing maintenance plans and programs, providing the appropriate education and training, establishing contracts with suppliers, among others. In addition, this latent variable is valued by medians with six items and they are described in Table 8.10, where all of them have values under four, but up to three. Also, the difference between the one with the lowest value and the one with a low value is slightly distant.

Moreover, the most important aspect for participants is that they consider that department supervisors in the company have a high commitment level and responsibility regarding maintenance plans and programs; In addition, within senior management, they are responsible to report the company's vision towards product quality, therefore, maintenance is an essential technique to guarantee it (Wahab et al. 2013).

Table 8.10 Descriptive analysis: management commitment

Item	Median	IR
The TPM responsibility from heads department	3.71	1.76
Management focuses on quality and maintenance	3.70	1.84
Leadership attitudes in the TPM performance by senior managers	3.68	1.71
Meetings between maintenance and production	3.68	1.88
Management involvement in maintenance projects	3.63	1.94
Promotion of workers participation in maintenance activities by managers	3.61	1.84

However, according to the penultimate item in this variable, a significant commitment and involvement from senior management in maintenance plans and programs is required, since their task does not end in designing and testing plans and programs, because they must be part of their execution processes as well (Alayón et al. 2017). In other words, industrial maintenance is not only the management department responsibility, but other departments where it is related, and senior management are responsible as well. Similarly, executive management is required to implement a serious leadership to promote maintenance culture, mechanic machines, and managed tools preservation, since they are the main wealth source for maquiladora companies.

8.2.9 Customers

Like other philosophies and production tools, TPM must be focused on the client and its integration in the plans and programs that are executed has to be considered; therefore, it is part of the lean manufacturing (Wahab et al. 2013). In addition, this research is integrated to the latent customer variable, which is evaluated by four observed variables, which are illustrated in Table 8.11 and according to its information, it is shown that all the items have ratings over three, but under four, which indicates that they are more than regularly performed.

The item with the highest median in this latent variable that is associated with the customers integration in the determination process, since they are who establish the acceptable quality and tolerance levels in products, besides they are willing to pay for it (Lin et al. 2017). Another essential aspect is the possibility to delay management activities in order to pay attention to production orders already established, and to fulfill the delivery scheduled time with customers (Chemweno et al. 2018).

Fortunately, respondents declare that there are few occasions (regularly) where there has been a production breach of orders that clients have arranged, which is good. Also, it is important to recall that TQM is a program associated with quality and may affect it in several ways; with delays due to decomposition, lack of compliance with established standards, among others (Ruschel et al. 2017).

Table 8.11 Descriptive analysis: customers

Item	Median	IR
Consider the customer's opinion on maintenance issues	3.59	1.97
Delaying maintenance activities to be satisfied with production	3.57	1.91
Costs for noncompliance of orders because of poor maintenance	3.34	1.97
Avoid noncompliance with customers due to lack of maintenance	3.11	2.12

8.3 TPM Benefits

As it was previously mentioned, companies do not implement techniques connected with production systems if they cannot get anything back. In addition, TPM is not the exception, it offers many benefits, which have been divided into three categories for further analysis. Also, in the following subsections, the descriptive analyzes that have been carried out with the items that integrate the latent variables are presented, which are ordered according to their median in a descending manner.

8.3.1 Descriptive Analysis: Company Benefits

This variable is valued by eight items and all of them are integrated into the analysis, since it was not necessary to delete any of them in the validation process; Table 8.12 illustrates the descriptive analysis of the benefits that are directly obtained for the company. In general, it is observed that seven out of the eight items have a median of up to four, which shows that they are almost always obtained by companies that implement TPM. However, the item with the lowest value is only two-tenths to reach the four value.

Moreover, it is clearly noticeable that through the TPM implementation, companies get better operations control, since it is possible to plan the machines maintenance, programs, production and orders deliveries, among others (Hedvall et al. 2016). Likewise, it allows to know the operative capacities from the supply chains in real time. Similarly, it is observed that another benefit is related to the final product quality, which had been already discussed previously in other investigations (Gouiaa-Mtibaa et al. 2018; Shrivastava et al. 2016; ArunKumar and Dillibabu 2016; Friedli et al. 2010).

However, the company needs to improve its social communication with other departments that are involved with maintenance to take better decisions or integrate the decisions of those who are involved (Adeyeri and Mpofu 2017; Callewaert et al.

Table 8.12 Descriptive analysis: company benefits

Item	Median	IR
Better operations control	4.21	1.52
Improves the work environment quality	4.12	1.56
Creates a responsibility culture, discipline, and respect towards standards	4.08	1.60
Permanent learning	4.06	1.65
Creates an environment where participation, collaboration, and creativity are a reality	4.05	1.63
Increases employee morale	4.01	1.64
Redes de comunicación eficaces	4.00	1.71
Appropriate staff schedule sizing	3.98	1.62

2018). Also, enterprises must carry out a better personnel dimensioning templates for maintenance work, since worker crews are often integrated with just two or three employees, but many more are required or sometimes many operators are assigned in a task that does not require them, and a lot of downtimes is incurred by some of the operators (Wang and Xia 2017).

8.3.2 *Descriptive Analysis: Productivity Benefits*

The present variable is integrated into eight items and it was not necessary to remove any item from the list during the validation process. In addition, the descriptive analysis of the observed variables is described in Table 8.13, where in general, it is detected that all medians are over four, which indicates that all the benefits are obtained when implementing TPM. Also, it is seen that the difference between the item with the highest median and the lowest median it is a very low value, which shows that there is a similar median in all the items, therefore, all the benefits are equally obtained.

Furthermore, the item on the top of the list with the highest median refers to the products quality, which makes sense, since with proper maintenance and calibration it can be guaranteed that machines are going to have an adequate perform (Gouiaa-Mtibaa et al. 2018; Shrivastava et al. 2016; ArunKumar and Dillibabu 2016), so they comply with the standards and technical specifications that have been already discussed in other paragraphs. Additionally, the second item with the highest median may refer to the equipment reliability and availability increasement, since technical lockouts related to decomposition are reduced, because they are in the good working state, which increases the team's efficiency indexes (Tang et al. 2017; Calixto 2016).

In addition, items with the lowest value in the median are still important according to the scale used, and they are dealing with the company response capacity toward its productive system, in order to make changes due to fluctuations in demand by customers or external factors. In the same way, it is discovered that

Table 8.13 Descriptive analysis: productivity benefits

Item	Median	IR
Improve the final product quality	4.31	1.49
Improve equipment reliability and availability	4.24	1.46
Eliminate losses that affect plants productivity	4.20	1.52
Create competitive capabilities from the factory	4.19	1.56
Improve company's technology	4.16	1.62
Low financial costs for spare parts	4.14	1.59
Increase responsiveness to market movements	4.14	1.55
Maintenance costs reduction	4.12	1.55

maintenance costs are an opportunity area where the company must work, which include aspects associated with labor, spare parts and training (Nguyen and Chou 2018; Dui et al. 2017). Also, it is relevant to mention that this financial aspect is manifested in the second last item, which connects the cost due to fast changes when supply chains move from one product to another (Brito et al. 2017).

8.3.3 Descriptive Analysis: Security Benefits

The current latent variable is integrated by six items or observed variables, since it was not necessary to remove any item during the validation process; Table 8.14 shows that all variables have a median over four, which indicates that maquiladora industries are almost always obtaining those benefits associated with security when implementing TPM. Also, in this category or latent variable, it is observed that the contrast between the item with the highest and the lowest value, is relatively different, which indicates that they are achieved in an equivalent way.

In first place, there is the prevention and elimination of potential accidents causes, since decomposition problems from moving equipment are far away from where it is possible that components may cause harm to the operators, or events associated with explosions may happen, among others (Okoh and Haugen 2014; Vinnem et al. 2016). In the same way, equipment that is not properly calibrated is likely to contaminate or emit more pollutants to the environment, therefore, companies through maintenance and proper equipment adjustment may avoid this type of events (Hosono et al. 2010). In addition, that first item according to its median brings consequently the third item, which introduces the environmental conditions improvement where maintenance tasks are carried out, since there are fewer potential accidents causes. Also, regarding the items with the lowest median, it is shown in the penultimate place the prevention of negative events for health, which are associated with the first and third items that are closely related. In addition, in the last place, there is the rules understanding and regulations linked to security, which should be associated with the culture in the regions where the companies are located.

Table 8.14 Descriptive analysis: security benefits

Item	Median	IR
Prevention and elimination of potential accidents causes	4.19	1.54
Radically eliminate pollution sources	4.15	1.55
Improve environmental conditions	4.09	1.58
Increase in the capacity to identify potential problems and search for corrective actions	4.09	1.52
Preventive culture of negative events for health	4.07	1.58
Acknowledge why there are certain rules, rather than how to perform a task	4.04	1.55

References

Adeyeri MK, Mpofu K (2017) Development of system decision support tools for behavioral trends monitoring of machinery maintenance in a competitive environment. J Ind Eng Int 13(2):249–264. https://doi.org/10.1007/s40092-017-0184-z

Alayón C, Säfsten K, Johansson G (2017) Conceptual sustainable production principles in practice: do they reflect what companies do? J Clean Prod 141:693–701. https://doi.org/10.1016/j.jclepro.2016.09.079

Antosz K, Ratnayake RMC (2016) Classification of spare parts as the element of a proper realization of the machine maintenance process and logistics—case study. IFAC-PapersOnLine 49(12):1389–1393. https://doi.org/10.1016/j.ifacol.2016.07.760

Arnaiz A, Konde E, Alarcón J (2013) Continuous improvement on information and on-line maintenance technologies for increased cost-effectiveness. Proc CIRP 11:193–198. https://doi.org/10.1016/j.procir.2013.07.038

ArunKumar G, Dillibabu R (2016) Design and application of new quality improvement model: Kano Lean six sigma for software maintenance project. Arab J Sci Eng 41(3):997–1014. https://doi.org/10.1007/s13369-015-1933-1

Atieh AM, Kaylani H, Al-abdallat Y, Qaderi A, Ghoul L, Jaradat L, Hdairis I (2016) Performance improvement of inventory management system processes by an automated warehouse management system. Proc CIRP 41:568–572. https://doi.org/10.1016/j.procir.2015.12.122

Barberá L, Crespo A, Viveros P, Stegmaier R (2012) Advanced model for maintenance management in a continuous improvement cycle: integration into the business strategy. Int J Syst Assur Eng Manag 3(1):47–63. https://doi.org/10.1007/s13198-012-0092-y

Brito M, Ramos AL, Carneiro P, Gonçalves MA (2017) Combining SMED methodology and ergonomics for reduction of setup in a turning production area. Proc Manuf 13:1112–1119. https://doi.org/10.1016/j.promfg.2017.09.172

Calixto E (2016) Chapter 3—Reliability and maintenance. In: Calixto E (ed) Gas and oil reliability engineering, 2nd edn. Gulf Professional Publishing, Boston, pp 159–267. https://doi.org/10.1016/B978-0-12-805427-7.00003-8

Callewaert P, Verhagen WJC, Curran R (2018) Integrating maintenance work progress monitoring into aircraft maintenance planning decision support. Transp Res Proc 29:58–69. https://doi.org/10.1016/j.trpro.2018.02.006

Caridade R, Pereira T, Pinto Ferreira L, Silva FJG (2017) Analysis and optimisation of a logistic warehouse in the automotive industry. Proc Manuf 13:1096–1103. https://doi.org/10.1016/j.promfg.2017.09.170

Cumbler E, Kneeland P, Hagman J (2016) Motivation of participants in an interprofessional quality improvement leadership team. J Interprof Educ Pract 3:5–7. https://doi.org/10.1016/j.xjep.2016.03.005

Chand G, Shirvani B (2000) Implementation of TPM in cellular manufacture. J Mater Process Technol 103(1):149–154. https://doi.org/10.1016/S0924-0136(00)00407-6

Chemweno P, Pintelon L, Muchiri PN, Van Horenbeek A (2018) Risk assessment methodologies in maintenance decision making: a review of dependability modelling approaches. Reliab Eng Syst Safety 173:64–77. https://doi.org/10.1016/j.ress.2018.01.011

Dui H, Si S, Yam RCM (2017) A cost-based integrated importance measure of system components for preventive maintenance. Reliab Eng Syst Safety 168:98–104. https://doi.org/10.1016/j.ress.2017.05.025

Eti MC, Ogaji SOT, Probert SD (2006a) Impact of corporate culture on plant maintenance in the Nigerian electric-power industry. Appl Energy 83(4):299–310. https://doi.org/10.1016/j.apenergy.2005.03.002

Eti MC, Ogaji SOT, Probert SD (2006b) Reducing the cost of preventive maintenance (PM) through adopting a proactive reliability-focused culture. Appl Energy 83(11):1235–1248. https://doi.org/10.1016/j.apenergy.2006.01.002

Friedli T, Goetzfried M, Basu P (2010) Analysis of the implementation of total productive maintenance, total quality management, and just-in-time in pharmaceutical manufacturing. J Pharmaceutical Innov 5(4):181–192. https://doi.org/10.1007/s12247-010-9095-x

García-Alcaraz JL, Maldonado-Macías AA, Hernández-Hernández SI, Hernández-Arellano JL, Blanco-Fernández J, Sáenz Díez-Muro JC (2016) New product development and innovation in the maquiladora industry: a causal model. Sustainability (Switzerland) 8(8). https://doi.org/10.3390/su8080707

Gouiaa-Mtibaa A, Dellagi S, Achour Z, Erray W (2018) Integrated maintenance-quality policy with rework process under improved imperfect preventive maintenance. Reliab Eng Syst Safety 173:1–11. https://doi.org/10.1016/j.ress.2017.12.020

Hadjimarcou J, Brouthers LE, McNicol JP, Michie DE (2013) Maquiladoras in the 21st century: six strategies for success. Bus Horiz 56(2):207–217. https://doi.org/10.1016/j.bushor.2012.11.005

Hedvall K, Dubois A, Lind F (2016) Analysing an activity in context: a case study of the conditions for vehicle maintenance. Ind Mark Manage 58:69–82. https://doi.org/10.1016/j.indmarman.2016.05.016

Hipkin I (1996) Evaluating maintenance management information systems. Eur J Inf Syst 5 (4):261–272. https://doi.org/10.1057/ejis.1996.31

Hosono T, Su C-C, Siringan F, Amano A, S-i Onodera (2010) Effects of environmental regulations on heavy metal pollution decline in core sediments from Manila Bay. Mar Pollut Bull 60 (5):780–785. https://doi.org/10.1016/j.marpolbul.2010.03.005

Jasiulewicz-Kaczmarek M (2016) SWOT analysis for planned maintenance strategy—a case study. IFAC-PapersOnLine 49(12):674–679. https://doi.org/10.1016/j.ifacol.2016.07.788

Ju W-H (2016) Study on fire risk and disaster reducing factors of cotton logistics warehouse based on event and fault tree analysis. Proc Eng 135:418–426. https://doi.org/10.1016/j.proeng.2016.01.150

Jugraj Singh R, Inderpreet Singh A (2017) 5S—a quality improvement tool for sustainable performance: literature review and directions. Int J Qual Reliab Manag 34(3):334–361. https://doi.org/10.1108/IJQRM-03-2015-0045

Kamath NH, Rodrigues LLR (2016) Simultaneous consideration of TQM and TPM influence on production performance: a case study on multicolor offset machine using SD Model. Perspect Sci 8:16–18. https://doi.org/10.1016/j.pisc.2016.01.005

Kinney S (2006) 3—Overview of the TPM architecture. In: Kinney S (ed) Trusted platform module basics. Newnes, Burlington, pp 21–30. https://doi.org/10.1016/B978-075067960-2/50004-X

Kobayashi K, Fisher R, Gapp R (2008) Business improvement strategy or useful tool? Analysis of the application of the 5S concept in Japan, the UK and the US. Total Qual Manag 19(3):245–262

Kock N, Verville J, Danesh-Pajou A, DeLuca D (2009) Communication flow orientation in business process modeling and its effect on redesign success: results from a field study. Decis Support Syst 46(2):562–575. https://doi.org/10.1016/j.dss.2008.10.002

Kumar Sharma R, Gopal Sharma R (2014) Integrating six sigma culture and TPM framework to improve manufacturing performance in SMEs. Qual Reliab Eng Int 30(5):745–765. https://doi.org/10.1002/qre.1525

Lin L, Luo B, Zhong S (2017) Development and application of maintenance decision-making support system for aircraft fleet. Adv Eng Softw 114:192–207. https://doi.org/10.1016/j.advengsoft.2017.07.001

Makaci M, Reaidy P, Evrard-Samuel K, Botta-Genoulaz V, Monteiro T (2017) Pooled warehouse management: an empirical study. Comput Ind Eng 112:526–536. https://doi.org/10.1016/j.cie.2017.03.005

Mourtzis D, Zogopoulos V, Vlachou E (2017) Augmented reality application to support remote maintenance as a service in the robotics industry. Proc CIRP 63:46–51. https://doi.org/10.1016/j.procir.2017.03.154

Nazeri A, Naderikia R (2017) A new fuzzy approach to identify the critical risk factors in maintenance management. Int J Adv Manuf Technol. https://doi.org/10.1007/s00170-017-0222-4

Nguyen TAT, Chou S-Y (2018) Maintenance strategy selection for improving cost-effectiveness of offshore wind systems. Energy Convers Manag 157:86–95. https://doi.org/10.1016/j.enconman.2017.11.090

Nouri Gharahasanlou A, Ataei M, Khalokakaie R, Barabadi A, Einian V (2017) Risk based maintenance strategy: a quantitative approach based on time-to-failure model. Int J Syst Assur Eng Manag. https://doi.org/10.1007/s13198-017-0607-7

Okoh P, Haugen S (2014) A study of maintenance-related major accident cases in the 21st century. Process Saf Environ Prot 92(4):346–356. https://doi.org/10.1016/j.psep.2014.03.001

Park S-C, Lee S-C, Suárez-Barraza MF, Ramis-Pujol J (2012) An exploratory study of 5S: a multiple case study of multinational organizations in Mexico. Asian J Qual 13(1):77–99

Ringsberg H (2016) Bar coding for product traceability. In: Reference module in food science. Elsevier. https://doi.org/10.1016/B978-0-08-100596-5.03165-6

Rodrigues M, Hatakeyama K (2006) Analysis of the fall of TPM in companies. J Mater Process Technol 179(1):276–279. https://doi.org/10.1016/j.jmatprotec.2006.03.102

Ruschel E, Santos EAP, Loures EdFR (2017) Industrial maintenance decision-making: a systematic literature review. J Manuf Syst 45:180–194. https://doi.org/10.1016/j.jmsy.2017.09.003

Seth D, Tripathi D (2006) A critical study of TQM and TPM approaches on business performance of Indian manufacturing industry. Total Qual Manag Bus Excellence 17(7):811–824. https://doi.org/10.1080/14783360600595203

Shen CC (2015) Discussion on key successful factors of TPM in enterprises. J Appl Res Technol 13(3):425–427. https://doi.org/10.1016/j.jart.2015.05.002

Shrivastava D, Kulkarni MS, Vrat P (2016) Integrated design of preventive maintenance and quality control policy parameters with CUSUM chart. Int J Adv Manuf Technol 82(9):2101–2112. https://doi.org/10.1007/s00170-015-7502-7

Singh R, Gohil AM, Shah DB, Desai S (2013) Total productive maintenance (TPM) implementation in a machine shop: a case study. Proc Eng 51:592–599. https://doi.org/10.1016/j.proeng.2013.01.084

Tang Y, Liu Q, Jing J, Yang Y, Zou Z (2017) A framework for identification of maintenance significant items in reliability centered maintenance. Energy 118:1295–1303. https://doi.org/10.1016/j.energy.2016.11.011

Tjosvold D, Tjosvold M (2015) Leadership for teamwork, teamwork for leadership. In: Building the team organization: how to open minds, resolve conflict, and ensure cooperation. Palgrave Macmillan, UK, London, pp 65–79. https://doi.org/10.1057/9781137479938_5

Utar H, Ruiz LBT (2013) International competition and industrial evolution: evidence from the impact of Chinese competition on Mexican maquiladoras. J Dev Econ 105:267–287. https://doi.org/10.1016/j.jdeveco.2013.08.004

Vinnem JE, Haugen S, Okoh P (2016) Maintenance of petroleum process plant systems as a source of major accidents? J Loss Prev Process Ind 40:348–356. https://doi.org/10.1016/j.jlp.2016.01.021

Wahab ANA, Mukhtar M, Sulaiman R (2013) A conceptual model of lean manufacturing dimensions. Proc Technol 11:1292–1298. https://doi.org/10.1016/j.protcy.2013.12.327

Wang B, Xia X (2017) A preliminary study on the robustness of grouping based maintenance plan optimization in building retrofitting. Energy Proc 105:3308–3313. https://doi.org/10.1016/j.egypro.2017.03.752

Wilson TD (2002) The masculinization of the Mexican maquiladoras. Rev Radical Polit Econ 34 (1):3–17. https://doi.org/10.1016/S0486-6134(01)00117-6

Willmott P, McCarthy D (2001) 9—TPM for equipment designers and suppliers. In: Total productivity maintenance, 2nd edn. Butterworth-Heinemann, Oxford, pp 181–192. https://doi.org/10.1016/B978-075064447-1/50012-6

Zhou W, Piramuthu S, Chu F, Chu C (2017) RFID-enabled flexible warehousing. Decis Support Syst 98:99–112. https://doi.org/10.1016/j.dss.2017.05.002

Part V
Estructural Equation Models

Chapter 9
Simple Models

Abstract In this chapter, simple models where only two latent variables intervene are described, in which the main objective is to find the relationship between them, as well as their dependence level. In addition, only three simple models from two variables are presented, therefore, some tables are developed including a summary that allows observing all the relationships from each of the bivariate latent variables, where the independent variable, the dependent variable, the beta value for each relationship, the *p*-value, and the *R*-square value as a measure of the explained variance are illustrated. Also, a series of industrial implications and conclusions are reported.

9.1 Simple Model A: Work Culture—Benefits for the Organization

9.1.1 Hypothesis—Model A

The proposed model is presented in Fig. 9.1, where the *Work culture* variable is illustrated as an independent variable and the *Benefits for the organization* as dependent variable.

Furthermore, the relationship between these two latent variables has not been studied directly, although there are reports by Dangayach and Deshmukh (2000), where *Work culture* is considered a fundamental benefits source associated with the company, and since TPM is a technique based on people, it is considered a competitive strategy to obtain quality in products. Similarly, Hana (2010) mentions that in TPM it is important to consider people customs and traditions when implementing TPM since these can influence on the results obtained. Also, Ahmadzadeh and Bengtsson (2017) recommend that maintenance activities should always be prioritized according to the different human factors because this motivates operators when they are considered in the company plans and programs.

In addition, among the simplest activities when carrying out maintenance planning, it should consider the operators' level of competence and their skills

J. R. Díaz-Reza et al., *Impact Analysis of Total Productive Maintenance*,
https://doi.org/10.1007/978-3-030-01725-5_9

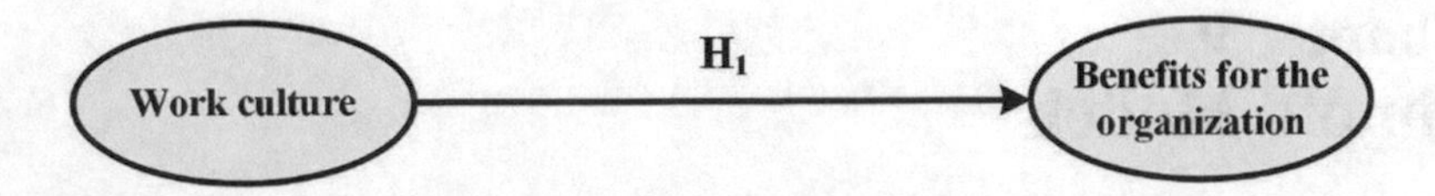

Fig. 9.1 Proposed simple model A

(Kachalov et al. 2015). Also, companies that do not consider the cultural aspects of their plans and programs, will surely have a poor execution and are very likely to change (Collins et al. 2006). According to the previous information, the following research hypotheses is proposed:

H_1 The *Work culture* in the *TPM implementation* process has a direct and positive effect on the *Benefits for the organization*.

9.1.2 Validation of the Variables—Model A

This simple model integrates as an independent variable the *Work culture* and as a dependent variable the *Benefits for the organization*. In addition, the validity indexes from the variables are shown in Table 9.1, although they had already been described previously, it is worth saying that both variables achieved the expected requirements, as a result, they are not discussed in this section.

9.1.3 Efficiency Indexes—Model A

As a matter of fact, before interpreting the model and its results, it is convenient to review its efficiency indexes, which are listed as follows:

- Average path coefficient (APC) = 0.563, $p < 0.001$
- Average R-squared (ARS) = 0.317, $p < 0.001$
- Average adjusted R-squared (AARS) = 0.315, $p < 0.001$
- Average block VIF (AVIF) not available
- Average full collinearity VIF (AFVIF) = 1.462, acceptable if ≤ 5, ideally ≤ 3.3

Table 9.1 Validation of latent variables in model A

Index	*Work culture*	*Benefits for the organization*
Compound reliability	0.865	0.950
Cronbach alpha	0.817	0.939
Average variance extracted	0.479	0.702

- Tenenhaus GoF (GoF) = 0.433, small $\geq$ 0.1, medium $\geq$ 0.25, large $\geq$ 0.36
- Sympson's paradox ratio (SPR) = 1.000, acceptable if $\geq$ 0.7, ideally = 1
- *R*-squared contribution ratio (RSCR) = 1.000, acceptable if $\geq$ 0.9, ideally = 1
- Statistical suppression ratio (SSR) = 1.000, acceptable if $\geq$ 0.7
- Nonlinear bivariate causality direction ratio (NLBCDR) = 1.000, acceptable if $\geq$ 0.7.

According to the previous information, it can be concluded that there is an adequate relationship between the two latent variables because of the APC index, but regarding the *R*-squared function and the adjusted *R*-squared, it is observed that a predictive validity exists. In the same way, the VIF index indicates the absence of collinearity while the GoF indicates an adequate data adjustment to the model. Finally, according to the other indexes, it is observed that there are no problems in the hypothesis perspective.

9.1.4 Results and Conclusion—Model A

Figure 9.2 illustrates the evaluated model along with the respective beta value as a dependence measure, the *p*-value, and the *R*-square as a measure of the explained variance between the two variables.

Based on the information established in Fig. 9.2, the following can be concluded regarding the proposed hypothesis:

H_1 There is enough statistical evidence to state that the *Work culture* in the *TPM implementation* process has a direct and positive effect on the *Benefits for the organization*, since when the first latent variable increases its standard deviation by one unit, the second increases in 0.563 units, which can explain 31.7% of its variability.

Moreover, this relationship between the *Work culture* and *Benefits for the organization* variables in a *TPM implementation* environment has been implicit in other researches, for instance, Shen (2015) discusses widely the critical success factors of TPM and declares that this *Work culture* is vital to guarantee the benefits offered by this technique. On the other hand, Mwanza and Mbohwa (2015) claim that the operator's cultural aspects must be taken into account since the *TPM implementation* is planned because they must be integrated during the plans and programs design. Finally, Kelly (2006) declares that if these cultural aspects are not

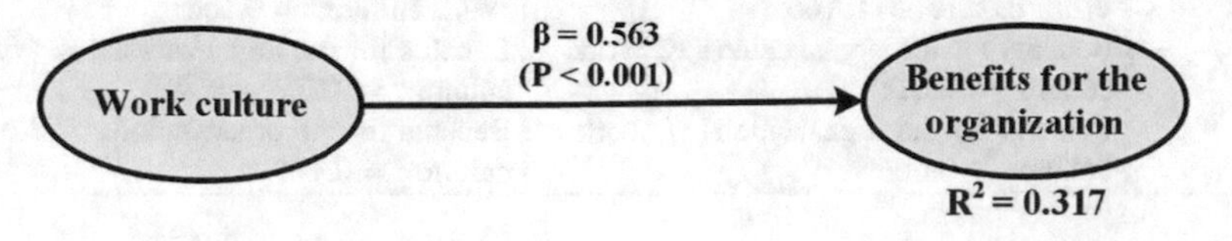

Fig. 9.2 Evaluated simple model A

integrated, there may be a limitation at the TPM plans execution moment, in that case, they should be integrated into its planning stage. However, another relevant cultural aspect is that they should be associated with leadership in this technique implementation (Tjosvold and Tjosvold 2015).

Therefore, it is concluded that the *Work culture* is a key factor to achieve *Benefits for the organization* in the maquiladora industry when TPM is implemented, as a result, its managers must focus on understanding workers habits and customs in order to integrate them into maintenance plans and programs, as well as respecting them as much as possible. Nevertheless, Eti et al. (2006b) declared that not only the *Work culture* from employees and people in charge of maintenance must be analyzed directly, but also from managers.

9.1.5 Sensitivity Analysis—Simple Model A

In the sensitivity analysis, the probabilities that the latent variables in different scenarios or status, specifically high and low levels are explained. Therefore, in the *Work culture* case which is related to the *Benefits for the organization*, there are four scenarios illustrated in Table 9.2.

According to data analysis in Table 9.2, a series of interesting conclusions can be obtained, which are given as follows:

- When analyzing the scenario where *Benefits for the organization* and *Work culture* are found at their high levels, it is observed that the probabilities for both are 0.163 and 0.160, respectively, which indicates that it is unlikely that independently those two scenarios occur. However, the probability that they occur simultaneously is also a relatively low value; only 0.063. In addition, the most

Table 9.2 Sensitivity analysis—simple model A

Work culture	*Benefits for the organization*	
	High	Low
High	Benefits for the organization+ = 0.163 Work culture+ = 0.160 Benefits for the organization+ *&* Work culture+ = 0.063 Benefits for the organization+ *if* Work culture+ = 0.390	Benefits for the organization− = 0.166 Work culture+ = 0.160 Benefits for the organization− *&* Work culture+ = 0.003 Benefits for the organization− *if* Work culture+ = 0.017
Low	Benefits for the organization+ = 0.163 Work culture− = 0.166 Benefits for the organization+ *&* Work culture− = 0.011 Benefits for the organization+ *if* Work culture− = 0.066	Benefits for the organization− = 0.166 Work culture− = 0.166 Benefits for the organization− *&* Work culture− = 0.079 Benefits for the organization− *if* Work culture− = 0.475

important aspect in this scenario, it is when the *Benefits for the organization* are high as well as the *Work Culture* with a probability of 0.390, which demonstrates the importance of the second variable. In other words, there is a probability of 0.227 (0.390 − 0.163) that these good benefits are due to the adequate *Work culture* in the productive systems focused on TPM.

- The second analyzed scenario is when there are *Benefits for the organization* at its low level and *Work culture* at its high level; it is observed that the probability of presenting the scenario independently for the first variable is 0.166, and management must pay special attention to it to prevent this from happening, however, the probability that this scenario is presented for the second variable is 0.160 only. Also, the probability that the two scenarios are presented simultaneously for the two variables is almost null with 0.003, finally, the probability of having low *Benefits for the organization* because there is a high *Work culture*, which is 0.017; it indicates that the *Work culture* has a positive impact on obtaining the *Benefits for the organization*.
- The third scenario may occur when there are *Benefits for the organization* at its high and *Work culture* at its low level. In addition, it is observed that the probability of having high *Benefits for the organization* independently is 0.163, and the probability of having a low *Work culture* is 0.166. Also, the probability that these two events are presented simultaneously is 0.011, and that the probability of having high *Benefits for the organization*, because there are low levels in the *Work culture* is 0.066. The previous data shows that the *Work culture* positively impacts on the *Benefits for the organization*.
- The fourth scenario that can occur when comparing these two variables is when the *Benefits for the organization* and the *Work culture* are at their low levels. In addition, the probability that the *Benefits for the organization* are low independently is 0.166; a very high-risk value that companies should consider, while the probability that the *Work culture* is low is 0.166; another high value. Also, the probability that both variables are presented in their low levels simultaneously is 0.079, therefore, the senior management must focus on generating that *Work culture*.

9.2 Simple Model B: Managerial Commitment—Safety Benefits

9.2.1 Hypothesis—Model B

One of the principal factors in any technique or tool implementation in the production process refers to the *Managerial commitment* and one of its consequences is the *Safety benefits* that can be obtained when talking about TPM, since the maintenance must be focused not only to help the quality and preserve machines and equipment but also operators' integrity. See Fig. 9.3.

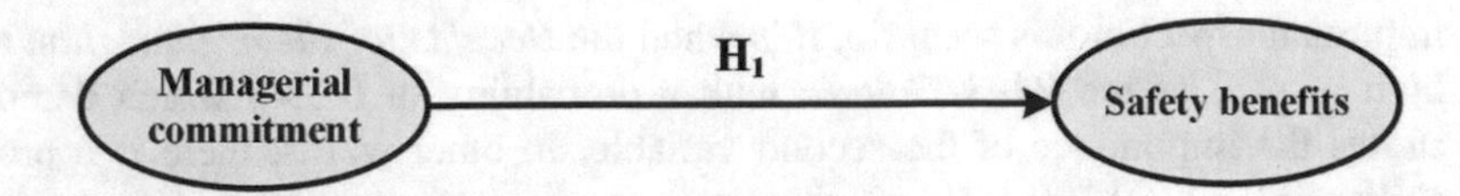

Fig. 9.3 Proposed simple model B

In a research, Shen (2015) declares that for workers, the most relevant aspect is their safety and that is why managers must assume that commitment to guarantee their physical integrity. However, in many production processes, maintenance should not be focused only on physical integrity, but to ensure adequate working atmospheric conditions, since aspects associated with noise, pollution in the environment, light, others, must be assisted as well (Singh et al. 2013).

In order to guarantee the workers safety and obtain the benefits that are a consequence from the adequate *TPM implementation*, senior management is responsible for establishing policies that must be followed, published, and provided to the employees by writing, especially those who are newly admitted. In fact, if new employees do not have the feeling that a company is safe, there is a risk of early dissertations, and that is one of the reasons why the *TPM implementation* fails since there is little continuity in programs (Rodrigues and Hatakeyama 2006).

In addition, when security is used as a working principle imposed by the senior management, a culture is created in the operators that helps to identify possible risk sources, and it is very common that there is a safety and hygiene committee integrated by syndicate unions, workers, doctors, and the senior management that is in charge of the oversight regulations compliance (Thomas et al. 2006).

According to the previous data, it is considered that the *Managerial commitment* is vital to obtain the *Safety benefits* offered by TPM, therefore, the following hypothesis is proposed, which is explained in Fig. 9.3.

H_1 The *Managerial commitment* has a direct and positive effect on the *Safety benefits* obtained when implementing TPM.

9.2.2 Validation of the Variables—Model B

This simple model integrates only two variables; the independent latent variable is the *Managerial commitment* and the dependent latent variable is the *Safety benefits*. In addition, Table 9.3 illustrates the validity indexes obtained from the variables in the model, which had been already discussed because they are integrated into more complex models. Also, according to the data in Table 9.3, it is concluded that the latent variables achieve the validity parameters, therefore, the model is evaluated.

Table 9.3 Validation of the latent variables in model B

Index	*Managerial commitment*	*Safety benefits*
Composite reliability	0.942	0.945
Cronbach alpha	0.926	0.93
Average variance extracted	0.729	0.740

9.2.3 Efficiency Indexes—Model B

In the indexes list presented below, it can be seen that model B complies with all the validity parameters to be interpreted, since predictive validity is observed through the ARS and AARS indexes, absence of collinearity according to the AVIF and AFVIF indexes, an adequate adjustment of the data to the model according to the Tenenhaus index, as a result, it is noticeable that there are no problems in the hypotheses perspective.

- Average path coefficient (APC) = 0.568, $p < 0.001$
- Average *R*-squared (ARS) = 0.322, $p < 0.001$
- Average adjusted *R*-squared (AARS) = 0.321, $p < 0.001$
- Average block VIF (AVIF) not available
- Average full collinearity VIF (AFVIF) = 1.447, acceptable if ≤ 5, ideally ≤ 3.3
- Tenenhaus GoF (GoF) = 0.487, small ≥ 0.1, medium ≥ 0.25, large ≥ 0.36
- Sympson's paradox ratio (SPR) = 1.000, acceptable if ≥ 0.7, ideally = 1
- *R*-squared contribution ratio (RSCR) = 1.000, acceptable if ≥ 0.9, ideally = 1
- Statistical suppression ratio (SSR) = 1.000, acceptable if ≥ 0.7
- Nonlinear bivariate causality direction ratio (NLBCDR) = 1.000, acceptable if ≥ 0.7

9.2.4 Results and Conclusion—Model B

Figure 9.4 illustrates the evaluated model B, where the Parameter β value is presented as a dependence measure between the variables, which is expressed in standard deviations, as well as the *R*-square value as a measure of the explained variance by the independent variable in the dependent variable.

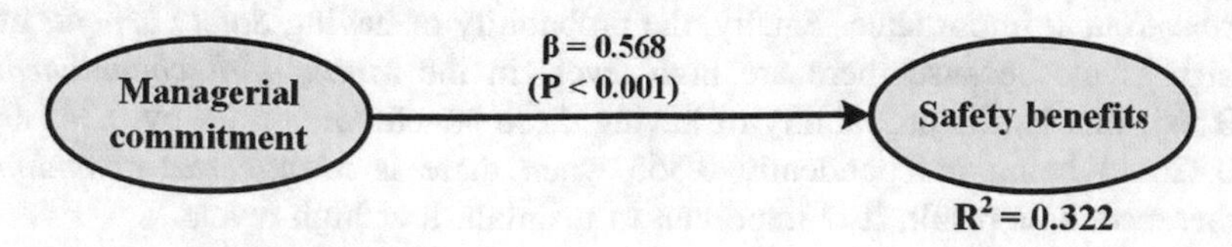

Fig. 9.4 Evaluated model B

According to the β value and the associated *p*-value that are presented in Fig. 9.4, it can be concluded regarding the proposed hypothesis in the following way:

H_1 There is enough statistical evidence to declare that the *Managerial commitment* has a direct and positive effect on the *Safety benefits* obtained when implementing TPM, since when the independent latent variable increases its standard deviation in one unit, the second increases in 0.568 units; in addition, it can explain up to 32.2% of its variability.

Moreover, this relationship between the *Managerial commitment* and the employees' safeness has already been reported in other investigations, although this is the first to quantify the dependency relationship that the variables have. For example, in a study conducted by Guariente et al. (2017) it is reported that when security cannot be guaranteed by the company in certain production processes, it must look for ways to automate them, and thus avoid the risks presence for operators.

In Morales et al. (2017) it is described that TPM is a competitiveness source, but the accidents and incidents presence in the production process increases the personnel absenteeism, insurance costs, and low production, thus, it can be concluded that management must take the initiative to provide security to their employees, although they are also responsible for their own actions in the workplace.

9.2.5 *Sensitivity Analysis—Simple Model B*

The Simple model B represents another case that is worth analyzing in its different scenarios since it links the *Managerial commitment* to the *TPM implementation* and the *Safety benefits* that can be obtained. As in the previous case, the two independent scenarios for each of the variables are analyzed, which results in a total of four combined scenarios about their low and high levels.

In addition, Table 9.4 shows a summary of the results obtained from these scenarios and the conclusions, which are the following:

- The first scenario occurs when the *Safety benefits* and the *Managerial commitment* towards TPM is high. Also, it is observed that the probability of occurrence for the first variable is 0.220 while for the second is 0.168, which happens independently. However, the probability that both variables have high values simultaneously is 0.095; a very low value that indicates the *Managerial commitment* importance. Finally, the probability of having *Safety benefits* at its high levels because there are high levels in the *Managerial commitment* is 0.565, that is, the probability of having these benefits increases by 0.345 from 0.220 to being independently 0.565 when there is *Managerial commitment* presence, as a result, it is important to maintain it at high levels.

Table 9.4 Sensitivity analysis—simple model B

Managerial commitment	Safety benefits	
	High	Low
High	Safety benefits+ = 0.220 Managerial commitment+ = 0.168 Safety benefits+ & Managerial commitment+ = 0.095 Safety benefits+ if Managerial commitment+ = 0.565	Safety benefits− = 0.166 Managerial commitment+ = 0.168 Safety benefits− & Managerial commitment+ = 0.000 Safety benefits− if Managerial commitment+ = 0.00
Low	Safety benefits+ = 0.220 Managerial commitment− = 0.158 Safety benefits+ & Managerial commitment− = 0.022 Safety benefits+ if Managerial commitment− = 0.138	Safety benefits− = 0.166 Managerial commitment− = 0.158 Safety benefits− & Managerial commitment− = 0.079 Safety benefits− if Managerial commitment− = 0.500

- The second scenario occurs when the *Managerial commitment* has high levels while there are low levels in the *Safety benefits*. In addition, there is a probability that the scenario for the first variable is presented in 0.166 and a probability that the scenario will be presented for the second is 0.168 (in an independent way). However, a very relevant aspect is presented here, since the probability of obtaining *Safety benefits* at its low level and high levels in the *Managerial commitment* simultaneously is zero or equal to zero. Also, the same occurs when analyzing the conditional probability of having low levels in the *Safety benefits* and high levels in the *Managerial commitment*. The previous data demonstrates that this *Managerial commitment* is efficient, and it is a great support to obtain adequate security levels when a maintenance program is implemented.
- The third scenario occurs when there are high *Safety benefits* and low levels in the *Managerial commitment*. In addition, the probability that the scenario occurs independently for the first variable is 0.220, but the probability of the scenario occurring independently for the second variable is 0.158; a very high value since it is unacceptable to have different values to zero in that variable. Also, the probability of simultaneous occurrence for the two scenarios regarding the two variables is very low; only 0.022, which indicates that management will almost always assume its commitment to TPM security.
- The most pessimistic scenario is when the *Safety benefits* and the *Managerial commitment* are at their low levels. In addition, the probability of finding that scenario independently for the first variable is 0.166 while for the second is 0.158. However, it is alarming to notice that there is a probability of 0.079 that these two variables are presented in this scenario simultaneously, and there is a probability of 0.500 to have low levels in the *Safety benefits*, because there are low levels in the *Managerial commitment*, which is something to be concerned about as well.

9.3 Simple Model C: TPM Implementation—Productivity Benefits

The present third and last simple model links two latent variables; the *TPM implementation* process as a productivity strategy and the *Productivity benefits* that are obtained in the company. In addition, the first variable has 12 items, while the second has 8 items; a graphic model representation is illustrated in Fig. 9.5.

9.3.1 Hypothesis—Model C

Undoubtedly, the methodology and procedures used to implement TPM in production systems affect the benefits in terms of productivity for the company. For example, before implementing TPM, a training process should be carried out on the basic concepts that this tool implies in a general way, but in a specific way, the operators of each and every machine must be trained, these are the first aspects about the optimal operating conditions, and in this way, it guarantees the production systems quality parameters (Seth and Tripathi 2006).

However, the supervisors and engineers' commitment in charge of the production process must be presented, since they are the ones who know the customer's demands and the production processes needs, therefore, they must play a leading role in the TPM process implementation. In addition, the absence of this leadership is understood as a high commitment from the superior managers to the machines and tools operators. In general, it is considered that the human resources education and training is vital in order to eliminate the losses that the company may have regarding to productivity when it comes to TPM.

In the same way, the integration of those machines operators in the *TPM implementation* process helps to have early warnings when they do not have the optimal production conditions, and warn about anomalies when the product quality could be at risk (Cua et al. 2001). Therefore, it is concluded that the communication that must exist between the operators and their supervisors is important to guarantee the productivity indexes; however, this communication does not end there, but it must exist between all the departments that integrate the company, since it is necessary to notify when a machine is at a technical stop in order to carry out planned or eventual repairs, and thus look forward to replace it, hire extra time on another machine or outsource production orders to other companies (Konecny and

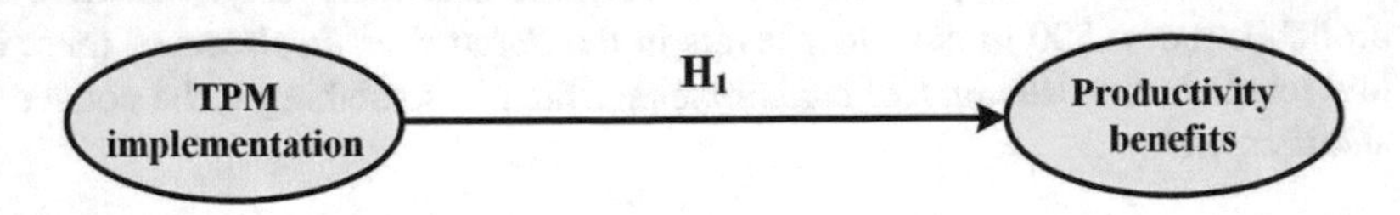

Fig. 9.5 Proposed model C

Thun 2011). Another significant aspect that must be considered in the *TPM implementation* is the machine components useful life, because it is required to keep a record about the operation times and dates of changes in order to perform technical stoppages planning due to component changes, which must be known by operators and employees working in the production system. Additionally, this requires that the machines output quality is constantly monitored (Ahmad et al. 2012).

According to the previous information, considering that the planning of the *TPM implementation* can affect the *Productivity benefits*, the following hypothesis is proposed:

H_1 The methodology and procedures for the *TPM implementation* has a direct and positive effect on the *Productivity benefits* that the company obtains.

9.3.2 Validation of the Variables—Model C

Table 9.5 illustrates the validity indexes from the latent variables that integrate the Simple model C. According to this information, it is observed that the two variables comply with the internal validity because the compound validity index and the alpha of Cronbach are over 0.7; minimum admitted value. In addition, enough convergent validity is observed, since the AVE is greater than 0.5, therefore, the model is interpreted.

9.3.3 Efficiency Indexes—Model C

Before interpreting the model, it is reviewed that it complies with all the validity specifications established in the methodology. In addition, the indexes obtained are illustrated in the section below.

- Average path coefficient (APC) = 0.575, $p < 0.001$
- Average *R*-squared (ARS) = 0.330, $p < 0.001$
- Average adjusted *R*-squared (AARS) = 0.328, $p < 0.001$
- Average block VIF (AVIF) not available

Table 9.5 Validation of the variables—model C

Index	*TPM implementation*	*Productivity benefits*
Composite reliability	0.944	0.945
Cronbach alpha	0.935	0.948
Average variance extracted	0.583	0.732

- Average full collinearity VIF (AFVIF) = 1.415, acceptable if ≤ 5, ideally ≤ 3.3
- Tenenhaus GoF (GoF) = 0.466, small ≥ 0.1, medium ≥ 0.25, large ≥ 0.36
- Sympson's paradox ratio (SPR) = 1.000, acceptable if ≥ 0.7, ideally = 1
- *R*-squared contribution ratio (RSCR) = 1.000, acceptable if ≥ 0.9, ideally = 1
- Statistical suppression ratio (SSR) = 1.000, acceptable if ≥ 0.7
- Nonlinear bivariate causality direction ratio (NLBCDR) = 1.000, acceptable if ≥ 0.7.

According to the information in the previous indexes list, it can be concluded that the model has enough predictive validity based on the ARS and AARS indexes, which are greater than 0.02. In the same way, it is observed that it does not exist a collinearity between the variables, since the VIF and AFVIF indexes under 3.3. Also, it is observed that the data have an adequate acceptability to the model since the Tenenhaus index is over 0.36. Finally, there are no directionality problems in the case of the dependency performed. Therefore, it is proceeding to interpret the model.

9.3.4 Results and Conclusions—Model C

Figure 9.6 illustrates the evaluated Model C, which illustrates the beta value as a measure of dependence between the two variables, and the *R*-square value as a measure of the explained variance by the independent variable in the dependent variable.

Based on the β value and the *p*-value that exists between the analyzed variables, regarding the hypothesis that has been initially proposed, the following can be concluded:

H_1 There is enough statistic evidence to declare that the methodology and procedures for the *TPM implementation* has a direct and positive effect on the *Productivity benefits* that the company obtains, since when the independent latent variable increases its standard deviation by one unit, the second one increases in 0.575 units, which can explain up to 33% of its variability.

The conclusion that has been reached from the statistical analysis of the variables associated with the *TPM implementation* and the *Productivity benefits* has been suggested by other researchers in other areas as well, however, in this research

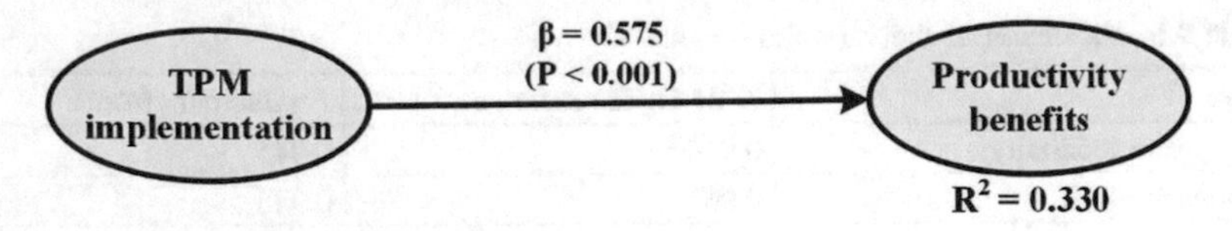

Fig. 9.6 Evaluated model C

conducted in the maquila industry field offers comparable results. For instance, there are tasks where certain *TPM implementation* activities are modeled, simulated, and associated with benefits, where they focus on determining areas for improvement opportunities (Pascale et al. 2012), which indicates that a TPM program is never complete and that it can always evolve and show improvement. In addition, failures that are obtained in the implementation process must have a feedback to the senior management in order that corrective actions are performed, and a new plan is generated (Willmott and McCarthy 2001b). In fact, some authors recommend associating TPM with other lean manufacturing techniques, such as Kaizen, where the processes and productivity continuous improvement is considered as the main purpose of its application (Prabhuswamy et al. 2013).

Some other authors consider that TPM should be associated with TQM since it is considered that these programs are efficient only if the processed products in the machines are able to meet the required quality standards. Thus, TQM and TPM are tools that must be in complete integration to guarantee *Productivity benefits* (Kamath and Rodrigues 2016). Also, other examples about the TPM integration with other tools are along with JIT (Cua et al. 2001), BPR (Hipkin and De Cock 2000), among others.

Therefore, it is concluded that TPM is another tool that should be integrated into production system, as well as with those that are already implemented because it should not be isolated, however, it is part of everyone's commitment that should be a focus on generating better productivity and quality.

As a matter of fact, because the analysis from all the simple models that can be generated with two variables is very high, their summaries are presented below, where only the β values, the *p*-value for the statistical test, the *R*-square as a measure of the explained variance, and the VIF as a measure of collinearity are presented. Also, in this analysis, the validation process of the variables is omitted, which can be consulted in a chapter dedicated specifically to it.

9.3.5 Sensitivity Analysis—Model C

In the Simple model C, the *TPM implementation* process is related to the *Productivity benefits* that a company can obtain. As in the previous cases, four combinations about their different high and low scenarios that these variables may have are analyzed.

In addition, Table 9.6 shows a summary of the values that the analyzed variables in their different scenarios may have and are described in the section below:

- The first scenario is when the *Productivity benefits* and the *TPM implementation* have high values. The probability of presenting the first variable in this scenario independently is 0.190, while for the second is 0.193. However, the probability that these two variables are presented simultaneously in their high levels is only 0.073, which is a very small probability. Finally, the probability that the

Table 9.6 Sensitivity analysis—simple model C

TPM implementation	*Productivity benefits*	
	High	Low
High	Productivity benefits+ = 0.190 TPM implementation+ = 0.193 Productivity benefits+ & TPM implementation+ = 0.073 Productivity benefits+ *if* TPM implementation+ = 0.380	Productivity benefits− = 0.155 TPM implementation+ = 0.193 Productivity benefits− & TPM implementation+ = 0.003 Productivity benefits− *if* TPM implementation+ = 0.014
Low	Productivity benefits+ = 0.190 TPM implementation− = 0.168 Productivity benefits+ & TPM implementation− = 0.027 Productivity benefits+ *if* TPM implementation− = 0.161	Productivity benefits− = 0.155 TPM implementation− = 0.168 Productivity benefits− & TPM implementation− = 0.076 Productivity benefits− *if* TPM implementation− = 0.452

Productivity benefits will be high because the *TPM implementation* is at its high levels is 0.380, which indicates that the probability of occurrence for these benefits is double to when it is found independently, hence the importance of a proper planning and execution in the implementation process.

- The second scenario for the analyzed variables is when the *Productivity benefits* are at its low level, but the *TPM implementation* process at its high level. The probability of that scenario occurring in the first variable is 0.155, while for the second is 0.193. However, the probability that these two variables occur simultaneously is 0.003; a very low probability, which has common sense, since TPM is aimed to increase productivity rates. Finally, the probability that the *Productivity benefits* are low because there is an implementation process at high levels is 0.014; a low probability of occurrence.
- The third scenario that may occur is when the *Productivity benefits* are high, but the *TPM implementation* process is low. The probability of having the scenario for the first variable independently is 0.190, while the probability of occurrence for the second variable is 0.168. However, the probability of these two scenarios being presented simultaneously is 0.027; a relatively low probability, but the probability of having *Productivity benefits* at its high level, and a *TPM implementation* process at its low levels is 0.161. The previous information indicates that these high productivity indexes may be due to other types of reasons, not only because of aspects associated with TPM.
- The fourth and last scenario can occur when there are *Productivity benefits* at its low levels, but also the *TPM implementation* process, which represents a pessimistic state. The probability that the first variable occurs independently is 0.155 while the probability for the second variable 0.168. However, the probability that both variables are in that scenario simultaneously is relatively high, with a value of 0.076. Finally, the probability of having *Productivity benefits* at its low level, because there is a *TPM implementation* process at its low level as

well, is 0.452; a very high value, which indicates that senior management must focus a lot of effort in this last process.

9.4 Simple Hypotheses—Work Culture

In this section, the models where the *Work culture* is related to other variables are analyzed, where it can play an independent latent variable or a latent dependent variable role. In addition, there are a total of eleven relationships or hypotheses, the variables that are related among them are the following: *Suppliers, PM implementation, TPM implementation, Technological status, Layout, Warehouse management, Managerial commitment,* and *Clients.*

The hypotheses that may arise from the relationships between *Work culture* and other variables when TPM is implemented in the company are the following:

H_1 The *Work culture* in a company has a direct and positive effect on the relationship that exists with the machines and equipment *Suppliers* in the production lines.

H_2 The *Work culture* in a company has a direct and positive effect on the *PM implementation* process in the production lines.

H_3 The *Work culture* in a company has a direct and positive effect on the *TPM implementation* process in the production lines.

H_4 The *Work culture* in a company has a direct and positive effect on the *Technological status* in the company.

H_5 The *Work culture* in a company has a direct and positive effect on the plant or *Layout* distribution in the productive system.

H_6 The *Work culture* in a company has a direct and positive effect on the machines and equipment parts and components *Warehouse management* processes in the production lines.

H_7 The *Managerial commitment* in the *TPM implementation* process in a company has a direct and positive effect on the *Work culture* towards this technique.

H_8 The *Work culture* in a company has a direct and positive effect on the company *Clients.*

H_9 The *Work culture* in a company has a direct and positive effect on the *Benefits for the organization.*

H_{10} The *Work culture* in a company has a direct and positive effect on the *Productivity benefits.*

H_{11} The *Work culture* in a company has a direct and positive effect on the *Safety benefits.*

9.4.1 *Results*—Work Culture

Table 9.7 illustrates the results obtained when the *Work culture* is related to other variables. In addition, the first column indicates the hypothesis that it represents, the second column indicates the variables that are related with, and from the third to the sixth column the relationship parameters are presented.

Based on the information from Table 9.7, the following can be concluded on the hypotheses previously stated:

H_1 There is enough statistical evidence to declare that *Work culture* in a company has a direct and positive effect on the relationship that exists with the machines and equipment *Suppliers* in the production lines, since when the first variable increases its standard deviation in one unit, the second increases in 0.625 units, and it can explain up to 39.1% of its variability.

H_2 There is enough statistical evidence to declare that the *Work culture* in a company has a direct and positive effect on the *PM implementation* process in the production lines, since when the first variable increases its standard deviation in one unit, the second increases in 0.600 units, and it can explain up to 36% of its variability.

H_3 There is enough statistical evidence to declare that the *Work culture* in a company has a direct and positive effect on the *TPM implementation* process in the production lines, since when the first variable increases its standard deviation in one unit, the second increases in 0.653 units, and it can explain up to 42.6% of its variability.

H_4 There is enough statistical evidence to declare that the *Work culture* in a company has a direct and positive effect on the *Technological status* in the company, since when the first variable increases its standard deviation in one

Table 9.7 Relationships between *Work culture* and other variables

Hypotheses	Variable	Parameter			
		B	*p*-value	R^2	VIF
H_1	*Suppliers*	0.625	$p < 0.001$	0.391	1.637
H_2	*PM implementation*	0.600	$p < 0.001$	0.360	1.557
H_3	*TPM implementation*	0.653	$p < 0.001$	0.426	1.711
H_4	*Technological status*	0.569	$p < 0.001$	0.324	1.475
H_5	*Layout*	0.559	$p < 0.001$	0.312	1.453
H_6	*Warehouse management*	0.602	$p < 0.001$	0.363	1.556
H_7	*Managerial commitment*	0.599	$p < 0.001$	0.359	1.551
H_8	*Clients*	0.362	$p < 0.001$	0.131	1.149
H_9	*Benefits for the organization*	0.563	$p < 0.001$	0.317	1.462
H_{10}	*Productivity benefits*	0.536	$p < 0.001$	0.286	1.379
H_{11}	*Safety benefits*	0.541	$p < 0.001$	0.293	1.403

unit, the second variable increases in 0.659 units, and it can explain up to 32.4% of its variability.

H_5 There is enough statistical evidence to declare that the *Work culture* in a company has a direct and positive effect on the plant or *Layout* distribution in the productive system, since when the first variable increases its standard deviation by one unit, the second increases in 0.559 units, and it can explain 31.2% of its variability.

H_6 There is enough statistical evidence to declare that the *Work culture* in a company has a direct and positive effect on the machines and equipment parts and components *Warehouse management* processes in the production lines, since when the first variable increases its standard deviation in one unit, the second one increases in 0.602 units, and it can explain up to 36.3% of its variability.

H_7 There is enough statistical evidence to declarc that the *Managerial commitment* in the *TPM implementation* process in a company has a direct and positive effect on the *Work culture* towards this technique, since when the first variable increases its standard deviation in one unit, the second increases in 0.599 units, and it can explain up to 35.9% of its variability.

H_8 There is enough statistical evidence to declare that the *Work culture* in a company has a direct and positive effect on the company *Clients*, since when the first variable increases its standard deviation in one unit, the second variable increases in 0.362 units, and it can explain up to 13.1% of its variability.

H_9 There is enough statistical evidence to declare that the *Work culture* in a company has a direct and positive effect on the *Benefits for the organization*, since when the first variable increases its standard deviation in one unit, the second increases in 0.363 units, and it can explain up to 31.7% of its variability.

H_{10} There is enough statistical evidence to declare that the *Work culture* in a company has a direct and positive effect on the *Productivity benefits*, since when the first variable increases its standard deviation in one unit, the second variable increases in 0.536 units, and it can explain 28.6% of its variability.

H_{11} There is enough statistical evidence to declare that the *Work culture* in a company has a direct and positive effect on the *Safety benefits*, since when the first variable increases its standard deviation in one unit, the second variable increases in 0.541 units, and it can explain 29.3% of its variability.

9.4.2 Conclusions and Industrial Implications—Work Culture

In general terms, regarding the relationship that *Work culture* has as a variable that is related to other variables that intervene during the *TPM implementation* process, the following can be concluded:

- All the effects from the *Work culture* variable along with other variables are statistically significant because the *p*-values are under 0.05. The highest beta value is the related to the plant *Layout* or distribution while the lowest is related to the *Clients*. In addition, the first is basically because the *Work culture* and the *Layout* are internal aspects whereas *Clients* are external to the company.
- The *R*-square value is very high; therefore, it is concluded that there is explanatory capacity and predictive validity in the models, except for the relationship with *Clients*, because of the information that was already explained above.
- The *R*-square value is variable but significant in all the relationships, although it is observed that the strongest relationships are found between the *TPM implementation* and the plant *Layout* or distribution.
- There are no problems of collinearity between the *Work culture* and the other variables which it is related with.
- However, these relationships have a series of industrial implications and based on the results shown in Table 9.7, the following suggestions are indicated:
- As mentioned in the previous chapters, TPM has a series of pillars that are indispensable to guarantee its success, such as the 5 s technique, that's why placing the tools in their section after using them, keep the work environment clean and organized is part of the *Work culture* that must be implemented. However, those attitudes of cleanliness and order should not be limited to a single area, but to the entire company. In fact, cleanliness and order in the company are part of the social image that it can project.
- A record should always be kept about the changes and adjustments on the machines through a maintenance logbook, which must be visible and accessible to employees. In the same way, that log must have future activities to be performed, but, in addition, at the time of making the report, the operators must be considered, since they are the ones who know the machine operation.
- These future activities should be executed along with all involved departments. Also, the communication between departments is part of the *Work culture* that should be strengthened by the senior management.
- The senior management must focus on generating a *Work culture* that considers the machines and tools maintenance and preservation that they have in the productive system since they are the generating wealth means (Eti et al. 2006b). A machine is an income source when it is operating under optimal conditions, but it is an expense source when it is under repair or a technical stoppage.

- The *Work culture* is a very broad aspect, which is why aspects associated with the monitoring of plans and programs must be integrated, which facilitates the MP and *TPM implementation* in the productive systems, as result, it is required that the senior management take the leadership and supervision to create that culture (Sani et al. 2012).
- The *Work culture* should focus on aspects that are associated with employees safety and integrity, forcing them to report on the environment that is not healthy or that entails a risk for them, which motivates to generate new plant distributions or *Layout* that consider these aspects and not only the costs (Davila and Elvira 2012). Also, employees must be the first to be able to offer preventive maintenance to machines.
- The *Work culture* promotion should not be limited only to aspects associated with the production systems, but it should be expanded to the warehouses where machines parts and components are preserved, since rules must be followed for their accommodation, location, distribution, as well as maintaining an adequate work environment (Collins et al. 2006; Sani et al. 2012).

9.5 Simple Hypotheses—Suppliers

In this section, the simple models where the *Suppliers* variable is related to other variables are analyzed. Logically, the relationship along with the *Work culture* variable has already been analyzed, therefore, there are only 10 variables which it is analyzed with; generating the same number of hypotheses. In addition, the variables that it is related with are the following: *PM implementation*, *TPM implementation*, *Technological status*, *Layout*, *Warehouse management*, *Managerial commitment*, *Clients*, *Benefits for the organization*, *Productivity benefits*, and *Safety benefits*.

The hypotheses that can be generated from the relationships between The *Suppliers* of machinery, equipment, and components during the *TPM implementation* process are the following:

H_1 The *Suppliers* of machinery, equipment, and components in a company have a direct and positive effect on the *PM implementation* process in the production lines.

H_2 The *Suppliers* of machinery, equipment, and components in a company have a direct and positive effect on the *TPM implementation* process in the production lines.

H_3 The *Suppliers* of machinery, equipment, and components in a company have a direct and positive effect on the *Technological status* in the company.

H_4 The *Suppliers* of machinery, equipment, and components in a company have a direct and positive effect on the plant distribution or *Layout* in the production system.

H_5 The *Suppliers* of machinery, equipment, and components in a company have a direct and positive effect on the machines and equipment parts and elements in the *Warehouse management* process in production lines.

H_6 The *Managerial commitment* that a company has toward the *TPM implementation* has a direct and positive effect on the *Suppliers* of machinery, equipment, and components required.

H_7 The *Suppliers* of machinery, equipment, and components in a company have a direct and positive effect on the company *Clients*.

H_8 The *Suppliers* of machinery, equipment, and components in a company have a direct and positive effect on the *Benefits for the organization*.

H_9 *Suppliers* of machinery, equipment, and components in a company have a direct and positive effect on Productivity benefits.

H_{10} *Suppliers* of machinery, equipment, and components in a company have a direct and positive effect on *Safety benefits*.

9.5.1 *Results*—Suppliers

After evaluating the ten models to validate the hypotheses that have been previously proposed on the relationship that *Suppliers* of equipment, machinery, and components have, the results are obtained, which are illustrated in Table 9.8.

The information from Table 9.8 allows to conclude regarding the proposed hypotheses about the *Suppliers* of machinery, equipment, and components in maintenance; the conclusions are the following:

H_1 There is enough statistical evidence to declare that the *Suppliers* of machinery, equipment, and components in a company have a direct and positive effect on the *PM implementation* process in the production lines,

Table 9.8 Relationships between *Suppliers* and other variables

Hypotheses	Dependent variable	Parameter			
		β	*p*-value	R^2	VIF
H_1	*PM implementation*	0.595	$p < 0.001$	0.354	1.547
H_2	*TPM implementation*	0.666	$p < 0.001$	0.444	1.786
H_3	*Technological status*	0.600	$p < 0.001$	0.360	1.562
H_4	*Layout*	0.558	$p < 0.001$	0.312	1.448
H_5	*Warehouse management*	0.568	$p < 0.001$	0.323	1.476
H_6	*Managerial commitment*	0.587	$p < 0.001$	0.345	1.524
H_7	*Clients*	0.424	$p < 0.001$	0.180	1.203
H_8	*Benefits for the organization*	0.520	$p < 0.001$	0.270	1.340
H_9	*Productivity benefits*	0.512	$p < 0.001$	0.263	1.331
H_{10}	*Safety benefits*	0.509	$p < 0.001$	0.259	1.315

since when the first variable increases its standard deviation in one unit, the second increases in 0.595, and it can explain up to 35.4% of its variability.

H_2 There is enough statistical evidence to declare that the *Suppliers* of machinery, equipment, and components in a company have a direct and positive effect on the *TPM implementation* process in the production lines, since when the first variable increases its standard deviation in one unit, the second one increases in 0.666, and it can explain up to 44.4% of its variability.

H_3 There is enough statistical evidence to declare the *Suppliers* of machinery, equipment, and components in a company have a direct and positive effect on the *Technological status* in the company, since when the first variable increases its standard deviation in one unit, the second increases in 0.600, and it can explain up to 36.0% of its variability.

H_4 There is enough statistical evidence to declare that the *Suppliers* of machinery, equipment, and components in a company have a direct and positive effect on the plant distribution or *Layout* in the production system, since when the first variable increases its deviation standard in one unit, the second increases in 0.558, and it can explain up to 31.2% of its variability.

H_5 There is enough statistical evidence to declare that the *Suppliers* of machinery, equipment, and components in a company have a direct and positive effect on the machines and equipment parts and elements in the *Warehouse management* process in production lines, since when the first variable increases its standard deviation by one unit, the second increases in 0.568, and it can explain up to 32.3% of its variability.

H_6 There is enough statistical evidence to declare that the *Managerial commitment* that a company has toward the *TPM implementation* has a direct and positive effect on the *Suppliers* of machinery, equipment, and components required, since when the first variable increases its standard deviation in one unit, the second increases in 0.587, and it can explain up to 34.5% of its variability.

H_7 There is enough statistical evidence to declare that the *Suppliers* of machinery, equipment, and components in a company have a direct and positive effect on the company *Clients*, since when the first variable increases its standard deviation in one unit, the second increases in 0.424, and it can explain up to 18.0% of its variability.

H_8 There is enough statistical evidence to state that the *Suppliers* of machinery, equipment, and components in a company have a direct and positive effect on the *Benefits for the organization*, since when the first variable increases its standard deviation by one unit, the second variable increases in 0.520, and it can explain up to 27.0% of its variability.

H_9 There is enough statistical evidence to state that the *Suppliers* of machinery, equipment, and components in a company have a direct and positive effect on productivity benefits, since when the first variable increases its standard deviation in one unit, the second variable increases in 0.512, and it can explain up to 26.3% of its variability.

H_{10} There is sufficient statistical evidence to state that the *Suppliers* of machinery, equipment, and components in a company have a direct and positive effect on *Safety benefits*, since when the first variable increases its standard deviation in one unit, the second variable increases in 0.509, and it can explain up to 25.9% of its variability.

9.5.2 *Conclusions and Industrial Implications*—Suppliers

The results from Table 9.8 allow to make a series of conclusions in a general way regarding the parameters that have been obtained, which are the following:

- *Suppliers* are directly and positively related to all other variables since the *p*-values are under 0.05. In addition, the β coefficients are literally high in all relationships. Also, the highest value refers to the relationship with *the TPM implementation*, which is expected, while the lowest refers to the relationship with the *Clients*, since they are external entities to the company.
- The *R*-square value as a measure of the explained variance is very different when it is related to the other latent variables, but as well as in the regression coefficients case, the highest value is associated with the *TPM implementation*, and the lowest with the *Clients*, which is due to the causes already explained above.
- No problems of collinearity are observed in any of the relationships that the *Suppliers* have with the other latent variables.
- The results from above have a series of industrial implications in the maquiladora industry, which also allow making recommendations, such as the following:
- TPM is not a tool that should only integrate the staff who works in the company, but it should go further, integrating the *Suppliers* of the machinery, equipment, consumables, and components that are required. In addition, the previous statement will allow maintaining a continuous production system flow to comply with all every order (Willmott and McCarthy 2001c).
- Before starting a PM or *TPM implementation* process, the capabilities and commitments that can be contracted with the *Suppliers* should be analyzed. Also, it is the responsibility from the Senior Management to establish the collaboration agreements with the *Suppliers*, establishing aspects associated with the education and training that will be carried out after the purchase of a team, the clear description of the operation manuals where handling critical aspects are indicated. In the same way, emphasis should be placed on the guarantees along with the *Suppliers*, besides always create the most effective guarantees when necessary (Bourke and Roper 2016).

- It is also important that the companies provide the necessary information from the production processes that they have in operation because in this way the *Suppliers* will be able to provide the components and supplies that fit the equipment already installed. Sometimes, it is recommended that mutual meetings be held, and training programs established in the provider's facilities, as these often have adequate facilities for it (Alipour 2011; Deering et al. 2011).
- Preferably, *Suppliers* should carry out the equipment installation in the manufacturer supply chains, unless the company has a lot of experience in the equipment that it has acquired, and wants to perform it by itself. Sometimes, the equipment warranty is not valid when they are altered by people who are not authorized by the *Suppliers*, and frequently, the installation is part of the signed contract between both entities (Shen 2015).
- Equipment *Suppliers* must visit the production systems in the company to identify technological levels that are installed, therefore, they can propose alternatives sales that are already in the market or may arise innovative ideas, where aspects associated to productivity and safety are improved (Shen 2015).
- It is the *Suppliers* responsibility to provide, by writing or through labels on the products, the storage conditions that the elements and spare parts must have in order to avoid their degradation or which quality characteristics are going to be lost when they are used.
- In the relationship that the senior management establishes with the *Suppliers*, it must always be considered that the equipment can provide the quality standards that are established in the operating manuals, but, among those aspects, under which type of working conditions it is possible to have them (Konecny and Thun 2011; Seth and Tripathi 2006). In addition, it helps to avoid work overloads on machines that may result in fatigue or in a decreasement on its useful life and parts.
- Purchasing decisions by management must consider several aspects, not only those that refer to the lower operating and maintenance costs, but also aspects related to safety environmental, and the traders that operate them must be guaranteed, because otherwise, that may be one of the reasons why the TPM program received low support (Rodrigues and Hatakeyama 2006).

9.6 Simple Hypotheses—PM Implementation

In this section, the hypotheses that arise from the relationship between the *PM implementation* with other variables are presented since this variable has already been related to two other variables in previous sections, then only nine hypotheses are presented here. In addition, the variables which it is related with are the following: *TPM implementation, Technological status, Layout, Warehouse management, Managerial commitment, Clients, Benefits for the organization, Productivity benefits,* and *Safety benefits.*

The hypotheses obtained from the relationships between the *PM implementation* and other variables are the following:

H_1 The *PM implementation* process in a production system has a direct and positive impact on the *TPM implementation*.
H_2 The *PM implementation* process in a production system has a direct and positive impact on the company's *Technological status*.
H_3 The *PM implementation* process in a production system has a direct and positive impact on the plant distribution or *Layout*.
H_4 The *PM implementation* process in a production system has a direct and positive impact on the parts and components *Warehouse Management*.
H_5 The *Managerial commitment* that a company has during the *TPM implementation* has a direct and positive impact on the *PM implementation*.
H_6 The *PM implementation* process in a production system has a direct and positive impact on the *Clients* influence and participation with TPM.
H_7 The *PM implementation* process in a production system has a direct and positive impact on the *Benefits for the organization* that are obtained when implementing TPM.
H_8 The *PM implementation* process in a production system has a direct and positive impact on the *Productivity benefits* that are obtained when implementing TPM.
H_9 The *PM implementation* process in a production system has a direct and positive impact on the *Safety benefits* that are obtained when implementing TPM.

9.6.1 Results—PM Implementation *Process*

Table 9.9 shows the results obtained after executing the simple models, where the independent variable is the *PM implementation* process while the response variables are other variables. As in previous tables, the variable which it is related with, the beta value, the *p*-associated value, the *R*-squared value, and the VIF value as a measure of collinearity is illustrated.

Based on the results illustrated in Table 9.9, the following can be concluded regarding the stated hypotheses:

H_1 There is enough statistical evidence to declare that the *PM implementation* process in a production system has a direct and positive impact on the *TPM implementation*, since when the first latent variable increases its standard deviation by one unit, the second increases in 0.700 units, and it can explain up to 49.1% of its variability.
H_2 There is enough statistical evidence to declare that the *PM implementation* process in a production system has a direct and positive impact on the company *Technological status*, since when the first latent variable increases its

Table 9.9 Relationships between the *PM implementation* and other variables

Hypotheses	Dependent variable	Parameter			
		β	*p*-value	R^2	VIF
H_1	*TPM implementation*	0.700	$p < 0.001$	0.491	1.939
H_2	*Technological status*	0.632	$p < 0.001$	0.399	1.639
H_3	*Layout*	0.558	$p < 0.001$	0.311	1.378
H_4	*Warehouse management*	0.580	$p < 0.001$	0.336	1.441
H_5	*Managerial commitment*	0.638	$p < 0.001$	0.407	1.671
H_6	*Clients*	0.441	$p < 0.001$	0.194	1.213
H_7	*Benefits for the organization*	0.519	$p < 0.001$	0.269	1.349
H_8	*Productivity benefits*	0.465	$p < 0.001$	0.216	1.235
H_9	*Safety benefits*	0.479	$p < 0.001$	0.230	1.280

standard deviation in one unit, the second increases in 0.632 units, and it can explain up to 39.9% of its variability.

H_3 There is enough statistical evidence to declare that the *PM implementation* process in a production system has a direct and positive impact on the plant distribution or *Layout*, since when the first latent variable increases its standard deviation in one unit, the second increases in 0.558 units, and it can explain up to 31.1% of its variability.

H_4 There is enough statistical evidence to declare that the *PM implementation* process in a production system has a direct and positive impact on the parts and components *Warehouse Management*, since when the first latent variable increases its standard deviation in one unit, the second increases in 0.580 units, and it can explain up to 33.6% of its variability.

H_5 There is enough statistical evidence to declare that the *Managerial commitment* that a company has during the *TPM implementation* has a direct and positive impact on the *PM implementation*, since when the first latent variable increases its standard deviation in one unit, the second increases in 0.0638 units, and it can explain up to 40.7% of its variability.

H_6 There is enough statistical evidence to declare that the *PM implementation* process in a production system has a direct and positive impact on the *Clients* influence and participation with TPM, since when the first latent variable increases its standard deviation in one unit, the second increases in 0.441 units, and it can explain up to 19.4% of its variability.

H_7 There is enough statistical evidence to state that the *PM implementation* process in a production system has a direct and positive impact on the *Benefits for the organization* that are obtained when implementing TPM, since when the first latent variable increases its standard deviation in one unit, the second increases in 0.519 units, and it can explain up to 26.9% of its variability.

H_8 There is enough statistical evidence to declare that the *PM implementation* process in a production system has a direct and positive impact on the *Productivity benefits* that are obtained when implementing TPM, since when the first latent variable increases its standard deviation in one unit, the second increases in 0.465 units, and it can explain up to 21.6% of its variability.

H_9 There is enough statistical evidence to state that the *PM implementation* process in a production system has a direct and positive impact on the *Safety benefits* that are obtained when implementing TPM, since when the first latent variable increases its standard deviation in one unit, the second increases in 0.479 units, and it can explain up to 23.0% of its variability.

9.6.2 Conclusions and Industrial Implications—PM Implementation

Based on the data from Table 9.9, some general conclusions can be mentioned, such as the following:

- The *PM implementation* is a variable that has direct and positive effects on all the variables which it is related with since the *p*-value associated with β is over 0.05, therefore, this inference can be achieved with 95% of reliability.
- The β magnitude is significant since in this case, the maximum value is the one related to the *TPM implementation*, which indicates that it should always start with a preventive maintenance program to complete it with a TPM program. Similarly, the β lowest value is related to the *Clients*, as it has happened in previous relationships.
- Those high and low β values in some relationships lead to high and low *R*-square values. Thus, the highest values in this relationship are because of the *TPM implementation* and *Managerial commitment*, but the lowest values are found in *Clients*, *Productivity benefits*, and *Safety benefits*.
- There are no problems of collinearity in the latent variables analysis since all relationships have associated VIF values under 3.3.
- The previous results allow to make a series of conclusions and industrial implications in the sector where the study was carried out, such as the following:
- The preventive maintenance program existence will always be adequate for the total productive maintenance program expansion where the entire company is integrated (Anand et al. 2018).
- The *PM implementation* success depends on the *Managerial commitment* and other programs support; however, it is the management's responsibility to obtain parameters and indicators that allow identifying the causes why the planned goals are not achieved if there is a case (Kim et al. 2004). In addition, all the personnel in the company must be an organization that learns from mistakes and takes actions in order that these failures are not repeated, therefore, not reaching some indicators should not be enough reason for management to quit those projects and withdraw them their support, on the contrary, they should seek to improve the implementation processes and identify the errors.

- If a company strives to always have assigned days to perform the equipment preventive maintenance, then they will always be in optimal working conditions, and the quality and standards requested by the client can be guaranteed. In the same way, there will be a constant materials flow throughout the entire supply chain, where the storage process is reduced (Fumagalli et al. 2017).
- Preventive maintenance does not only depend on a specialized department but depends on everyone in the company, especially the operators who are the ones who know and observe the machines performance every day. Therefore, it is very important to look forward to train operators in order that they are the ones who keep a record about the conditions that the machines have under their responsibility. Likewise, these operators must be trained about the equipment use and management to avoid breakdowns due to inappropriate operations or poor performance (Eti et al. 2006c).
- Preventive maintenance should be an open program, with accessible information for everyone, hence, the operator must know when major repairs will be made on the machines that they operate. Also, the production department must be informed to assign a worker for another job if necessary. In addition, this coordination between departments and operator helps to have less losses for workers and idle machines (Shrivastava et al. 2016).
- One way to monitor the preventive maintenance programs success is to monitor the quality of products outcomes that are processed in each machine to perform corrective actions as soon as possible because it is not convenient to wait to finish whole batches of production to assess if they were manufactured according to the quality specifications required by the client (Eti et al. 2006a).
- Preventive maintenance should be focused on preventing failures from happening, but in case they already happened, a study must always be carried out to determine the main cause of the problems generated in the machines (Wu and Seddon 1994). Operators are a valuable source of information in this case.
- When a machine causes constant failures during the production process, the *Technological status* that it has with those already installed in the market must be reviewed, since it is possible that it is cheaper to buy and install a new one instead of maintaining one that is already damaged (Kim et al. 2004).
- It is important that the maintenance managers have the precise knowledge about the useful life of each machine in the production line.
- Those responsible for maintenance must know the useful life of the machines components, as well as keep a record about those that are used frequently, which will allow a better spare parts and accessories *Warehouse management*.
- The proper preventive maintenance planning facilitates the machines administration, the productive processes, and decreases the costs due to technical failures, which provides *Benefits for the organization*, which are associated with low administrative costs, repair speed, better relationships with *Suppliers*, among others. However, it should be mentioned that it is no coincidence that machines in good operating condition are safer for workers who operate them (Eti et al. 2006a).

9.7 Simple Hypotheses—TPM Implementation

In this section, one of the most important variables is analyzed, the *TPM implementation* process in the company's production system. In addition, some relationships between this variable have already been exposed as hypothesis in previous sections, consequently, only the relationships between the following variables are analyzed: *Technological status, Layout, Warehouse management, Managerial commitment, Clients, Benefits for the organization, Productivity benefits,* and *Safety benefits.* According to the previous information, eight hypotheses that can be proposed are the following:

H_1 The *TPM implementation* process in a productive system has a direct and positive effect on the company *Technological status*.

H_2 The *TPM implementation* process in a productive system has a direct and positive effect on the plant distribution or *Layout*.

H_3 The *TPM implementation* process in a productive system has a direct and positive effect on the company *Warehouse management* of parts, components, and consumables.

H_4 The *Managerial commitment* process during the TPM adoption process has a direct and positive effect on the *TPM implementation* process in a productive system.

H_5 The *TPM implementation* process in a productive system has a direct and positive effect on the company's *Clients* or customers.

H_6 The *TPM implementation* process in a productive system has a direct and positive effect on the *Benefits for the organization* obtained in the company.

H_7 The *TPM implementation* process in a productive system has a direct and positive effect on the *Productivity benefits* obtained in the company.

H_8 The *TPM implementation* process in a productive system has a direct and positive effect on the *Safety benefits* obtained in the company.

9.7.1 Results—TPM Implementation

Table 9.10 illustrates the results obtained after evaluating the relationships between the *TPM implementation* process with other latent variables. In addition, the variables which it is related with is presented, as well as the hypothesis that it represents, the β value, the associated p-value, the R-squared value as a measure of the explained variability, and the VIF as a measure of collinearity.

Considering the values illustrated in Table 9.10, the following is concluded regarding the hypotheses proposed at the beginning of this section:

H_1 There is enough statistical evidence to state that the *TPM implementation* process in a productive system has a direct and positive effect on the company *Technological status*, since when the first variable increases its standard

Table 9.10 Relationships between the *TPM implementation* and other variables

Hypotheses	Dependent variable	Parameter			
		β	*p*-value	R^2	VIF
H_1	*Technological status*	0.742	$p < 0.001$	0.550	2.212
H_2	*Layout*	0.635	$p < 0.001$	0.403	1.645
H_3	*Warehouse management*	0.656	$p < 0.001$	0.430	1.671
H_4	*Managerial commitment*	0.768	$p < 0.001$	0.590	2.436
H_5	*Clients*	0.493	$p < 0.001$	0.244	1.299
H_6	*Benefits for the organization*	0.563	$p < 0.001$	0.317	1.427
H_7	*Productivity benefits*	0.575	$p < 0.001$	0.330	1.415
H_8	*Safety benefits*	0.564	$p < 0.001$	0.319	1.416

deviation in one unit, the second variable increases in 0.742 units, and it can explain up to 55.0% of its variability.

H_2 There is enough statistical evidence to declare that the *TPM implementation* process in a productive system has a direct and positive effect on the plant distribution or *Layout*, since when the first variable increases its standard deviation in one unit, the second variable increases in 0.635 units, and it can explain up to 40.3% of its variability.

H_3 There is enough statistical evidence to declare that the *TPM implementation* process in a productive system has a direct and positive effect on the company *Warehouse management* of parts, components, and consumables, since when the first variable increases its standard deviation in one unit, the second one increases in 0. 656 units, and it can explain up to 43.0% of its variability.

H_4 There is enough statistical evidence to state that the *Managerial commitment* process during the TPM adoption process has a direct and positive effect on the *TPM implementation* process in a productive system, since when the first variable increases its standard deviation in one unit, the second variable increases in 0.768 units, and it can explain up to 59.0% of its variability.

H_5 There is enough statistical evidence to declare that the *TPM implementation* process in a productive system has a direct and positive effect on the company's *Clients* or customers, since when the first variable increases its standard deviation in one unit, the second variable increases in 0.493 units, and it can explain up to 24.4% of its variability.

H_6 There is enough statistical evidence to declare that the *TPM implementation* process in a productive system has a direct and positive effect on the *Benefits for the organization* obtained in the company, since when the first variable increases its standard deviation in one unit, the second variable increases in 0.563 units, and it can explain up to 31.7% of its variability.

H_7 There is enough statistical evidence to declare that the *TPM implementation* process in a productive system has a direct and positive effect on the *Productivity benefits* obtained in the company, since when the first variable increases its standard deviation in one unit, the second increases in 0.575 units, and it can explain up to 33.0% of its variability.

H_8 There is enough statistical evidence to declare that the *TPM implementation* process in a productive system has a direct and positive effect on the *Safety benefits* obtained in the company, since when the first variable increases its standard deviation in one unit, the second increases in 0.564 units, and it can explain up to 31.9% of its variability.

9.7.2 *Conclusions and Industrial Implications*—PM Implementation

The general conclusions that can be obtained from the relationships between the *TPM implementation* process and other variables are the following:

- All the relationships that the *TPM process* has are statistically significant, because the *p*-value associated with the β value is under 0.05, as a result, all the assertions can be established with 95% of reliability.
- In this specific case, it is important to observe the relationship that exists along with the *Technological status* and *Managerial commitment* variables, since they are the highest values that have been generated until this moment, with 0.742 and 0.768, respectively. However, the lowest β value is found along with the *Clients*.
- These β high values influence that the *R*-squared value will be large as well. For example, the relationship with the *Managerial commitment* has a *R*-squared value = 0.590, which indicates that up to 59% of its variability can be explained.
- There are no problems of collinearity between the variables since the VIF values are under 3.3; the maximum value allowed.
- It is important to mention that according to the temporality of events principle, some of the relationships between variables exposed here may change their role, for instance, the *Managerial commitment* toward the *TPM implementation* process, as it is illustrated in hypothesis H_5.
- In the same way, the information in Table 9.10 allows to include a series of aspects related to the *TPM implementation* in the production lines, and these results have a series of industrial implications, as it is described below:
- The *TPM implementation* process does not take place by itself, but rather it is the company's human resources that plan and carry it out, for this reason, training and education programs must be included in this process (Hana 2010).
- It is the management and supervisors responsibility to follow up on the machines and equipment maintenance task, shutting down the initial plans, and take corrective actions when there are deviations from the objectives, which

must be informed to their operators; it promotes communication and integration in the productive systems (Ahmadzadeh and Bengtsson 2017; Okoh and Haugen 2015).

- The senior management and maintenance supervisors must demonstrate leadership in the planning and execution of TPM plans and programs, since, if employees do not identify that commitment, then they will not be motivated to fix it, and it may be a TPM project abandonment accusation. It is important to recall that workers are a reflection from the attitudes that higher commands have (Akroyd et al. 2009; Gong et al. 2009).
- The plans and programs communication during the *TPM implementation* process are vital to guarantee success, consequently, there must be an information flow in all the senses: horizontal, from the workers in a shift to another shift, vertical, from supervisors to operators and vice versa. In addition, it is important to mention that operators must know the routine maintenance programming tasks, which should be integrated into these activities or into another production line. Also, it is recommended that informative meetings be held periodically between the operators and supervisors from different departments to exchange maintenance information as well as production system needs (Deering et al. 2011; Rico and Cohen 2005).
- In order to avoid operational problems, during the training processes, it is relevant that operators are instructed on the critical machines aspects; in other words, they are the parameters that the equipment can perform without sacrificing its quality and integrity from the operators themselves (Alfonso and Mara 2016).
- The productive systems change constantly, therefore, it is very common that the generated knowledge is replaced by a newer one in other to have the possibility of investing in new training, new machinery, and new work tools. Also, each time a new technological investment is complied, adequate training should be provided to operators (McKone and Weiss 2000).
- When purchasing new equipment, management must always consider the aspects associated with maintenance, since many times highly specialized personnel are required to carry it out, therefore, there will always be a dependence on the supplier. In addition, those investments in new equipment are those that will define the *Technological status* that the company has against its competitors, and it will determine in a certain way, the level of competitiveness and the possibility of complying with technical specifications (Bourke and Roper 2016; Cardoso et al. 2012; McKone and Weiss 2000).
- The *TPM implementation* should always consider aspects associated with the plant distribution or *Layout*, since many aspects associated with safety are studied from this point of view; in other words, narrow corridors should be avoided, saddles should be integrated to protect equipment in operation, routes that pieces will follow should be established, as well as what automated system will perform the task. In addition, it is significant to remember that maintenance is directly linked to the *Safety benefits* that can be obtained, and having wide aisles is safer for the company, however, they require enough space, which will be idle most of the time (Pang et al. 2016).

- The *TPM implementation* should not be limited to production systems, because it should be extended to the warehouses where the components and spare parts are arranged, as a result, TPM does not focus on keeping only the machines of the production system, but on ensuring that those parts or components are preserved and in good condition when required, since not having the required parts at the right time may be a limitation for TPM, which cause delays, idle times, and breach of contracts (Attri et al. 2013).
- As an integral philosophy, TPM must consider the company *Clients* to carry out and execute their work plans. In addition, the standards that they must adjust and calibrate to machines in order to offer the required quality must be known. Also, demand trends must also be acknowledged to know if the installed production capacity is enough, analyze the purchase needs from the new equipment if required due to an increased demand, or lease part of the production process to an external manufacturer if the demand is only by season (Azizi 2015; Morales Méndez and Rodriguez 2017).
- In order to comply with the plans and programs during the *TPM implementation* process, it will undoubtedly bring a series of benefits, as a result, the company must make an effort to quantify them from different points of view, since it is the only way to justify investments in this type of programs (Tong et al. 2015; Willmott and McCarthy 2001a).

9.8 Simple Hypotheses—Technological Status

Undoubtedly, having a high *Technological status* in production lines has several advantages and disadvantages for maintenance personnel. First, highly specialized personnel are required for this activity, but also, these systems have many sensors that help to identify malfunction problems quickly, therefore, the productive systems are not idle.

Since the relationship between the *Technological status* with other variables has already been studied, this section only reports relationships between: *Layout, Warehouse management, Managerial commitment, Clients, Organizational benefits, Productivity benefits,* and *Safety benefits*; the hypotheses that can be proposed are the following:

H_1 The *Technological status* that a company has in its production lines has a direct and positive effect on the plant distribution or *Layout* during the *TPM implementation.*

H_2 The *Technological status* that a company has in its production lines has a direct and positive effect on the parts and consumables *Warehouse management.*

H_3 The *Management commitment* during the TPM adoption process has a direct and positive effect on the *Technological status* that a company has on its production lines.

H_4 The *Technological status* that a company has in its production lines has a direct and positive effect on the products and services *Clients* that the company has during the *TPM implementation.*

H_5 The *Technological status* that a company has in its production lines has a direct and positive effect on the *Benefits for the organization* that can be obtained during the *TPM implementation.*

H_6 The *Technological status* that a company has in its production lines has a direct and positive effect on the *Productivity benefits* that can be obtained during the *TPM implementation.*

H_7 The *Technological status* that a company has in its production lines has a direct and positive effect on the *Safety benefits* that can be obtained during the *TPM implementation.*

9.8.1 Results—Technological Status

Table 9.11 portrays the results from the relationship between the *Technological status* independent latent variable with other variables; as in previous sections, the dependent variables which the *Technological status* is related with, the beta regression parameter, and its respective *p*-value, the *R*-squared value as a measure of the explained variance, and the VIF as a measure of collinearity are illustrated.

The results presented in Table 9.11 allow to conclude the following regarding the initial hypotheses:

H_1 There is enough statistical evidence to declare that the *Technological status* that a company has in its production lines has a direct and positive effect on the plant distribution or *Layout* during the *TPM implementation*, since when the first variable increases its standard deviation in one unit, the second variable increases in 0.328, and can explain up to 39.5% of its variability.

H_2 There is enough statistical evidence to declare that the *Technological status* that a company has in its production lines has a direct and positive effect on the

Table 9.11 Relationships between *Technological status* and other variables

Hypotheses	Dependent variable	Parameter			
		β	*p*-value	R^2	VIF
H_1	*Layout*	0.628	$p < 0.001$	0.395	1.645
H_2	*Warehouse management*	0.631	$p < 0.001$	0.398	1.633
H_3	*Managerial commitment*	0.719	$p < 0.001$	0.517	2.072
H_4	*Clients*	0.477	$p < 0.001$	0.228	1.270
H_5	*Benefits for the organization*	0.546	$p < 0.001$	0.298	1.376
H_6	*Productivity benefits*	0.517	$p < 0.001$	0.268	1.320
H_7	*Safety benefits*	0.524	$p < 0.001$	0.275	1.344

parts and consumables *Warehouse management*, since when the first variable increases its standard deviation in one unit, the second variable increases in 0.631, and can explain up to 39.8% of its variability.

H_3 There is enough statistical evidence to declare that the *Management commitment* during the TPM adoption process has a direct and positive effect on the *Technological status* that a company has on its production lines, since when the first variable increases its standard deviation in one unit, the second variable increases in 0.719, and can explain up to 51.7% of its variability.

H_4 There is enough statistical evidence to declare that the *Technological status* that a company has in its production lines has a direct and positive effect on the products and services *Clients* that the company has during the *TPM implementation*, since when the first variable it increases its standard deviation in one unit, the second variable increases in 0.477, and can explain up to 22.8% of its variability.

H_5 There is enough statistical evidence to declare that the *Technological status* that a company has in its production lines has a direct and positive effect on the *Benefits for the organization* that can be obtained during the *TPM implementation*, since when the first variable increases its deviation standard in one unit, the second variable increases in 0.546, and can explain up to 29.8% of its variability.

H_6 There is enough statistical evidence to declare that the *Technological status* that a company has in its production lines has a direct and positive effect on the *Productivity benefits* that can be obtained during the *TPM implementation*, since when the first variable increases its standard deviation in one unit, the second variable increases in 0.517, and can explain up to 26.8% of its variability.

H_7 There is enough statistical evidence to declare that the *Technological status* that a company has in its production lines has a direct and positive effect on the *Safety benefits* that can be obtained during the *TPM implementation*, since when the first variable increases its standard deviation in one unit, the second variable increases in 0.527, and can explain up to 27.5% of its variability.

9.8.2 Conclusions and Industrial Implications— Technological Status

When reviewing the data in Table 9.11, the following can be concluded in general terms based on the models presented:

- The *Technological status* as an independent latent variable has a direct and positive effect on: *Layout, Warehouse management, Managerial commitment, Clients, Benefits for the organization, Productivity benefits,* and *Safety benefits* since the *p*-values associated with β parameters are always less than 0.05.

- The highest relationship that the *Technological status* variable has is with the *Managerial commitment*, with a β-value equal to 0.719, which has a broad common sense, even though the role of the variables in the analysis is interchanged, since the senior management makes the investment decisions in new equipment and machinery, therefore, it depends on the production systems technological level and modernity. Meanwhile, the lowest β-value is found in the relationship with *Clients*, as it is an external entity to the company.
- It can be said that the relationships that the *Technological status* variable has with the other variables have enough predictive validity because in all of them there is a R-squared value greater than 0.02.
- There are no collinearity problems between the *Technological status* variable with other variables since the VIF value is less than 3.3 in each of them.
- This allows to make a series of conclusions associated with the industrial implications that these results entail, such as the following:
- The company, through the senior management, must seek to keep up with the technological advances that arise in the market to update its production system. In addition, it implies that maintenance managers, as well as engineering and production managers, look for alternatives to find equipment in the market that is capable of generating greater production for the company, better safety levels for operators or they are simpler and occupy less volume in the plant distribution (Bourke and Roper 2016).
- Technology has a useful life cycle; therefore, a company should always be considering the new machinery and equipment generation that will be installed in production systems in the future. In addition, some government systems allow the equipment devaluation, which indicates that they can estimate taxes from investments in technologies and, therefore, after a certain time, that equipment is totally estimated, and a new one can be acquired. Also, devaluation times may vary greatly depending on the types of equipment, for example, computer systems deteriorate themselves much faster than a production equipment (Cardoso et al. 2012).
- Management must be committed to making the company a technological leader, since, otherwise, in this globalized era it is easier for other companies to move and occupy the niche market that has conquered. Consequently, the best way to know the technological level that a company has is to compare itself with national and international competitors (Fulton and Hon 2010).
- It is common that the company requires more production capacity than it has installed, however, the new equipment purchase is expensive, as a result, it is recommended that the option of modifying and upgrading technology should be always considered when it is already installed, which consists of certain components purchase that are added to the existing ones. In addition, some managers mention that often the productive capacity can be increased by making technological modifications with suggestions from the operators themselves, and with the maintenance department support (Aravindan and Punniyamoorthy 2002).

- It should be remembered that maintaining a high technological level within the company is not the goal, but rather having an economically productive company, therefore, before making any machines acquisition, it must be analyzed if they can offer a competitive advantage to the company when they are operating. Also, the new technologies justification must be an activity that is performed with care (Koc and Bozdag 2009).
- In the highly technological equipment, the maintenance software provided should always be used by the *Suppliers* to perform diagnostics and identify failures quickly in machines and equipment. It is important to remember that many machines with a high technological level have many integrated sensors that may help to monitor their operational status (Realyvásquez-Vargas et al. 2014).
- When a company has a high *Technological status*, it is required to have solid relationships with their equipment maintenance services *Suppliers*, in case they do not have them in their own plant. However, something that should be discussed with *Suppliers* is the storage level that there should be in the stores regarding components and consumables (Spanos and Voudouris 2009).
- Similarly, the company must maintain direct relationships with its *Clients*, know the future products demand, and, based on this, plan the technological capacity that must be installed. Also, it is not advisable to make heavy economic investments in machinery and equipment that is idle due to the high demand (Choe 2004).
- When purchasing a new machine, the size and automation level in the company must be considered, since currently small machines in physical dimensions can perform several activities if they are reprogrammed by an expert, and by adding components, functions can be added as well (Fulton and Hon 2010).

9.9 Simple Hypotheses—Layout

A good plant distribution that undoubtedly helps the *TPM implementation* process, since many times that aspect does not facilitate this task and even, frequently, they are a source of risks. Although the relationship between the *Layout* with other variables has been studied previously, in this section the relationship with the following latent variables is analyzed: *Warehouse management, Managerial commitment, Clients, Benefits for the organization, Productivity benefits,* and *Safety benefits*.

The hypotheses that can be considered where the *Layout* is the independent latent variable and is related to the others are the following:

H_1 The plant distribution or *Layout* that a company has in its production lines has a direct and positive effect on the parts and consumables *Warehouse Management* during the *TPM implementation*.

H_2 The *Managerial commitment* during the *TPM implementation* has a direct and positive effect on the productive system plant distribution or *Layout*.

H_3 The plant distribution or *Layout* that a company has in its production lines has a direct and positive effect on the *Clients* that it has for its goods and services during the *TPM implementation*.

H_4 The plant distribution or *Layout* that a company has in its production lines has a direct and positive effect on the *Benefits for the organization* that can be obtained during *TPM implementation*.

H_5 The plant distribution or *Layout* that a company has in its production lines has a direct and positive effect on the H_5 on the *Productivity benefits* that can be obtained during the *TPM implementation*.

H_6 The plant distribution or *Layout* that a company has in its production lines has a direct and positive effect on the *Safety benefits* that can be obtained during the *TPM implementation*.

9.9.1 *Results*—Layout

After evaluating the relationships that are defined in the previous models, the results in Table 9.12 are obtained, where, as in previous cases, the name of the dependent latent variable, the β-value and the p-value associated for the significance statistical test, the R-squared value as a measure of predictive validity, and the VIF as a measure of collinearity are illustrated.

Based on the data from Table 9.12, the following can be concluded regarding the hypotheses initially proposed:

H_1 There is enough statistical evidence to declare that the plant distribution or *Layout* that a company has in its production lines has a direct and positive effect on the parts and consumables *Warehouse management* during the *TPM implementation*, since when the first latent variable increases its standard deviation in one unit, the second variable increases in 0.706, and it can explain up to 49.8% of its variability.

Table 9.12 Relationships between *Technological status* and other variables

Hypotheses	Dependent variable	Parameter			
		β	p-value	R^2	VIF
H_1	*Warehouse management*	0.706	$p < 0.001$	0.498	1.958
H_2	*Managerial commitment*	0.644	$p < 0.001$	0.414	1.704
H_3	*Clients*	0.534	$p < 0.001$	0.286	1.381
H_4	*Benefits for the organization*	0.513	$p < 0.001$	0.263	1.37
H_5	*Productivity benefits*	0.505	$p < 0.001$	0.255	1.334
H_6	*Safety benefits*	0.473	$p < 0.001$	0.224	1.281

H_2 There is enough statistical evidence to declare that the *Managerial commitment* during the *TPM implementation* has a direct and positive effect on the productive system plant distribution or *Layout*, since when the first latent variable increases its standard deviation in one unit, the second variable increases in 0.644, and it can explain up to 41.4% of its variability.

H_3 There is enough statistical evidence to declare that the plant distribution or *Layout* that a company has in its production lines has a direct and positive effect on the *Clients* that it has for its goods and services during the *TPM implementation*, since when the first latent variable increases its standard deviation in one unit, the second variable increases in 0.534, and it can explain up to 28.6% of its variability.

H_4 There is enough statistical evidence to state that the plant distribution or *Layout* that a company has in its production lines has a direct and positive effect on the *Benefits for the organization* that be obtained during *TPM implementation*, since when the first variable latent it increases its standard deviation in one unit, the second variable increases in 0.513, and it can explain up to 26.3% of its variability.

H_5 There is enough statistical evidence to state that the plant distribution or *Layout* that a company has in its production lines has a direct and positive effect on the on the *Productivity benefits* that can be obtained during the *TPM implementation*, since when the first latent variable increases its standard deviation in one unit, the second variable increases in 0.505, and it can explain up to 25.5% of its variability.

H_6 There is enough statistical evidence to state that the plant distribution or *Layout* that a company has in its production lines has a direct and positive effect on the *Safety benefits* that can be obtained during the *TPM implementation*, since when the first latent variable increases its standard deviation in one unit, the second one increases in 0.473, and it can explain up to 22.4% of its variability.

9.9.2 *Conclusions and Industrial Implications*—Layout

After executing the models related to the plant distribution or *Layout* with other variables, as well as analyzing the data in Table 9.12, the following is concluded:

- The plant distribution or *Layout* has a direct and positive effect on the other variables since the p-value associated with β is less than 0.05, consequently, inferences can be presented with 95% of reliability.
- The highest relationship that the *Layou*t has is along with the *Warehouse management*, with a value of 0.706, and interestingly, the lowest β-value is along with the *Safety benefits*, when it is well known and reported in the literature review that one of the *Layout* principles is to improve that aspect.

- The models have enough predictive validity since in all the relationships the *R*-square value was over 0.02.
- There is no collinearity between the variables that intervene in the model.
- However, the most important of the analysis from the previous relationships, are the industrial implications that these may have in the industrial field, therefore, that is described below.
- As a matter of fact, moving items from one place to another costs, and it does not add value to the product, therefore, it is convenient that the plant distributions or *Layout* is focused on avoiding unnecessary costs and risks in the production systems. According to the previous information, classifying products and machines in groups have been successful, such as technology groups (Attri et al. 2013).
- The machines distribution should not only focus on operating costs, but also on the ease of providing the maintenance that will be required in the future (usually called maintainability). For instance, access with huge machinery to move heavy parts, which requires wide corridors (Willmott and McCarthy 2001b).
- An important *Layout* aspect is to have the appropriate signs in all facilities, indicate the positions of emergency equipment, such as fire extinguishers, hoses with hydrants, first aid kits, danger places by movement of parts, meeting places, among others (McCarthy and Rich 2015).
- An appropriate plant distribution or *Layout* must be focused on generating a series of productive benefits, such as less circulation and materials flow, decreasement of losses by quality, keep equipment and machines in operational and optimal conditions to decrease machines and operators time idle, among others (Cua et al. 2001).
- The *Layout* should also focus on the operator safety, as a result, it should focus on establishing safe operation rules, lines establishment, and safety places in the production system, provide escape routes if necessary, and brigades to face contingencies, such as fires, floods, among others (Borkowski et al. 2014).

9.10 Simple Hypotheses—Warehouse Management

Regarding the *TPM implementation*, a vital aspect is the consumables and spare parts *Warehouse Management*, since there may be adequate maintenance plans and programs, but if the replacement is not available when required, then there will be a lot of idle time for machines and operators. In this section, the relationships between the *Warehouse management* variable and other variables are analyzed such as *Managerial commitment, Clients, Benefits for organization, Productivity benefits,* and *Safety benefits.*

The hypotheses that can be proposed where the *Warehouse management* is the independent variable are the following:

H_1 The *Managerial commitment* during the *TPM implementation* process has a direct and positive effect on the *Warehouse management* of spare parts and consumables.

H_2 The adequate *Warehouse management* of spare parts and consumables has a direct and positive effect on the *Clients* that the company has during the *TPM implementation* process.

H_3 The adequate *Warehouse management* of spare parts and consumables has a direct and positive effect on the *Benefits for the organization* that the company has during the *TPM implementation* process.

H_4 The adequate *Warehouse management* of spare parts and consumables has a direct and positive effect on the *Productivity benefits* that the company has during the *TPM implementation* process.

H_5 The adequate *Warehouse management* of spare parts and consumables has a direct and positive effect on the *Safety benefits* that the company has during the *TPM implementation* process.

9.10.1 *Results*—Warehouse Management

After evaluating the relationships between the *Warehouse management* and other variables; the obtained results are illustrated in Table 9.13, where the name of the variable which it is related with, the beta value, the *p*-value for the significance statistical test, the *R*-squared value as a measure of the explained variability, and the VIF as a measure of collinearity between the variables are portrayed.

The conclusions that can be declared regarding the hypotheses after reviewing the data in Table 9.13 are indicated below:

H_1 There is enough statistical evidence to declare that the *Managerial commitment* during the *TPM implementation* process has a direct and positive effect on the *Warehouse management* of spare parts and consumables, since when the first latent variable increases its deviation standard in one unit, the second variable increases in 0.600 units, and it can explain up to 36.0% of its variability.

Table 9.13 Relationships between *Warehouse management* and other variables

Hypotheses	Dependent variable	Parameter			
		β	*p*-value	R^2	VIF
H_1	*Managerial commitment*	0.600	$p < 0.001$	0.360	1.534
H_2	*Clients*	0.465	$p < 0.001$	0.216	1.238
H_3	*Benefits for the organization*	0.536	$p < 0.001$	0.287	1.396
H_4	*Productivity benefits*	0.516	$p < 0.001$	0.267	1.363
H_5	*Safety benefits*	0.493	$p < 0.001$	0.243	1.316

H_2 There is enough statistical evidence to declare that the adequate *Warehouse management* of spare parts and consumables has a direct and positive effect on the *Clients* that the company has during the *TPM implementation* process, since when the first latent variable increases its standard deviation in one unit, the second variable increases in 0.465 units, and it can explain up to 21.6% of its variability.

H_3 There is enough statistical evidence to state that the adequate *Warehouse management* of spare parts and consumables has a direct and positive effect on the *Benefits for the organization* that the company has during the *TPM implementation* process, since when the first variable latent increases its standard deviation in one unit, the second variable one in 0.536 units, and it can explain up to 28.7% of its variability.

H_4 There is enough statistical evidence to state that the adequate *Warehouse management* of spare parts and consumables has a direct and positive effect on the *Productivity benefits* that the company has during the *TPM implementation* process, since when the first latent variable increases its standard deviation in one unit, the second variable increases in 0.516 units, and it can explain up to 26.7% of its variability.

H_5 There is enough statistical evidence to declare that the adequate *Warehouse management* of spare parts and consumables has a direct and positive effect on the *Safety benefits* that the company has during the *TPM implementation* process, since when the first latent variable increases its standard deviation in one unit, the second variable increases in 0.493 unit, and it can explain up to 24.3% of its variability.

9.10.2 *Conclusions and Industrial Implications—* Warehouse Management

In general terms, it can be observed that the simple models where the *Warehouse management* is related have the following characteristics:

- The five relationships are statistically significant since the *p*-values are less than 0.05, therefore, inferences can be made with 95% of reliability.
- It is observed that the largest relationship with the *Warehouse management* is along with the *Managerial commitment*, which makes sense since they are administrative aspects. Similarly, the lowest relationship is along with *Clients*, as has happened with other variables, which may be because they are external to the company.
- There are no collinearity problems among the variables analyzed, since the VIF values are under 3.3.

- There is enough predictive validity in the models and among the variables that are related since the *R*-square values are greater than 0.02. According to the previous information, the models are interpreted.
- The results obtained previously have a series of industrial implications for the manufacturing sector, which are explained below:
- The senior management is directly in charge of the *Warehouse management*, as well as of the consumables and spare parts storage levels, therefore, along with those responsible for the maintenance and production programs, they must generate lists of products that are critical to the machines operation, also they must monitor how much idle equipment passes through high administration associated with warehouses (Laureani and Antony 2017; Tjosvold and Tjosvold 2015).
- Although there is a direct relationship with *Clients,* the *Warehouse management* will undoubtedly have other indirect relationships; this relationship implies that the spare parts availability should not cause delays in deliveries or delivery incomplete orders (Grout and Christy 1999; Morlock et al. 2014).
- If there is an adequate *Warehouse management*, then there will be a series of benefits, among them the following can be mentioned: it avoids having to get items unexpectedly, purchases are planned and are based on the components useful life, the idle time is reduced in machines and operators, the quality standards expected by the client are complied, and orders are delivered on time (Fandel and Trockel 2016).

9.11 Simple Hypotheses—Managerial Commitment

The *Managerial commitment* is one of the principal factors in the *TPM implementation* success since it is responsible for generating maintenance plans and programs, keep a track of them as well as taking corrective actions if necessary. In this section, the *Managerial commitment* is related to the following variables: *Clients, Benefits for the organization, Productivity benefits*, and *Safety benefits*.

The hypotheses that can be established are the following:

H_1 The *Managerial commitment* during the *TPM implementation* process has a direct and positive effect on the company *Clients*.

H_2 The *Managerial commitment* during the *TPM implementation* process has a direct and positive effect on the *Benefits for the organization* that the company can obtain.

H_3 The *Managerial commitment* during the *TPM implementation* process has a direct and positive effect on the *Productivity benefits* that the company can obtain.

H_4 The *Managerial commitment* during the *TPM implementation* process has a direct and positive effect on the *Safety benefits* that the company can obtain for the operators and the environment.

9.11.1 Results—Managerial Commitment *Process*

When evaluating the simple structural equation models to validate the proposed hypotheses, the results obtained are presented in Table 9.14, where the dependent variable, the *p*-value associated, the *R*-squared value, and the VIF values are illustrated.

After evaluating the relationships indicated above in the hypotheses, the following conclusions are reached:

H_1 There is enough statistical evidence to declare that the *Managerial commitment* during the *TPM implementation* process has a direct and positive effect on the company *Clients*, since when the first latent variable increases its standard deviation in one unit, the second variable increases in 0.566, and it can explain up to 32.0% of its variability.

H_2 There is enough statistical evidence to state that the *Managerial commitment* during the *TPM implementation* process has a direct and positive effect on the *Benefits for the organization* that the company can obtain, since when the first latent variable increases its standard deviation by one unit, the second variable increases in 0.585, and it can explain up to 34.2% of its variability.

H_3 There is enough statistical evidence to declare that the *Managerial commitment* during the *TPM implementation* process has a direct and positive effect on the *Productivity benefits* that the company can obtain, since when the first latent variable increases its standard deviation in one unit, the second variable increases in 0.561 and can explain up to 31.4% of its variability.

H_4 There is enough statistical evidence to declare that the *Managerial commitment* during the *TPM implementation* process has a direct and positive effect on the *Safety benefits* that the company can obtain for the operators and the environment, since when the first latent variable increases its standard deviation in one unit, the second variable increases in 0.568, and it can explain up to 32.2% of its variability.

Table 9.14 Relationships between *Managerial commitment* and other variables

Hypotheses	Dependent variable	Parameter			
		β	*p*-value	R^2	VIF
H_1	*Clients*	0.566	$p < 0.001$	0.320	1.413
H_2	*Benefits for the organization*	0.585	$p < 0.001$	0.342	1.482
H_3	*Productivity benefits*	0.561	$p < 0.001$	0.314	1.404
H_4	*Safety benefits*	0.568	$p < 0.001$	0.322	1.447

9.11.2 *Conclusions and Industrial Implications—* **Managerial Commitment**

The *Managerial commitment* is one of the most important aspects in the *TPM implementation* success in the production lines; it is a tool that should be seen as a competitive advantage that allows companies to comply with production orders in shape and time, and although many relationships have been analyzed previously, regarding its models, the following can be concluded:

- The *Managerial commitment* has a direct and positive effect on the other variables, since the *p*-value associated with the β value is less than 0.05, which indicates that all inferences can be made with at least 95% of reliability.
- The *Managerial commitment* variable that has the highest β value is along with the *Clients*, which indicates that, in the end, TPM is only a tool that helps the management to have a better fulfillment of the requirements and demands of this variable. In other words, the management must be highly focused on *Clients*.
- There are no collinearity problems in the simple models that relate the *Managerial commitment* with other variables since the VIF value is less than 3.3 in all relationships.
- The simple models where the *Managerial commitment* relates to other variables have enough predictive validity.
- The results obtained in Table 9.14 have a series of industrial implications regarding *Managerial commitment*, as it is indicated below:
- It is crucial that managers and department heads accept their responsibility with the *TPM implementation* process, since they are responsible for making decisions in case there are deviations and they focus them to comply with established production orders (Seth and Tripathi 2006).
- The senior management must demonstrate leadership in the *TPM implementation* process, but, in addition, it must seek to integrate this tool, its plans, and programs with other techniques already implemented, such as Kaizen, SMED, 5 s, among others. Also, the management department is the link between the different kind of people responsible for applying these techniques, as well as for maintaining the greatest possible coordination among everyone (Ahmad et al. 2012; Konecny and Thun 2011; Seth and Tripathi 2006).
- As part of the senior management and TPM commitment, meetings must be held periodically in order that the production and maintenance departments reach agreements and know their respective plans. In addition, this coordination between both departments is vital to avoid misunderstandings and generate corrective and preventive actions in a coordinated way (Cua et al. 2001).
- TPM is an integral production philosophy that is not the responsibility of a single department, however, the senior management and the supervisors must integrate the operators, since they are the ones who know the machine's operation, consequently, they must keep informed about them. Also, it is important to mention that sometimes TPM fails because the operators' opinion is not taken

into account, and because they are not informed either (Kamath and Rodrigues 2016).

- The senior management must generate a quality policy where TPM is a central part of it. In addition, do not forget that TPM is only a tool that helps to generate quality products as well as provide delivery production orders on time for the customer (Konecny and Thun 2011).
- TPM plans and programs that are performed by the senior management and by people in charge of taking decisions are not enough since it requires a personal involvement from managers in their execution processes.
- The senior management is responsible for maintaining a working environment that is safe to perform the work by the operators. Therefore, the security aspect must be a central part of maintenance plans and programs. Also, the high insecurity rates are reflected in the number of accidents and fees that have to be paid to insurance companies, besides the absenteeism and abandonment rates in the company (Pinjala et al. 2006).

9.12 Simple Hypotheses—Clients

Although *Clients* are an external entity to the company, they are the main reason that the company stands out, therefore, all the programs that are implemented must be focused on improving the service towards them. Although in the previous sections *Clients* have been related to many variables, however, it is not yet related to the benefits that can be obtained. In this section, it is related to *Benefits for the organization, Productivity benefits*, and *Safety benefits.*

The hypotheses that are proposed are the following:

H_1 The *Clients* that a company has during the *TPM implementation* have a direct and positive effect on the *Benefits for the organization* that are obtained.

H_2 The *Clients* that a company has during the *TPM implementation* have a direct and positive effect on the *Productivity benefits* that are received.

H_3 The *Clients* that a company has during the *TPM implementation* have a direct and positive effect on the *Safety benefits* that are obtained.

9.12.1 Results—Clients

After evaluating the models that represent the relationships proposed in the previous hypotheses, the results are obtained, which are shown in Table 9.15, where the latent dependent variables (in this case, only the benefits), the β values, the *p*-value associated to the statistical significance tests, the *R*-squared values for the predictive validity, and the VIF value as a measure of collinearity are illustrated.

Table 9.15 Relationships between *Clients* and other variables

Hypotheses	Dependent variable	Parameter			
		β	*p*-value	R^2	VIF
H_1	*Benefits for the organization*	0.396	$p < 0.001$	0.157	1.179
H_2	*Productivity benefits*	0.348	$p < 0.001$	0.121	1.126
H_3	*Safety benefits*	0.346	$p < 0.001$	0.120	1.131

When analyzing the information in Table 9.15, the following can be concluded regarding the hypotheses initially established:

H_1 There is enough statistical evidence to declare that the *Clients* that a company has during the *TPM implementation* have a direct and positive effect on the *Benefits for the organization* that are obtained, since when the first latent variable increases its standard deviation in one unit, the second variable increases in 0.396, and it can explain up to 15.7% of its variability.

H_2 The *Clients* that a company has during the *TPM implementation* have a direct and positive effect on the *Productivity benefits* that are received, since when the first variable increases its standard deviation in one unit, the second variable increases in 0.348 units, and it can explain up to 12.1% of its variability.

H_3 The *Clients* that a company has during the *TPM implementation* have a direct and positive effect on the *Safety benefits* that are obtained, since when the first variable increases its standard deviation in one unit, the second variable increases in 0.346 units, and it can explain up to 12.0% of its variability.

9.12.2 Conclusions and Industrial Implications—Clients

Although *Clients* are an external factor to the company where there is very little control on, in this section, the impact that they have on the *TPM implementation* process, as well as the benefits obtained is analyzed. In addition, the evaluated models allow to conclude the following in a generic way:

- *Clients* have direct and positive effects on the benefits obtained from the *TPM implementation*, since the *p*-values associated with the β values are less than 0.05, therefore, the inferences are made with 95% of reliability.
- When analyzing the β magnitude in the relationships that *Clients* have with other variables, it is observed that these values are the smallest (all are lower than 0.4), although they are statistically significant. In addition, the highest value related to the *Benefits for the organization*, while the other relationships are similar with differences expressed in thousandths.
- There are no problems of collinearity between the latent variables, since the VIF value is less than 3.3 in all the analyzed relationships.

- The models have predictive validity since the *R*-square value is greater than 0.02 in all relationships.
- In the same way, the relationships that *Clients* have along with the benefits obtained have a series of industrial implications that must be considered by those responsible for the decision-making process, such as the following:
- *Clients* must be part of the decision-making process before making the new equipment purchase since they must know the demand before adjusting or even buying new equipment. Also, it would not be appropriate to invest large amounts of money in equipment if the trend of demand and the design variations that products may have in a close future is unknown, since it is possible that the new equipment is useless in a short period of time (Bourke and Roper 2016).
- The management and supervisors must consider the established delivery times with *Clients* and seek to fulfill them, preventing maintenance from being a cause of non-compliance in deliveries. Also, a lack of delivery time is enough to stain the *Clients* and society image that the company has (Fandel and Trockel 2016).
- The product quality must never be sacrificed in order to comply with a delivery when maintenance is not performed; it is a fact that maintenance is important, but the quality and the image that the *Clients* have about the company is more relevant, since that quality is what the customer looks for and values (Hohan et al. 2015).
- When for any reason there is a breach, as far as possible, management must quantify the cost of the same to create awareness within the company for such errors (Eti et al. 2006c; Willmott and McCarthy 2001a).
- In the relationship with *Clients*, it must be very clear in the technical specifications that are required in the product in order that the maintenance department along with the production department generates a plan that seeks to always maintain the equipment in optimal work conditions associated with calibrations and adjustments (Sabet et al. 2016).
- Knowing delivery times helps the company plan production and coordinate the activities to achieve them, which includes the maintenance department. The previous information will allow to avoid unexpected events that decrease the system productive efficiency (Samaranayake and Laosirihongthong 2016).
- Acknowledging the *Clients* work orders allows a better production planning and avoids that in order to comply with them, operators handle equipment in poor conditions, which are a source of risk for them. In the same way, avoid paying extra time to operators and generating production orders when they are exhausted or with high levels of neglection and concentration (Gosavi 2006; Moradkhani et al. 2015; Nouri Gharahasanlou et al. 2017).

References

Ahmad MF, Zakuan N, Jusoh A, Takala J (2012) Relationship of TQM and business performance with mediators of SPC, lean production and TPM. Procedia Soc Behav Sci 65:186–191. https://doi.org/10.1016/j.sbspro.2012.11.109

Ahmadzadeh F, Bengtsson M (2017) Using evidential reasoning approach for prioritization of maintenance-related waste caused by human factors—a case study. Int J Adv Manufact Technol 90(9):2761–2775. https://doi.org/10.1007/s00170-016-9377-7

Akroyd D, Legg J, Jackowski MB, Adams RD (2009) The impact of selected organizational variables and managerial leadership on radiation therapists' organizational commitment. Radiography 15(2):113–120. https://doi.org/10.1016/j.radi.2008.05.004

Alfonso JG, Mara M (2016) Rewards for continuous training: a learning organisation perspective. Ind Commercial Train 48(5):257–264. https://doi.org/10.1108/ICT-11-2015-0076

Alipour FH (2011) The relationship between organizational climate and communication skills of managers of the Iranian physical education organization. Procedia Soc Behav Sci 30:421–428. https://doi.org/10.1016/j.sbspro.2011.10.083

Anand A, Singhal S, Panwar S, Singh O (2018) Optimal price and warranty length for profit determination: an evaluation based on preventive maintenance. In: Kapur PK, Kumar U, Verma AK (eds) Quality, IT and business operations: modeling and optimization. Springer, Singapore, pp 265–277

Aravindan P, Punniyamoorthy M (2002) Justification of advanced manufacturing technologies (AMT). Int J Adv Manufact Technol 19(2):151–156. https://doi.org/10.1007/s001700200008

Attri R, Grover S, Dev N, Kumar D (2013) Analysis of barriers of total productive maintenance (TPM). Int J Syst Assur Eng Manage 4(4):365–377. https://doi.org/10.1007/s13198-012-0122-9

Azizi A (2015) Evaluation improvement of production productivity performance using statistical process control, overall equipment efficiency, and autonomous maintenance. Procedia Manufact 2:186–190. https://doi.org/10.1016/j.promfg.2015.07.032

Borkowski S, Czajkowska A, Stasiak-Betlejewska R, Borade AB (2014) Application of TPM indicators for analyzing work time of machines used in the pressure die casting. J Ind Eng Int 10(2):55. https://doi.org/10.1007/s40092-014-0055-9

Bourke J, Roper S (2016) AMT adoption and innovation: an investigation of dynamic and complementary effects. Technovation 55–56:42–55. https://doi.org/10.1016/j.technovation.2016.05.003

Cardoso RdR, Pinheiro de Lima E, Gouvea da Costa SE (2012) Identifying organizational requirements for the implementation of advanced manufacturing technologies (AMT). J Manufact Syst 31(3):367–378. http://dx.doi.org/10.1016/j.jmsy.2012.04.003

Choe J-M (2004) Impact of management accounting information and AMT on organizational performance. J Inf Technol 19(3):203–214. https://doi.org/10.1057/palgrave.jit.2000013

Collins H, Gordon C, Terra JC (2006) Chapter 9—The culture and change dimension. In: Collins H, Gordon C, Terra JC (eds) Winning at collaboration commerce. Butterworth-Heinemann, Boston, pp 157–181

Cua KO, McKone KE, Schroeder RG (2001) Relationships between implementation of TQM, JIT, and TPM and manufacturing performance. J Oper Manage 19(6):675–694. https://doi.org/10.1016/S0272-6963(01)00066-3

Dangayach GS, Deshmukh SG (2000) Manufacturing strategy: experiences from select Indian organizations. J Manufact Syst 19(2):134–148. https://doi.org/10.1016/S0278-6125(00)80006-0

Davila A, Elvira MM (2012) Humanistic leadership: lessons from Latin America. J World Bus 47 (4):548–554. https://doi.org/10.1016/j.jwb.2012.01.008

Deering S, Johnston LC, Colacchio K (2011) Multidisciplinary teamwork and communication training. Semin Perinatol 35(2):89–96. https://doi.org/10.1053/j.semperi.2011.01.009

Eti MC, Ogaji SOT, Probert SD (2006a) Development and implementation of preventive-maintenance practices in Nigerian industries. Appl Energy 83(10):1163–1179. https://doi.org/10.1016/j.apenergy.2006.01.001

Eti MC, Ogaji SOT, Probert SD (2006b) Impact of corporate culture on plant maintenance in the Nigerian electric-power industry. Appl Energy 83(4):299–310. https://doi.org/10.1016/j.apenergy.2005.03.002

Eti MC, Ogaji SOT, Probert SD (2006c) Reducing the cost of preventive maintenance (PM) through adopting a proactive reliability-focused culture. Appl Energy 83(11):1235–1248. https://doi.org/10.1016/j.apenergy.2006.01.002

Fandel G, Trockel J (2016) Investment and lot size planning in a supply chain: coordinating a just-in-time-delivery with a Harris- or a Wagner/Whitin-solution. J Bus Econ 86(1):173–195. https://doi.org/10.1007/s11573-015-0800-6

Fulton M, Hon B (2010) Managing advanced manufacturing technology (AMT) implementation in manufacturing SMEs. Int J Prod Perform Manage 59(4):351–371. https://doi.org/10.1108/17410401011038900

Fumagalli L, Macchi M, Giacomin A (2017) Orchestration of preventive maintenance interventions. IFAC-PapersOnLine 50(1):13976–13981. https://doi.org/10.1016/j.ifacol.2017.08.2417

Gong Y, Law KS, Chang S, Xin KR (2009) Human resources management and firm performance: the differential role of managerial affective and continuance commitment. J Appl Psychol 94 (1):263–275. https://doi.org/10.1037/a0013116

Gosavi A (2006) A risk-sensitive approach to total productive maintenance. Automatica 42 (8):1321–1330. https://doi.org/10.1016/j.automatica.2006.02.006

Grout JR, Christy DP (1999) A model of supplier responses to just-in-time delivery requirements. Group Decis Negot 8(2):139–156. https://doi.org/10.1023/a:1008634008759

Guariente P, Antoniolli I, Ferreira LP, Pereira T, Silva FJG (2017) Implementing autonomous maintenance in an automotive components manufacturer. Procedia Manufact 13:1128–1134. https://doi.org/10.1016/j.promfg.2017.09.174

Hana P (2010) Human reliability in maintenance task. Front Mech Eng China 5(2):184–188. https://doi.org/10.1007/s11465-010-0002-4

Hipkin IB, De Cock C (2000) TQM and BPR: lessons for maintenance management. Omega 28 (3):277–292. https://doi.org/10.1016/S0305-0483(99)00043-2

Hohan AI, Olaru M, Pirnea IC (2015) Assessment and continuous improvement of information security based on TQM and business excellence principles. Procedia Econ Finance 32:352–359. https://doi.org/10.1016/S2212-5671(15)01404-5

Kachalov N, Kornienko A, Kvesko R, Nikitina Y, Kvesko S, Bukharina Z (2015) Integrated nature of professional competence. Procedia Soc Behav Sci 206:459–463. https://doi.org/10.1016/j.sbspro.2015.10.083

Kamath NH, Rodrigues LLR (2016) Simultaneous consideration of TQM and TPM influence on production performance: a case study on multicolor offset machine using SD Model. Perspect Sci 8:16–18. https://doi.org/10.1016/j.pisc.2016.01.005

Kelly A (2006) 13—Total productive maintenance: its uses and limitations. In: Kelly A (ed) Plant maintenance management set. Butterworth-Heinemann, Oxford, pp 247–265

Kim CS, Djamaludin I, Murthy DNP (2004) Warranty and discrete preventive maintenance. Reliab Eng Syst Saf 84(3):301–309. https://doi.org/10.1016/j.ress.2003.12.001

Koc T, Bozdag E (2009) The impact of AMT practices on firm performance in manufacturing SMEs. Robot Comput-Integr Manufact 25(2):303–313. https://doi.org/10.1016/j.rcim.2007.12.004

Konecny PA, Thun J-H (2011) Do it separately or simultaneously—an empirical analysis of a conjoint implementation of TQM and TPM on plant performance. Int J Prod Econ 133(2):496–507. https://doi.org/10.1016/j.ijpe.2010.12.009

Laureani A, Antony J (2017) Leadership characteristics for lean six sigma. Total Qual Manage Bus Excellence 28(3/4):405–426. https://doi.org/10.1080/14783363.2015.1090291

McCarthy D, Rich N (2015) Chapter Two—The lean TPM master plan. In: Lean TPM, 2nd edn. Butterworth-Heinemann, Oxford, pp 27–54

McKone KE, Weiss EN (2000) Analysis of investments in autonomous maintenance activities. IIE Trans 32(9):849–859. https://doi.org/10.1023/a:1007647329259

Moradkhani A, Haghifam MR, Abedi SM (2015) Risk-based maintenance scheduling in the presence of reward penalty scheme. Electric Power Syst Res 121:126–133. https://doi.org/10.1016/j.epsr.2014.12.006

Morales Méndez JD, Rodriguez RS (2017) Total productive maintenance (TPM) as a tool for improving productivity: a case study of application in the bottleneck of an auto-parts machining line. Int J Adv Manufact Technol 92(1):1013–1026. https://doi.org/10.1007/s00170-017-0052-4

Morlock F, Dorka T, Meier H (2014) Concept for a performance measurement method for the organization of the IPS2 delivery. Procedia CIRP 16:56–61. https://doi.org/10.1016/j.procir.2014.03.003

Mwanza BG, Mbohwa C (2015) Design of a total productive maintenance model for effective implementation: case study of a chemical manufacturing company. Procedia Manufact 4:461–470. https://doi.org/10.1016/j.promfg.2015.11.063

Nouri Gharahasanlou A, Ataei M, Khalokakaie R, Barabadi A, Einian V (2017) Risk based maintenance strategy: a quantitative approach based on time-to-failure model. Int J Syst Assur Eng Manage. https://doi.org/10.1007/s13198-017-0607-7

Okoh P, Haugen S (2015) Improving the robustness and resilience properties of maintenance. Process Saf Environ Prot 94:212–226. https://doi.org/10.1016/j.psep.2014.06.014

Pang S, Jia Y, Liu X, Deng Y (2016) Study on simulation modeling and evaluation of equipment maintenance. J Shanghai Jiaotong Univ (Sci) 21(5):594–599. https://doi.org/10.1007/s12204-016-1768-2

Pascale L, Mainea M, Patic PC, Duta L (2012) Mathematical decision model to improve TPM indicators. IFAC Proc Vol 45(6):934–939. https://doi.org/10.3182/20120523-3-RO-2023.00303

Pinjala SK, Pintelon L, Vereecke A (2006) An empirical investigation on the relationship between business and maintenance strategies. Int J Prod Econ 104(1):214–229. https://doi.org/10.1016/j.ijpe.2004.12.024

Prabhuswamy MS, Ravikumar KP, Nagesh P (2013) Implementation of Kaizen techniques in TPM. IUP J Mech Eng 6(3):38

Realyvásquez-Vargas A, Maldonado-Macías AA, García-Alcaraz JL, Alvarado-Iniesta A (2014) Expert System development using fuzzy if–then rules for ergonomic compatibility of AMT for lean environments. In: García-Alcaraz JL, Maldonado-Macías AA, Cortes-Robles G (eds) Lean manufacturing in the developing world: methodology, case studies and trends from Latin America. Springer International Publishing, Cham, pp 347–369

Rico R, Cohen SG (2005) Effects of task interdependence and type of communication on performance in virtual teams. J Manag Psychol 20(3/4):261–274. https://doi.org/10.1108/02683940510589046

Rodrigues M, Hatakeyama K (2006) Analysis of the fall of TPM in companies. J Mater Process Technol 179(1):276–279. https://doi.org/10.1016/j.jmatprotec.2006.03.102

Sabet E, Adams E, Yazdani B (2016) Quality management in heavy duty manufacturing industry: TQM vs. six sigma. Total Qual Manage Bus Excellence 27(1/2):215–225. https://doi.org/10.1080/14783363.2014.972626

Samaranayake P, Laosirihongthong T (2016) Configuration of supply chain integration and delivery performance: unitary structure model and fuzzy approach. J Modell Manage 11(1):43–74. https://doi.org/10.1108/JM2-01-2014-0005

Sani SIA, Mohammed AH, Misnan MS, Awang M (2012) Determinant factors in development of maintenance culture in managing public asset and facilities. Procedia Soc Behav Sci 65:827–832. https://doi.org/10.1016/j.sbspro.2012.11.206

Seth D, Tripathi D (2006) A critical study of TQM and TPM approaches on business performance of Indian manufacturing industry. Total Qual Manage Bus Excellence 17(7):811–824. https://doi.org/10.1080/14783360600595203

Shen CC (2015) Discussion on key successful factors of TPM in enterprises. J Appl Res Technol 13(3):425–427. https://doi.org/10.1016/j.jart.2015.05.002

Shrivastava D, Kulkarni MS, Vrat P (2016) Integrated design of preventive maintenance and quality control policy parameters with CUSUM chart. Int J Adv Manufact Technol 82 (9):2101–2112. https://doi.org/10.1007/s00170-015-7502-7

Singh R, Gohil AM, Shah DB, Desai S (2013) Total productive maintenance (TPM) implementation in a machine shop: a case study. Procedia Eng 51:592–599. https://doi.org/10.1016/j.proeng.2013.01.084

Spanos YE, Voudouris I (2009) Antecedents and trajectories of AMT adoption: the case of Greek manufacturing SMEs. Res Policy 38(1):144–155. https://doi.org/10.1016/j.respol.2008.09.006

Thomas AJ, Jones GR, Vidales P (2006) An integrated approach to TPM and six sigma development in the castings industry. In: Intelligent production machines and systems. Elsevier Science Ltd., Oxford, pp 620–625

Tjosvold D, Tjosvold M (2015) Leadership for teamwork, teamwork for leadership. Building the team organization: how to open minds, resolve conflict, and ensure cooperation. Palgrave Macmillan UK, London, pp 65–79

Tong DYK, Rasiah D, Tong XF, Lai KP (2015) Leadership empowerment behaviour on safety officer and safety teamwork in manufacturing industry. Saf Sci 72:190–198. https://doi.org/10.1016/j.ssci.2014.09.009

Willmott P, McCarthy D (2001a) 2—Assessing the true costs and benefits of TPM. In: Total productivity maintenance, 2nd edn. Butterworth-Heinemann, Oxford, pp 17–22

Willmott P, McCarthy D (2001b) 6—Applying the TPM improvement plan. In: Total productivity maintenance, 2nd edn. Butterworth-Heinemann, Oxford, pp 116–144

Willmott P, McCarthy D (2001c) 9—TPM for equipment designers and suppliers. In: Total productivity maintenance, 2nd edn. Butterworth-Heinemann, Oxford, pp 181–192

Wu B, Seddon JJM (1994) An anthropocentric approach to knowledge-based preventive maintenance. J Intell Manufact 5(6):389–397. https://doi.org/10.1007/bf00123658

Chapter 10
Structural Equation Models: Human Factor—Part I

Abstract In this chapter, four structural equation models are presented, where variables that integrate the human factor category appeared during the TPM implementation process, which operate as independent variables and are related to variables from the operational factor categories as well as from the obtained benefits. In addition, the models are validated and evaluated, and also direct, indirect, and total effects are obtained as well as their effect sizes. Finally, a series of industrial implications are discussed and displayed based on the results.

10.1 Complex Model 1

This model intertwined four latent variables: *Clients, Suppliers, Managerial commitment,* and *Work culture*. In this model, it is assumed that the Company's *Clients* are the most important aspect for them, and that according to their demands, relationships are established with machinery and equipment *Suppliers*, which undoubtedly requires a high *Managerial commitment*, since they are the ones who take the purchasing decisions. Therefore, it is stated that *Clients* are an independent variable, whereas *Work culture* is the dependent variable from all the previous variables, while *Suppliers* and *Managerial commitment* are the median variables. In other words, this model determines the way *Work culture* is built focusing on TPM, in terms of external factors: *Clients* and *Suppliers* where the internal factor is *Managerial commitment*.

In addition, a set of six hypotheses are generated to link the variables with each other in this model, which are discussed in the section below.

10.1.1 Hypotheses—Complex Model 1

It is known that great relationships with *Suppliers* may offer many opportunities for companies, since they have become innovation resources and generate competitive

J. R. Díaz-Reza et al., *Impact Analysis of Total Productive Maintenance*,
https://doi.org/10.1007/978-3-030-01725-5_10

advantages, but in order to promote that potential, certain specific competences must be developed in a partnership between the customer and the supplier, where the manufacturer is only the intermediary between these two entities, one before and one after the manufacturing process (Riedel et al. 2010). Additionally, the supplier may get benefits from the innovative activities of his client, such as technological invention as well as production efficiency, and this effect is not only driven by the increasement in the *Clients*' profitability from the new innovations but due to the knowledge that is transferred from one to the other, with the manufacturer being the transfer vehicle (Li 2018). In addition, if *Clients* seek opportunities for sustainable growth, they are also willing to support and collaborate with their *Suppliers* during the creating value process, which may increase the competitive advantages in both parts of the machinery market in production systems (Krolikowski and Yuan 2017) and those *Suppliers* must always pay attention to market trends in order to generate equipment that can manufacture products with these qualities. Therefore, the following hypothesis is proposed:

H_1: *Clients* that have a company have a direct and positive effect on machinery *Suppliers* and equipment that is already installed.

Furthermore, the competition driven by the customer service has forced that managers from industrial manufacturing companies realize that the concept of service influences the long-term survival of their companies, and that it should be considered at strategic and operational levels in their business (Sade et al. 2015), since it will determine whether the strategies pursued are successful or not (Sade et al. 2015). In addition, in the TPM implementation field, *Clients* must be a significant aspect when generating the organization, strategic plans, and programs, because their productive capacity depends on the *Suppliers* capacities to provide equipment that can generate products according to the *Clients*' needs. Also, depending on the required indicators, quality and quantity in the production orders, senior management must plan the new equipment purchase or acquisition, as well as the preservation of existing ones. In addition, it is important to remember that the production plans are created according to the identified demand in the *Clients*. Also, knowing the *Clients* demands will avoid delays in deliveries and possible administrative sanctions. Therefore, the following hypothesis is proposed:

H_2: *Clients* from a company have a direct and positive impact on the *Managerial commitment* in the TPM implementation process.

Moreover, senior management must always be attentive to the technological changes that may be generated in machines and equipment; they must be in regular contact with *Suppliers* to know the updated trends, the modern technologies generation, quality, and safety standards. In fact, Bourke and Roper (2016) state that *Suppliers* must visit the manufacturer's industrial factories to make suggestions about the sale of new equipment to improve already installed processes; however, in these meetings the production and maintenance managers must be included in order to take better decisions and analyze if the investments are justified, as well as personnel sales, who know the outlooks and trends from products. Also, another aspect that is important in the relationship with *Suppliers* is the work contracts establishment, which include aspects related to training and equipment installation, among others (McKone and Weiss 2000; Pinjala et al. 2006). Therefore, the following hypothesis is proposed:

H_3: Machinery *Suppliers* and equipment from a company have a direct and positive impact on the *Managerial commitment* toward the TPM implementation.

Furthermore, the market requirements change rapidly, and these changes require improvements in the performance of the company to focus on cost reduction, increased productivity and quality levels, and time order deliveries to satisfy *Clients* (Abhishek et al. 2014). In order to thrive in the current economic environment, any organization must dedicate itself to endless improvements as well as more efficient ways to acquire products or services that fulfill the customer needs on a daily basis, since globalization has forced manufacturing organizations engineers and managers to produce their products with high quality at a lower cost (Mandeep et al. 2012).

However, quality and maintenance functions are vital factors for achieving sustainability in a manufacturing organization and for satisfying *Clients* (Mandeep et al. 2012). In the TPM area, industrial maintenance plays a crucial role in improving employee productivity, unit value added, and competitiveness under international increasing pressures (Damiana and Gianni 2010). Also, the biggest challenge for organizations is to be able to carry out a cultural transformation to ensure general employees participation toward maintaining and improving manufacturing performance through TPM initiatives (Attri et al. 2013). In order to achieve the previous information, senior management must motivate its employees to generate an organizational and *Work culture* focused on raising awareness among employees about the true potential of TPM and let them know about its contributions (Ahuja and Khamba 2008).

Nowadays, globalized markets are ruled by quality standards, which are imposed by the client, and as a result, they must have an impact on the organization culture (Sani et al. 2012). In this sense, the following hypothesis can be proposed:

H_4: *Clients* who have a company for their own products have a direct and positive impact on the *Work culture* that is inside its relationship with TPM.

The production equipment *suppliers* are forced to provide their operation manuals and look forward to promoting the habit that the operators read and understand them. In addition, a special emphasis toward safeness, handling, and operation in machines as well as in optimum levels that operating parameters must have, such as temperature and humidity, among others, have to be pointed out (Tong et al. 2015). Also, *Suppliers* must promote improvement groups that can be organized within the company, since safety is an aspect that is almost always addressed, as well as an aspect linked to productive efficiency, where it must support the management in order to generate a *Work culture* that allows to improve the equipment that it manufactures and distributes based on the operators opinions (Tong et al. 2015; James et al. 2014). Consequently, the following hypothesis is proposed:

H_5: Machinery *Suppliers* and the equipment from a company have a direct and positive impact on its *Work culture*.

Moreover, the biggest challenge for the organization is to make a radical transformation in the organization culture, in order to ensure the general employee's participation toward the maintenance and improvement of the manufacturing performance through TPM initiatives (Attri et al. 2013). In addition, concentrated and concerted efforts should be organized by senior management to motivate the organization culture by raising employees' awareness about the potential of TPM and let them know about its contributions (Ahuja and Khamba 2008). In the same way, senior management must demonstrate leadership and responsibility in the TPM implementation in order to these good habits to be transmitted to operators, as well as a *Work culture* to be generated toward the preservation of production means. In this sense, the following hypothesis can be proposed:

H_6: The *Managerial commitment* in the TPM implementation has a direct and positive impact on the company's *Work culture* related to maintenance.

Figure 10.1 shows the relationships between the variables, which have been established as hypotheses. Also, the validity indexes of this model have been presented in previous chapters, so they are not discussed in this section.

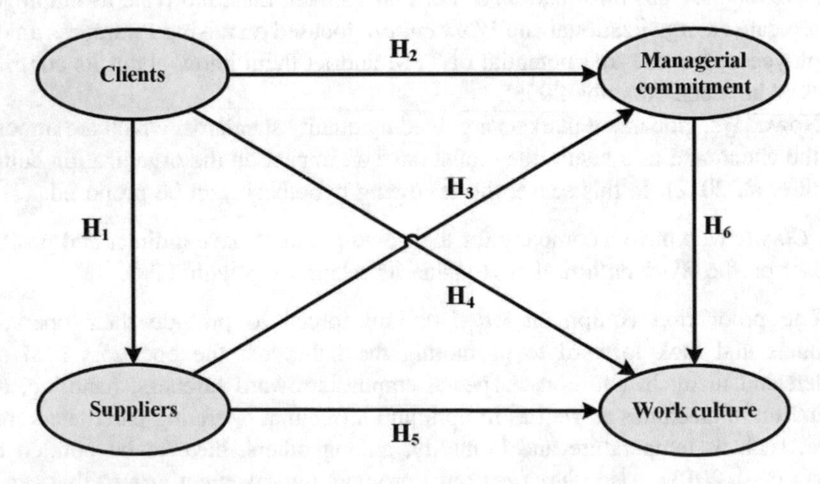

Fig. 10.1 Proposed hypotheses from Complex Model 1

10.1.2 *Efficiency Indexes—Complex Model 1*

The model is evaluated according to the methodology previously described and a series of results about the relationships between the variables are obtained, but before interpreting it, the efficiency indexes of the model are analyzed, which are illustrated below:

- Average path coefficient (APC) = 0.334, $P < 0.001$;
- Average R-squared (ARS) = 0.371, $P < 0.001$;
- Average adjusted R-squared (AARS) = 0.368, $P < 0.001$;
- Average block VIF (AVIF) = 1.431, acceptable if ≤ 5, ideally ≤ 3.3;
- Average full collinearity VIF (AFVIF) = 1.817, acceptable if ≤ 5, ideally ≤ 3.3;
- Tenenhaus GoF (GoF) = 0.464, small ≥ 0.1, medium ≥ 0.25, large ≥ 0.36;
- Sympson's paradox ratio (SPR) = 0.833, acceptable if > = 0.7, ideally = 1;
- R-squared contribution ratio (RSCR) = 0.999, acceptable if ≥ 0.9, ideally = 1;
- Statistical suppression ratio (SSR) = 1.000, acceptable if ≥ 0.7; and
- Nonlinear bivariate causality direction ratio (NLBCDR) = 1.000, acceptable if ≥ 0.7.

Based on the ARS and AARS values, it can be concluded that there is enough predictive validity in the model, while the AVIF and AFVIF values indicate the absence of collinearity problems between variables. In the same way, the Tenenhaus index indicates an adequate adjustment from the data to the model, and there are any issues in the relationships between variables or hypotheses that have been set. According to the previous information, it can be proceeding to interpret the model.

10.1.3 *Results—Complex Model 1*

Figure 10.2 displays the obtained results from the evolution of the model presented in Fig. 10.1, which refers to the Complex Model 1. In addition, a value for the β is illustrated as well as the p-value for its significance test, and an R-squared value in dependent latent variables.

10.1.3.1 Direct Effects—Complex Model 1

According to the direct effects, the hypotheses previously presented are validated and the following conclusions are reached:

H_1: There is enough statistical evidence to declare that *Clients* who have a company for their own products have a direct and positive effect on the machinery *Suppliers*

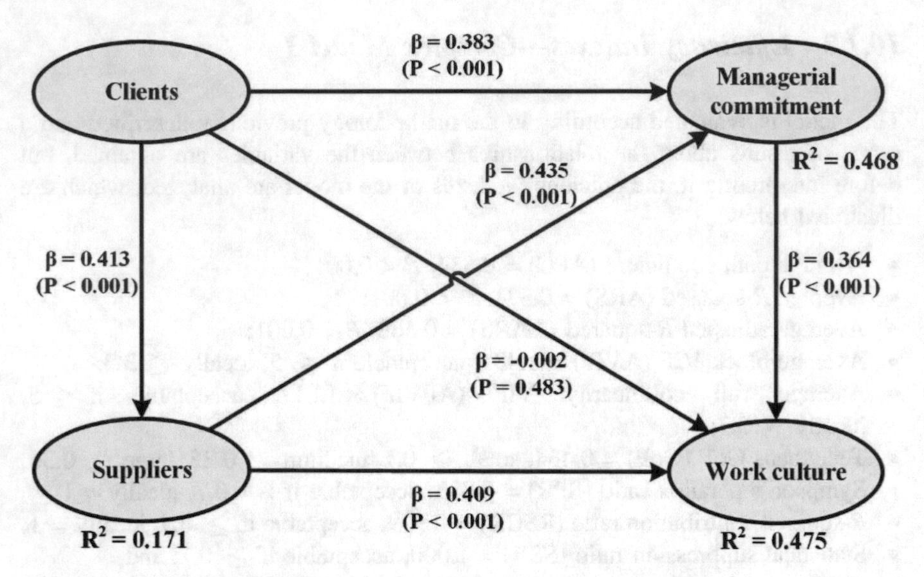

Fig. 10.2 Evaluated Complex Model 1

and equipment that is already installed, since when the first variable increases its standard deviation in one unit, the second one does it in 0.413 units.

H_2: There is enough statistical evidence to declare that *Clients* from a company have a direct and positive impact on the *Managerial commitment* in the TPM implementation process, since when the first variable increases its standard deviation in one unit, the second one goes up in 0.383 units.

H_3: There is enough statistical evidence to claim that machinery *Suppliers* and equipment from a company have a direct and positive impact on the *Managerial commitment* toward the TPM implementation, because when the first variable increases its standard deviation by one unit, the second one does it in 0.435 units.

H_4: There is not enough statistical evidence to declare that *Clients* who have a company for their own products have a direct and positive impact on the *Work culture* that is inside its relationship with TPM, since the *p*-value associated to the β is lower than 0.05; therefore, the relationship is statistically nonsignificant.

H_5: There is enough statistical evidence to state that machinery *Suppliers* and the equipment from a company have a direct and positive impact on its *Work culture*, since when the first variable increases its standard deviation in one unit, the second variable grows up in 0.409 units.

H_6: There is enough statistical evidence to declare that the *Managerial commitment* in the TPM implementation has a direct and positive impact on the company's *Work culture* related to maintenance, since when the first variable increases its standard deviation in a unit, the second one does it in 0.364 units.

Table 10.1 *R*-squared contribution—Complex Model 1

Dependent variable	Independent variable			R^2
	Suppliers	*Managerial commitment*	*Clients*	
Work culture	0.255	0.221	−0.001	0.475
Suppliers			0.171	0.171
Managerial commitment	0.255		0.213	0.468

10.1.3.2 Effects Size—Complex Model 1

Some of the latent variables dependent from the Complex Model 1 are explained by more than one independent latent variable, and as a result, Table 10.1 presents the amount of variance (*R*-squared) that corresponds to each one. In the last column, the *R*-squared value is added, which is the total effects size.

According to the data in Table 10.1, the following can be concluded:

- The *Work culture* variable is explained in 47.5% by three independent variables, but *Clients* are insignificant in their contribution. In this case, if a company desires to achieve a *Work culture* focused on TPM, it should pay special attention to the equipment and tools *Suppliers*, since it can explain 25.5%, but also to *Managerial commitment*, since it can explain 22.1% of its variability.
- In order to achieve an adequate *Managerial commitment*, 25.5% is explained in this model because of the relationship and commitment established with the *Suppliers* and 21.3% is related to the *Clients,* which gives an *R*-squared value = 0.468.
- Based on the independent variable contributions from the *Suppliers* along with other dependent variables, it is said that senior management and maintenance supervisors should pay special attention to this item to achieve an appropriate *Work culture* focused on TPM.

10.1.3.3 Total Direct and Indirect Effects—Complex Model 1

When there are mediating variables between latent and dependent variables, the indirect effects are generated. Additionally, Table 10.2 shows the total indirect effects, where the *p*-value is exposed, because of the statistical significance test and the effect size or variability explained.

According to the data in Table 10.2, the following can be concluded:

- When analyzing the relationship between *Clients* and *Work culture* through hypothesis H_4, it is observed that there is no significant direct relationship. However, when the *Suppliers* and *Managerial commitment* are mediating variables in that relationship, there is a total of indirect effects by 0.374, the greatest of all. The previous data is a result mainly since the *Clients* are an external entity and only managers or senior management has a relationship with them.

Table 10.2 Total indirect effects—Complex Model 1

Dependent variable	Independent variable	
	Suppliers	*Clients*
Work culture	0.158 ($P < 0.001$) ES = 0.099	0.374 ($P < 0.001$) ES = 0.135
Managerial commitment		0.179 ($P < 0.001$) ES = 0.100

- *Suppliers* promote *Work culture* indirectly through the *Managerial commitment*, which demonstrates the importance of management in the development of habits focused on machines and equipment preservation, but always following the recommendations made by the *Suppliers* through their training courses, training, and operation manuals.
- An external entity such as *Clients* also has an indirect effect on *Managerial commitment* through the mediating variable of *Suppliers*, because they are who determine the technological capabilities in the production system and finally, *Clients* only receive what *Suppliers* and the manufacturers can produce.

10.1.3.4 Total Effects—Complex Model 1

Table 10.3 shows the total effects from the Complex Model 1, which represents the total direct and indirect effects. Also, the effect value, the associated *p*-value, and the effect size are illustrated as a measure of the explained variance.

It is noticeable that in Table 10.3 all total effects are statistically significant, since the associated *p*-value is over 0.05. In addition, depending on the effect size, it is observed that the relationship between *Suppliers* and *Work culture* is the highest, followed by the relationship between *Clients* and *Managerial commitment*, which leads to a series of industrial implications that are exposed in the section below.

Table 10.3 Total effects—Complex Model 1

Dependent variables	Independent variables		
	Suppliers	*Managerial commitment*	*Clients*
Work culture	0.567 ($P < 0.001$) ES = 0.354	0.364 ($P < 0.001$) ES = 0.221	0.372 ($P < 0.001$) ES = 0.134
Suppliers			0.413 ($P < 0.001$) ES = 0.171
Managerial commitment	0.435 ($P < 0.001$) ES = 0.255		0.562 ($P < 0.001$) ES = 0.313

10.1.4 *Industrial Conclusions and Implications—Complex Model 1*

The Complex Model 1 is intended to quantify how the external entities, *Suppliers* and *Clients*, using the *Managerial commitment* as a mediating variable can support the *Work culture* that is established in a company during the TPM implementation process. In addition, according to the outcomes from the model evaluation, it is possible to conclude and give a series of recommendations to the industrial sector, such as the following:

- Due to the direct relationship with the highest value in the β between the *Suppliers* and the *Managerial commitment*, it is stated that managers must be attentive to the suggestions from new technologies applied to TPM, the operating manuals, and other guides that the first ones provide for their usage and management, since only in this way they can promote *Work culture*.
- *Clients* have a strong effect on *Suppliers* directly, which implies that they must pay attention to their *Clients*' needs to know if they are able to offer the technology that can be produced or the requested products, also if they can start innovation processes on already existing ones.
- It is interesting to observe the values that *Suppliers* have along with *Work culture*, which is 0.409, higher than the one that *Managerial commitment* has with only 0.364. The previous statement may be due to the fact that the maintenance and the equipment operation rules are established by the *Suppliers* and managers, and as a result, it only supports to the released and established rules through the machines operating manuals.
- *Clients* do not directly influence the *Work culture*, but they do it indirectly through *Suppliers* and *Managerial commitment*, which implies that management must be attentive to its *Clients*' needs.

10.1.5 *Sensitivity Analysis—Complex Model 1*

In Complex Model 1, there are six relationships that have been presented as hypotheses; therefore, in this section, a series of sensitivity analyzes are described to discuss the probability of occurrence in different scenarios from the latent variables that integrate it.

10.1.5.1 Relationship Between *Clients* and *Suppliers* (H_1)—Complex Model 1

The present relationship is essential in TPM, since they are two external entities to the company. In addition, it is assumed that *Clients* exist in first place and based on

Table 10.4 Sensitivity analysis: *Clients* and *Suppliers* (H_1)—Complex Model 1

Clients	*Suppliers*	
	High	Low
High	Suppliers+ = 0.144 Clients+ = 0.209 Suppliers+ *&* Clients+ = 0.052 Suppliers+ *if* Clients+ = 0.247	Suppliers− = 0.168 Clients+ = 0.209 Suppliers− *&* Clients+ = 0.008 Suppliers− *if* Clients+ = 0.039
Low	Suppliers+ = 0.144 Clients− = 0.144 Suppliers+ *&* Clients− = 0.016 Suppliers+ *if* Clients− = 0.113	Suppliers− = 0.168 Clients− = 0.144 Suppliers− *&* Clients− = 0.057 Suppliers− *if* Clients− = 0.396

it, machines and equipment purchases are done through the *Suppliers*. Also, four different scenarios are analyzed for each variable, which combined retrieve a total of four variables, which are shown in Table 10.4. Some general conclusions that can be acknowledged are the following:

- There is a scenario that is especially impressive, which is presented when there is optimism, where *Suppliers* and *Clients* have elevated levels. The probability that the first variable happens in this scenario independently is 0.144, while the second variable is 0.209 which indicates that management must work on the increment of these probabilities, since it is the mediating entity between both variables. Also, the probability that both variables have scenarios with prominent levels simultaneously is only 0.052, whereas the probability that *Suppliers* have elevated levels because there are high levels in *Clients* is 0.247. In other words, it goes up from 0.144 independently to 0.247 with the presence of those *Clients* at a high level, which indicates a difference of 0.103.
- The second scenario refers to the situation where *Suppliers* have low levels and *Clients* high levels. The probability that the scenario from the first variable occurring independently is 0.168, and the probability that the scenario of the second variable occurring is 0.209. Also, the probability that the two scenarios occur simultaneously in both variables is 0.008, which shows that management actually seeks to be a mediating entity that regulates relationships with both variables. Finally, the probability that *Suppliers* have low levels, because *Clients* have high levels, is 0.168, which is a work opportunity area for management, since this value should tend to be zero.
- The third scenario is concern about the situation where *Suppliers* have a high level and *Clients* a low level of participation in the TPM implementation. The probability that the scenario occurring independently from the first variable is 0.144, and that it occurs from the second variable is 0.144. However, the probability that they occur simultaneously is 0.016, a low value which indicates that management is always mediating between both entities. Finally, if the probability that *Suppliers* have a high level because *Clients* have a low one is 0.113.

- Finally, the pessimistic scenario of the relationship between these variables is presented when *Suppliers* and *Clients* have low levels. The probability of occurrence from the first variable in this scenario independently is 0.168, while the second is 0.144. Also, the probability that both variables are presented in their low values is only 0.057, and management should look forward to reducing that probability to zero. Likewise, the probability that *Suppliers* have low levels because *Clients* have low values is 0.396; therefore, TPM management and managers must seek to reduce those scenarios.

10.1.5.2 Relationship Between *Clients* and *Managerial Commitment* (H_2)—Complex Model 1

In this section, the sensitivity analysis for the relationship between *Clients* and *Managerial commitment* is assessed, which is represented by H_2. In addition, only the high and low scenarios are analyzed for each of the variables, which combined result in a total of four conjunctions.

Table 10.5 illustrates a summary of the probabilities to find the variables in their different combinations, and they are detailed below:

- The first scenario occurs when the *Managerial commitment* and *Clients* are at their high levels. The probability that these two scenarios occur independently from the first latent variable is 0.168, while from the second variable is 0.209. However, the probability of presenting both scenarios simultaneously in the two variables is 0.063, a very low value, and as a result, management should seek to increase these values. Finally, the probability of having a high *Managerial commitment* because there are high values in the *Clients* is 0.299, which indicates that the presence of this second variable represents an increasement in the probability by 0.131.

Table 10.5 Sensitivity analysis: *Clients* and *Managerial commitment* (H_2)—Complex Model 1

Clients	*Managerial commitment*	
	High	Low
High	Managerial commitment+ = 0.168 Clients+ = 0.209 Managerial commitment+ & Clients + = 0.063 Managerial commitment+ *if* Clients + = 0.299	Managerial commitment− = 0.158 Clients+ = 0.209 Managerial commitment− & Clients + = 0.008 Managerial commitment− *if* Clients + = 0.039
Low	Managerial commitment+ = 0.168 Clients− = 0.144 Managerial commitment+ & Clients − = 0.003 Managerial commitment+ *if* Clients − = 0.019	Managerial commitment− = 0.158 Clients− = 0.144 Managerial commitment− & Clients − = 0.076 Managerial commitment− *if* Clients − = 0.528

- The relationship where *Managerial commitment* is high, and *Clients* have a low level, seems to be a normal situation, but it is observed that the probability of presenting the scenario for the first variable, because it is pre-sentenced the scenario for the second variable, is only 0.019; a very low probability, because when a management action is desired, an adequate action from the *Clients* is expected as well.
- In a pessimistic scenario, when the *Managerial commitment* is low, and *Clients* have a low role in the participation of TPM projects, the probability of occurrence for the first variable is 0.158, whereas the second is 0.144. However, the probability of having a low *Managerial commitment* if there is a low level of client integration is 0.528, a risk situation that the management should never get, since the probability that *Managerial commitment* goes from 0.158 to 0.538, which represents an increasement of 0.370.

10.1.5.3 Relationship Between *Suppliers* and *Managerial Commitment* (H_3)—Complex Model 1

The sensitivity analysis of this relationship aims to determine the scenarios that *Managerial commitment* may have as a response to an integration by the *Suppliers* in the TPM implementation process. In addition, Table 10.6 illustrates the four scenarios that exist when they are at their low and high levels. Also, from a quick analysis of these probabilities, the following is concluded:

- In an optimistic scenario, where there is a wide integration of *Suppliers* in the TPM implementation and, there is a high *Managerial commitment*, it is observed that the first variable has a probability to happen in 0.168, but the second at 0.144, which it is something to worry about because it is when the probability that both variables are occurring simultaneously is analyzed, since it

Table 10.6 Sensitivity analysis: *Suppliers* and *Managerial commitment* (H_3)—Complex Model 1

Suppliers	*Managerial commitment*	
	High	Low
High	Managerial commitment+ = 0.168 Suppliers+ = 0.144 Managerial commitment+ & Suppliers+ = 0.068 Managerial commitment+ *if* Suppliers+ = 0.472	Managerial commitment− = 0.158 Suppliers+ = 0.144 Managerial commitment− & Suppliers+ = 0.011 Managerial commitment− *if* Suppliers + = 0.075
Low	Managerial commitment+ = 0.168 Suppliers− = 0.168 Managerial commitment+ & Suppliers− = 0.000 Managerial commitment+ *if* Suppliers− = 0.000	Managerial commitment− = 0.158 Suppliers− = 0.168 Managerial commitment− & Suppliers− = 0.071 Managerial commitment− *if* Suppliers− = 0.419

is only 0.068, which indicates that effort must be made in order to this situation does not go further. Finally, the probability of having high levels on *Managerial commitment* because there is a broad integration of *Suppliers* is 0.472; it represents an increasement of 0.302, which indicates that management needs to be committed when the supplier is integrated into the TPM implementation process.

- The previous information is demonstrated when *Managerial commitment* is having a high integration under the *Suppliers*, because the probability from the first occurrence, because the second has already happened, is 0.075.
- In the worst situation where *Managerial commitment* is low during the TPM process and others, *Suppliers* have little interest, and it is noted that the probability that they occur simultaneously is low, with only 0.071, which indicates that, in general terms, *Managerial commitment* is high. But there is a considerable risk that if there is no integration by *Suppliers*, *Managerial commitment* will be very low because the first scenario has been presented, since the probability is 0.419.

10.1.5.4 Relationship Between *Work Culture* and *Clients* (H_4)—Complex Model 1

Table 10.7 presents the possible scenarios where *Work culture* can be found because integrating *Clients* in the TPM implementation process. Also, it is important to mention that this analysis is complementary to the analyzed direct effects, which were statistically not significant.

In addition, four scenarios are analyzed, high and low levels' combinations in each of the variables as well. According to the data in Table 10.7, the following can be concluded:

- In an optimistic scenario, when *Work culture* has high levels and *Clients* are properly integrated into the TPM integration, there is a probability of 0.160 that the first variable is presented in that scenario independently, while the second

Table 10.7 Sensitivity analysis: *Clients* and *Work culture* (H_4)—Complex Model 1

Clients	*Work culture*	
	High	Low
High	Work culture+ = 0.160 Clients+ = 0.209 Work culture+ *&* Clients+ = 0.065 Work culture+ *if* Clients+ = 0.312	Work culture− = 0.166 Clients+ = 0.209 Work culture− *&* Clients+ = 0.011 Work culture− *if* Clients+ = 0.052
Low	Work culture+ = 0.160 Clients− = 0.144 Work culture+ *&* Clients− = 0.014 Work culture+ *if* Clients− = 0.094	Work culture− = 0.166 Clients− = 0.144 Work culture− *&* Clients− = 0.041 Work culture− *if* Clients− = 0.231

variable has a probability of 0.209. In other words, *Clients* always seem to be portrayed with notifications of technical specifications. However, the probability that these two variables are presented optimistically is only 0.065. Also, another interesting situation occurs when it is observed that there is a 0.312 probability of having a high *Work culture*, since *Clients* are also at high levels, which indicates that *Clients* are relevant in the creation of *Work culture.*

- The previous information is demonstrated while analyzing the probability when a low *Work culture* is found, but with very committed *Clients* to TPM, which is 0.052, a low probability that indicates that those *Clients* directly impact *Work culture* in a positive way.
- The pessimistic scenario in this relationship is found when *Work culture* is low, and *Clients* have a low commitment to TPM. Also, the scenario of the first variable can be presented with a probability of 0.166, while for the second is 0.144. In addition, the probability that both scenarios are happening simultaneously is very low, with only 0.041, and the probability that there a lack of *Work culture* because there is a low commitment from *Clients* toward TPM is 0.231; therefore, management should always look forward to avoiding this situation.

10.1.5.5 Relationship Between *Work Culture* and *Suppliers* (H_5)—Complex Model 1

Undoubtedly, machinery and tools *Suppliers* have a high influence in *Work culture* when implementing TPM, since they are the ones that generate the operation manuals and carry out the training processes. In this section, the scenarios where these two variables are at their high and low levels are analyzed. In addition, Table 10.8 displays a summary where according to this data the following is concluded:

- In the optimistic scenario, when *Work culture* has high levels and *Suppliers* a high commitment to TPM, it is observed that the probability of occurrence

Table 10.8 Sensitivity analysis: *Suppliers* and *Work culture* (H_5)—Complex Model 1

Suppliers	*Work culture*	
	High	Low
High	Work culture+ = 0.160 Suppliers+ = 0.144 Work culture+ & Suppliers+ = 0.071 Work culture+ *if* Suppliers+ = 0.491	Work culture− = 0.166 Suppliers+ = 0.144 Work culture− & Suppliers+ = 0.003 Work culture− *if* Suppliers+ = 0.019
Low	Work culture+ = 0.160 Suppliers− = 0.168 Work culture+ & Suppliers− = 0.003 Work culture+ *if* Suppliers− = 0.016	Work culture− = 0.166 Suppliers− = 0.144 Work culture− & Suppliers− = 0.073 Work culture− *if* Suppliers− = 0.435

regarding that scenario for the first variable is 0.160 while for the second is 0.144. However, measures must be taken in order to the probability of occurrence from these scenarios to be simultaneous, since it is only 0.071. Finally, the probability of having a high *Work culture*, because there is a high commitment from *Suppliers,* is 0.491. For this reason, management should strive to ensure that these *Suppliers* are integrated into the TPM implementation process, since the results will be shown in the *Work culture* toward TPM.

- Fortunately, the previous data is clearly visible on the stage when the probability of finding a *Work culture* is low because there is a high commitment from the *Suppliers*, which is only 0.019; it indicates that the role of the second variable influences the occurrence of the first variable. Similarly, when the roles and levels of the variables are inverted, the probability of occurrence is 0.016.
- In the pessimistic scenario, *Work culture* and *Suppliers* have low levels. In this case, the probability of finding the first variable at its low level is 0.166, while for the second is 0.144. However, fortunately the probability of finding these two variables at their low levels simultaneously is 0.073, which can be considered a low risk; but the probability of having a *Work culture* at its low level when *Suppliers* have a low integration is 0.435, which may be worse for the company. Consequently, the management and people in charge of machines and tools purchases must ensure that they always fulfill the operating standards established in the manuals and that they receive adequate training.

10.1.5.6 Relationship Between *Work Culture* and *Managerial Commitment* (H_6)—Complex Model 1

This is one of the most interesting relationships since it seeks to determine the effect that *Managerial commitment* has on the generation of a *Work culture* in the TPM implementation process. In addition, the two variables are analyzed in their high and low scenarios, and a probability of occurrence summary in the mixed levels is shown in Table 10.9. The conclusions from the data analysis are the following:

- In its optimistic state, when *the Managerial commitment* and *Work culture* are at their high level, there is a probability of 0.160 that the scenario for the first latent variable will occur independently, while for the second it is 0.168, while the probability that these two variables are pre-sentenced at their high levels simultaneously is 0.084, which is a low value and it should be expected to be high. Finally, the probability of having a *Work culture* at its high levels, because there is a *Managerial commitment* at its high level, is 0.500, which indicates that the presence of the *Managerial commitment* makes it go from 0.160 to 0.500 in the *Work culture*, which indicates an increasement of 0.3; it indicates the role that management has in the creation of work habits regarding TPM.
- The previous statement can be clearly observed when analyzing the scenario where *Managerial commitment* is high, and *Work culture* is low, since the probability that these scenarios occur simultaneously is very low, with only

Table 10.9 Sensitivity analysis: *Managerial commitment* and *Work culture*—Complex Model 1

Managerial commitment	*Work culture*	
	High	Low
High	Work culture+ = 0.160 Managerial commitment+ = 0.168 Work culture+ & Managerial commitment+ = 0.084 Work culture+ *if* Managerial commitment+ = 0.500	Work culture− = 0.166 Managerial commitment+ = 0.168 Work culture− & Managerial commitment+ = 0.005 Work culture− *if* Managerial commitment+ = 0.032
Low	Work culture+ = 0.160 Managerial commitment− = 0.158 Work culture+ & Managerial commitment− = 0.014 Work culture+ *if* Managerial commitment− = 0.086	Work culture− = 0.166 Managerial commitment− = 0.158 Work culture− & Managerial commitment− = 0.071 Work culture− *if* Managerial commitment− = 0.448

0.005. Also, the probability of a low *Work culture* being present, because a high *Managerial commitment* has been shown, is only 0.032. In other words, whenever *Managerial commitment* exists, there will be an adequate *Work culture*. In case that the roles and scenarios of the variables are inverse, there is a probability of 0.086, which indicates that there will be no *Work culture* if there is no *Managerial commitment.*

- In the pessimistic case, when there are low levels in the *Work culture* and in the *Managerial commitment*, the probability that the first variable is presented independently is 0.166 while for the second is 0.158, but the probability that they are presented simultaneously is low, since it is only 0.071, which is a risk for the administration. Similarly, the probability of having a low *Work culture* because there is a low *Managerial commitment* is 0.448, which represents an area of opportunity to be improved by people in charge of maintenance.

10.2 Complex Model 2

The Complex Model 2 integrates a total of four latent variables, which refer to the *Managerial commitment*, *Technological status*, *Work culture,* and *TPM implementation*. In addition, it is assumed that *Managerial commitment* is the independent variable that the others depend on; for this reason, it is placed in the upper left section. Also, the dependent variable in this model is the *TPM implementation*, where *Work culture* and *technological status* are mediated variables.

Therefore, the objective of this model is to demonstrate that when an adequate *Managerial commitment* is present, an appropriate *TPM implementation* can be achieved, as well as when the support from an adequate *Work culture* of all human resources and *Technological status* is presented in the machinery and equipment that are installed in the supply chains. Additionally, the hypotheses that relate these variables are described in the section below.

10.2.1 Hypotheses

In this model, the relationship between two latent variables that integrate it, they have already been analyzed in the Complex Model 1, refers to the hypothesis six H_6 where it links itself to the *Managerial commitment* with *Work culture*, that is why, in this section it is not justified only simply exposed. However, the presence of other variables influences on the regression coefficients from the relationships to be different, that is, the value of the β will surely be not the same, since the relationship is evaluated in a different context. The hypothesis is expressed in the following sense:

H_1: The *Managerial commitment* in the TPM implementation process has a direct and positive impact on the *Work culture* toward this technique.

One of the most important aspects that is required to ensure the success of TPM is to guarantee the support and participation from the senior management, making this support visible to the entire organization at organizational, financial, technological, and effective communication levels (Park and Han 2001). In that sense, the effective application of modern technology can only be achieved through people in charge of that technology in the production systems, such as operators and maintainers to hence the success of TPM as the enabling tool to maximize the equipment effectiveness as well as establishing and maintaining the optimal relationship between people and their machines (Bon and Lim Ping 2011).

In addition, senior management is always the one that takes the decisions related to the acquisition of new technologies, and it is responsible for planning the company technological future and taking these types of decisions (Bourke and Roper 2016), since that technological capacity will determine the innovation capacity that the company has to produce new products and services, which allows them to remain in the globalized market (Cardoso et al. 2012). As a result, the following hypothesis is presented:

H_2: *Managerial commitment* has a direct and positive impact on the company *Technological status*.

In this globalization era, the manufacturing strategy is changing rapidly, the innovative strategies that became vital for each manufacturing organization want to adopt advanced manufacturing technologies, where their role in the organization is to produce high-quality products at low cost in the shortest delivery time (Singh and Kumar 2013). In addition, the implementation of these technologies brings positive results, such as better quality and flexibility, improves competitive advantages, and increases productivity (Singh and Kumar 2013; Goyal and Grover 2012).

In order to increase the successful probability of implementation in advanced manufacturing technology, management must focus on different operational aspects in the organization where the organizational culture stands out (Singh et al. 2007). Also, a company whose culture is more flexible may be more likely to experience success with advanced manufacturing technologies than one that does not have this flexibility (McDermott and Stock 1999). In fact, some authors indicate that the low acceptance of technological changes is one of the principal barriers to overcome when implementing new technologies, since these systems are automated and generally represent organizational restructuring (Waldeck and Leffakis 2007). However, when there is a progressive *Work culture*, the implementing process for these new technologies is facilitated (Swink and Nair 2007; Okure et al. 2006). According to the information above, the following hypothesis can be proposed:

H_3: The *Work culture* in a company toward TPM has a direct and positive impact on its *Technological status*.

Moreover, it should be understood that a TPM implementation program is not achieved overnight and it requires a reasonable period of holistic interventions, ranging from 3 to 5 years to realize the true potential of TPM (Ahuja 2009). Additionally, the lack of long-term vision related to the results that TPM can offer is one of the greatest limitations that any technique gives when it is being implemented, since traditionally, managers want to obtain results quickly (Lodgaard et al. 2016). For this reason, it is needed an adequate planning and a TPM implementation plan focused and assisted by senior management, through the absorption of improvements in the organizational culture during a considerable period of time to achieve significant improvements in the manufacturing performance from the TPM implementation (Ahuja 2009). However, some authors recommend that investments in new technologies should be done in a gradual manner to avoid rejection by operators (Zammuto and O'Connor 1992).

In addition, other authors claim that many techniques, including TPM, can offer favorable results and benefits, but it is often difficult to associate them. In other words, it is not known whether the success obtained is due to the technique implemented or because of another fortuitous situation. For this reason, a responsibility from senior management is to associate the obtained results with the applied techniques and the machines that are being used in the production process (Yamada et al. 2013; Attri et al. 2013). In this way, the following hypothesis can be proposed:

H_4: The *Managerial commitment* toward TPM has a direct and positive impact on the *TPM implementation* in the supply chains.

Furthermore, by managing the changes in the operating strategy during and before the *TPM implementation*, the company needs to achieve a change in culture at all employees levels (Ng et al. 2011). In addition, the organizational culture is a fundamental force that changes the way employees perform their daily routine as well as how they take decisions that play a fundamental role in guaranteeing long-term success in the current competitive market (Park and Han 2001), since it is the biggest challenge that the organization has in order to be able to make a radical transformation in the company culture, in order to ensure that all employees participate in maintenance and manufacturing performance through the TPM implementation (Ahuja and Khamba 2007). Also, the support and participation from the management is crucial to ensure that the TPM program does not lose its implementation momentum, the staff can feel the support and full participation from the senior management, and it will improve the success chances of TPM in the organization (Ng et al. 2011). However, this management requires the support from each company organizational level, including machines operators; consequently, a progressive *Work culture* is required. Therefore, the following hypothesis can be proposed:

H_5: The *Work culture* toward TPM has a direct and positive impact on the *TPM implementation* in the supply chains.

Furthermore, the technology choices must match the business requirements from the company (Park and IIan 2001), since technology provides advantages to the manufacturing companies' managers in terms of flexibility, quality, reduced delivery times, and global competitiveness (Goyal and Grover 2012), because in a dynamic and highly challenging environment, reliable manufacturing equipment is considered the main contributor to the performance and profitability of manufacturing systems (Kutucuoglu et al. 2001). In addition, its importance is increasing in the growing application of advanced manufacturing technology stages (Maggard and Rhyne 1992); therefore, equipment maintenance is an indispensable function in a manufacturing company (Shamsuddin et al. 2005).

Nowadays, some modern machines and equipment are currently equipped with sensors that allow the operator to know the operational status that they have, sending alarm signals when a critical situation is reached in a parameter, such as the temperature, among others (Fulton and Hon 2010). Logically, these signals allow quick reactions to fix problems; however, those situations cannot be possible in old machines, and decisions will be taken until the equipment has a generalized failure or the production process stops (Lewis and Boyer 2002). In other words, more technologically advanced equipment should be easily maintained in optimal working conditions. According to the previous information, the following hypothesis can be proposed:

H_6: The *Technological status* in the supply chains of the company has a direct and positive effect on the *TPM implementation*.

Figure 10.3 shows the relationships that have been previously justified and that are expressed as hypotheses. Also, it is relevant to recall that the validation of the latent variables that integrate this model has already been exposed in a previous chapter; therefore, it proceeds directly to evaluate the model and its performance as well as its efficiency indexes.

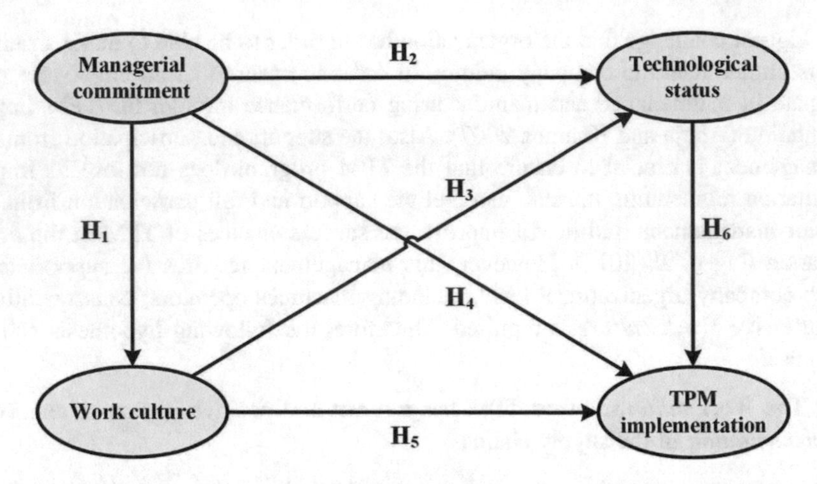

Fig. 10.3 Proposed hypotheses from Complex Model 2

10.2.2 Efficiency Indexes—Complex Model 2

The efficiency indexes from the Complex Model 2 are listed, where according to the ARS and AARS indexes, there is enough predictive value on average, since the *p*-associated values are under 0.05. Also, collinearity problems are not observed, since the AVIF and AFVIF values are less than 3.3. In the same way, it is noticeable that the data has an adequate adjustment to the model, since the Tenenhaus index is over 0.36. Finally, there are any issues in the hypotheses direction. Therefore, the model interpretation is proceeded.

- Average path coefficient (APC) = 0.396, $P < 0.001$;
- Average *R*-squared (ARS) = 0.543, $P < 0.001$;
- Average adjusted *R*-squared (AARS) = 0.541, $P < 0.001$;
- Average block VIF (AVIF) = 1.871, acceptable if $\leq$ 5, ideally $\leq$ 3.3;
- Average full collinearity VIF (AFVIF) = 2.614, acceptable if $\leq$ 5, ideally $\leq$ 3.3;
- Tenenhaus GoF (GoF) = 0.580, small $\geq$ 0.1, medium $\geq$ 0.25, large $\geq$ 0.36;
- Sympson's paradox ratio (SPR) = 1.000, acceptable if $\geq$ 0.7, ideally = 1;
- *R*-squared contribution ratio (RSCR) = 1.000, acceptable if > = 0.9, ideally = 1;
- Statistical suppression ratio (SSR) = 1.000, acceptable if $\geq$ 0.7; and
- Nonlinear bivariate causality direction ratio (NLBCDR) = 1.000, acceptable if $\geq$ 0.7.

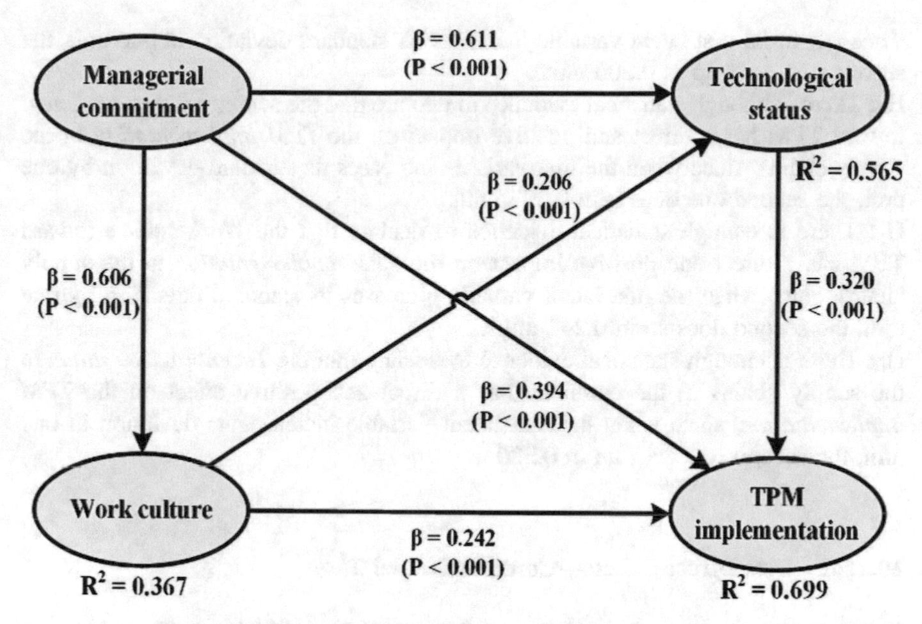

Fig. 10.4 Evaluated Complex Model 2

10.2.3 Results—Complex Model 2

Figure 10.4 illustrates the obtained results from the evaluation model presented in Fig. 10.3, where a value for the β is shown, the *p*-value for the statistical significance test and the dependent latent variables have an *R*-squared value as measure of dependency.

10.2.3.1 Direct Effects from Complex Model 2

The direct effects allow to validate the initial hypotheses, which are linked to the different latent variables analyzed. The conclusions are the following:

H_1: There is enough statistical evidence to declare that the *Managerial commitment* in the TPM implementation process has a direct and positive impact on the *Work culture* toward this technique, since when the first latent variable increases its standard deviation in one unit, the second does it in 0.606 units.

H_2: There is enough statistical evidence to declare that *Managerial commitment* has a direct and positive impact on the company *Technological status*, since when the first latent variable increases its standard deviation in one unit, the second one increases in 0.611 units.

H_3: There is enough statistical evidence to declare that the *Work culture* in a company toward TPM has a direct and positive impact on its *Technological status*,

since when the first latent variable increases its standard deviation in one unit, the second one goes up in 0.206 units.
H_4: There is enough statistical evidence to declare that the *Managerial commitment* toward TPM has a direct and positive impact on the *TPM implementation* in the supply chains, since when the first variable increases its standard deviation by one unit, the second one does it in 0.394 units.
H_5: There is enough statistical evidence to declare that the *Work culture* toward TPM has a direct and positive impact on the *TPM implementation* in the supply chains, since when the first latent variable increases its standard deviation by one unit, the second does so in 0.242 units.
H_6: There is enough statistical evidence to declare that the *Technological status* in the supply chains of the company has a direct and positive effect on the *TPM implementation*, since when the first latent variable increases its deviation in one unit, the second one goes up in 0.320 units.

10.2.3.2 Size Direct Effects—Complex Model 2

In this model, there are three dependent latent variables, which are affected by one or more independent latent variables; as a result, Table 10.10 displays the values where each one of these independent latent variables contributes in the *R*-squared value, which allows identifying the most significant independent variables for the dependent variables that are essential in order to obtain them.

According to the obtained values from Table 10.5, the next conclusions can be stated:

- The *Work culture* variable is explained only by the *Managerial commitment* in 36.7%, which indicates that to achieve the change and commitment culture from TPM, management must work through different sub-managements to direct that human resource attitude in the company.
- The *Technological status* is explained in 56.5% by two independent variables, in which 11.7% is from *Work culture* while 44.8% is from *Managerial*

Table 10.10 *R*-squared contribution—Complex Model 2

Dependent variable	Independent variable			R^2
	Work culture	*Managerial commitment*	*Technological status*	
Work culture		0.367		0.367
TMP implementation	0.158	0.304	0.237	0.699
Technological status	0.117	0.448		0.565

commitment. According to the previous information, it is concluded that to achieve an adequate *Technological status*, *Managerial commitment* is the most significant variable, which makes sense, since managers are responsible for making decisions to buy new machines and equipment.

- Finally, the TPM implementation variable is explained in 69.9% by three independent variables; where 15.8% is from *Work culture*, 30.4% from *Managerial commitment,* and 23.7% from the *Technological status*. In addition, the previous information indicates that *Managerial commitment* is the most significant variable to achieve the TPM implementation, since it has the greatest explanatory power, which is easy to understand, since management is responsible for the generation of plans as well as programs, training, and coaching for the implementation of this technique; although by the magnitude of the relation with the *Technological status*, it can be said that this data is of vital importance.

10.2.3.3 Total Indirect Effects—Complex Model 2

Table 10.11 illustrates the total effects that may arise between the latent variables analyzed when there are mediating variables. From this analysis, the following can be concluded:

- The *Work culture* has an indirect effect on the *TPM implementation*, where the *Technological status* is a mediating variable. Although this relationship is short, it is still statistically significant, since the *p*-value associated is still under 0.05.
- The *Managerial commitment* has two effects in two segments on the *TPM implementation* through the *Work culture* and *Technological status*, but also one out of three segments that are used simultaneously in those two variables, which in total are 0.381 units. In addition, it is important to observe that the direct effect is 0.394, two very similar values, which indicates that the presence of a *Work culture* and the *Technological status* are crucial supports for the *Managerial commitment* to achieve an adequate *TPM implementation*. Also, it is relevant to mention that this relationship between the two variables is the highest indirect effect.

Table 10.11 Total indirect effects—Complex Model 2

Dependent variable	Independent variable	
	Work culture	*Managerial commitment*
TMP implementation	0.066 ($P = 0.036$) ES = 0.043	0.381 ($P < 0.001$) ES = 0.294
Technological status		0.125 ($P < 0.001$) ES = 0.091

- Another important indirect effect is the one between the *Managerial commitment* and the *Technological status* variables, which is because through *Work culture*, which is very small regarding the direct effect.

10.2.3.4 Total Effects—Complex Model 2

Table 10.12 portrays the total effects that between the variables that have been analyzed in Complex Model 2. Additionally, it is known that the *Managerial commitment* variable has the largest total effects on the other variables, since it is the independent variable that has been placed at the top of the model.

According to the data in Table 10.12, the following conclusions and industrial implication can be described:

- The magnitude of these relationships allows researchers to conclude that *Managerial commitment* is the basis of the TPM implementation process, and that it requires the support of other variables, such as *Work culture* and an appropriate *Technological status* to facilitate this process. Likewise, *Managerial commitment* has a high effect on the *Work culture* of the company toward TPM.
- It is important that the company works on the creation of a *Work culture* toward TPM, since it has a high impact on the *TPM implementation* in a successful way. It is relevant to remember that, at the end of the day, people in charge of keeping the operational supply chains are the maintenance personnel and the operators, so it is convenient to create habits toward TPM.
- *Work culture* is something that must be taken into account when acquiring new technologies that allow to increase the *Technological status*, since, if this does not exist, a lot of resistance to change will be offered and it is possible that economic investments in machinery will be seen as a failure due to the lack of attitude from the personnel as well as the lack of commitment in the machinery preservation and handling.

Table 10.12 Total effects—Complex Model 2

Dependent variable	Independent variable		
	Work culture	*Managerial commitment*	*Technological status*
Work culture		0.606 ($P < 0.001$) ES = 0.367	
TMP implementation	0.307 ($P < 0.001$) ES = 0.201	0.776 ($P < 0.001$) ES = 0.598	0.320 ($P < 0.001$) ES = 0.237
Technological status	0.206 ($P < 0.001$) ES = 0.117	0.735 ($P < 0.001$) ES = 0.539	

10.2.4 Conclusions and Industrial Implications—Complex Model 2

According to the obtained data analysis in the Complex Model 2, a series of inferences and conclusions can be done, such as the following:

- *Managerial commitment* is the base where many other variables that support the appropriate *TPM implementation* depend on. In addition, when analyzing the direct effects of this *Managerial commitment*, it can be seen that the highest values are those linked to *Work culture* and *Technological status*, which indicates that management is responsible to promote the attitude toward equipment and machines change and preservation (Park 2016). Also, they are the ones who take the technological investments decisions.
- New technologies purchases for the productive system must be a discussed decision, where the benefits which justify that investment are analyzed, since it is well known that a bad decision can bankrupt a company, since AMT are expensive (Bai and Sarkis 2017). Frequently, machines are highly specialized, and it must be ensured that more flexibility is achieved in the production processes with the new AMT, since changes in the product designs occur quickly and the MAT must not be used for exclusive or specific way, but they must be easily adapted to develop several activities.
- Senior management should focus on achieving a *Work culture* toward machines and tools preservation, since these are the means of generating wealth. This culture must be focused mainly on operators, because they are the first source of information about the true conditions of a machine (Eti et al. 2006; Collins et al. 2006).
- The *TPM implementation process* requires many other support activities and its success is not fortuitous. For instance, investments in technology that offer greater productive capacity and automation capacity are required, since these types of equipment are repaired quickly, because they provide a lot of information on the operational conditions that they have (Singh et al. 2013).

10.2.5 Sensitivity Analysis—Complex Model 2

This model integrates six hypotheses or relationships between four variables. It is assumed that *Managerial commitment* is the independent variable and that the *TPM implementation* is the dependent variable while *Work culture* and *Technological status* are the mediating variables. In addition, for each relationship or hypothesis, a Sensitivity analysis is carried out, where four possible scenarios are analyzed in a connected way.

10.2.5.1 Relationship Between *Managerial Commitment* and *Work Culture* (H_1)—Complex Model 2

This relationship is one of the most relevant to analyze, since it is traditionally assumed that senior management is the most committed to the *TPM implementation* and, therefore, it must create a *Work culture* in order to have the best conditions for the adoption of the technique. However, it is important to remember that this relationship was analyzed in Complex Model 1, where it appears as H_6, so its analysis and interpretation are not described in this section.

10.2.5.2 Relationship Between *Managerial Commitment* and *Technological Status* (H_2)—Complex Model 2

Investments in modern technologies and machinery for the production process depend on the *Managerial commitment* to maintain a competitive *Technological status*. In this section, the scenarios where it is possible to find the relationship between these two variables, for their high and low levels, are analyzed. Also, summary of the results is illustrated in Table 10.13, and some of the conclusions and inferences that are obtained are the following:

- The most optimistic scenario where the relationship between *Managerial commitment* and *Technological status* have high levels of probability to happen. In this case, that scenario for the first latent variable is 0.163 and for the second is 0.168, which is independent. However, when analyzing, the probability that these two variables are presented simultaneously is only 0.092, which indicates that it is an area of opportunity and that this value is always required to be high. Finally, when it is observed that the probability of having a high *Technological*

Table 10.13 Sensitivity analysis: *Managerial commitment* and *Technological status*—Complex Model 2

Managerial commitment	*Technological status*	
	High	Low
High	Technological status+ = 0.163 Managerial commitment+ = 0.168 Technological status+ & Managerial commitment+ = 0.092 Technological status+ *if* Managerial commitment+ = 0.548	Technological status− = 0.182 Managerial commitment+ = 0.168 Technological status− & Managerial commitment+ = 0.003 Technological status− *if* Managerial commitment+ = 0.016
Low	Technological status+ = 0.163 Managerial commitment− = 0.158 Technological status+ & Managerial commitment− = 0.003 Technological status+ *if* Managerial commitment− = 0.017	Technological status− = 0.182 Managerial commitment− = 0.158 Technological status− & Managerial commitment− = 0.103 Technological status− *if* Managerial commitment− = 0.655

status in machines and tools, since there is a high *Managerial commitment*, it is 0.548, which indicates that there are other factors that affect the decision process to develop technological investments and that not everything relies on management, although it does have a strong influence.

- The previous data is easily demonstrated by observing the case where *Managerial commitment* is high, but the *Technological status* is low, since the probability that these two variables are presented simultaneously in that scenario is 0.003, which is almost invalid; that is, the first variable is influenced by the second. In addition, if the probability of having a low *Technological status*, because that there is a high *Managerial commitment*, is only 0.016, it indicates that whenever there is a *Managerial commitment*, there will be an adequate technological level in the machines and tools.
- In the pessimistic scenario for the *Technological status* and *Managerial commitment* variables, it is found when both have low values. Also, in an independent sense, the first one has a probability of being in this scenario by 0.182 while the second is 0.158, but the probability that they are presented simultaneously is 0.103, which represents a risk in the TPM implementation. Finally, the probability that the *Technological status* is low, because there is a low *Managerial commitment*, is 0.655, a high-risk value that managers must analyze before starting the TPM implementation process.

10.2.5.3 Relationship Between *Work Culture* and *Technological Status* (H_3)—Complex Model 2

Undoubtedly, management would commit a little in making investments in cutting-edge technology, if there is no *Work culture* that helps to preserve it. In this section, the Sensitivity analysis for *Work culture* and *Technological status* is performed, so Table 10.14 illustrates a summary in two scenarios: highs and lows. From the analysis of this data, the following is concluded:

Table 10.14 Sensitivity analysis: *Work culture* and *Technological status*—Complex Model 2

Work culture	*Technological status*	
	High	Low
High	Technological status+ = 0.182 Work culture+ = 0.158 Technological status+ & Work culture+ = 0.103 Technological status+ *if* Work culture+ = 0.655	Technological status− = 0.182 Work culture+ = 0.160 Technological status− & Work culture+ = 0.011 Technological status− *if* Work culture += 0.068
Low	Technological status+ = 0.163 Work culture− = 0.166 Technological status+ & Work culture− = 0.005 Technological status+ *if* Work culture− = 0.033	Technological status− = 0.182 Work culture− = 0.166 Technological status− & Work culture− = 0.073 Technological status− *if* Work culture− = 0.443

- The most optimistic scenario in the relationship between the *Technological status* and *Work culture* variables is when both variables have their high levels. In an independent way, this scenario has a probability of occurrence of 0.182 and the second one has it in 0.158. Also, the probability that these two favorable scenarios for companies are presented is only 0.103, a value that managers must take into account to seek to increase. Finally, the probability of having a good *Technological status*, because there is a good *Work culture* toward TPM, is 0.655; therefore, managers along with *Suppliers* should look forward for the *Work culture* to increase. In other words, the *Work culture* is a guarantee to implement machines and equipment with high *Technological status*.
- The previous data is easily demonstrated when analyzing the situation where the *Technological status* is low, but *Work culture* is high, in which it is observed that the probability of simultaneous occurrence of those two scenarios is only 0.011, a very low probability; but also, the probability of having a *Technological status*, because there is a high *Work culture*, is only 0.068, which indicates that these variables are strongly linked. Likewise, if the situation where the variables have inverse levels is analyzed, it is observed that the probability that there are situations where there is a *Technological status* with a low *Work culture* is 0.005, almost zero, which allows to conclude that these two variables are related.
- In the most pessimistic scenario, these two variables happen when the *Technological status* and *Work culture* are low. The probability that the first variable is presented independently is 0.182, while for the second it is 0.166, and there is a high risk of 0.073 that they occur simultaneously. However, the probability of having a low *Technological status* level, since there is a *Work culture* at its low level, is very high, since it has a probability of occurrence of 0.443 and that is not favorable for TPM.

10.2.5.4 Relationship Between *Managerial Commitment* and *TPM Implementation* (H_4)—Complex Model 2

As a matter of fact, without the *Managerial commitment* as collateral, no manager would dare to propose the *TPM implementation*. In this section, the probabilities of occurrence of these two variables in the scenarios where they have high and low levels independently are analyzed, but also, there is an analysis of four scenarios when they are combined. In addition, Table 10.15 illustrates the summary of these scenarios, and based on its information, the following can be concluded:

- In the most optimistic scenario, the variables of *TPM implementation* and *Managerial commitment* have high levels and, in this case, the probability of finding that scenario independently for the first variable is 0.193 while for the second is 0.168. However, the probability of finding these two scenarios

Table 10.15 Sensitivity analysis: *Managerial commitment* and *TPM implementation*—Complex Model 2

Managerial commitment	*TPM implementation*	
	High	Low
High	TPM implementation+ = 0.193 Managerial commitment+ = 0.168 TPM implementation+ & Managerial commitment+ = 0.103 TPM implementation+ *if* Managerial commitment+ = 0.613	TPM implementation− = 0.168 Managerial commitment+ = 0.168 TPM implementation− & Managerial commitment+ = 0.003 TPM implementation− *if* Managerial commitment+ = 0.016
Low	TPM implementation+ = 0.193 Managerial commitment− = 0.158 TPM implementation+ & Managerial commitment− = 0.005 TPM implementation+ *if* Managerial commitment− = 0.034	TPM implementation− = 0.168 Managerial commitment− = 0.158 TPM implementation− & Managerial commitment− = 0.101 TPM implementation− *if* Managerial commitment− = 0.638

simultaneously is 0.103; a low value, since it is expected to be high. Finally, the probability of having a high *TPM implementation* process when *Managerial commitment* is also high is 0.613, and as a result, it is concluded that maintenance managers should seek to guarantee that *Managerial commitment* among other aspects first.

- When analyzing the case where there is a low *TPM implementation* and a high *Managerial commitment* simultaneously, it is observed that the probability of occurrence is only 0.003, a very low value, which validates the conclusion from the previous section. In addition, the probability of a low *TPM implementation* occurring because there is a high *Managerial commitment* is 0.016, which indicates that management has a high impact on the success of TPM. In the same way, if the opposite situation is analyzed, where the *TPM implementation* is high and *Managerial commitment* low, they have a probability of occurrence of only 0.005, and the probability of the first situation due that the second occurred has happened by 0.034, which shows that these two variables have a high relationship.
- The pessimistic scenario can occur when the *TPM implementation* and *Managerial commitment* have low values. In addition, independently the scenario of the first variable has a probability of occurrence by 0.168 while the second has it by 0.158, but the probability of occurrence in a simultaneous way is 0.101; a high-risk value, and even worse it may be the situation where there is a low TPM implementation level because there is a low level of support or *Managerial commitment*, since it is a probability of 0.638, and the maintenance managers must do an effort in order to this, will not happen.

10.2.5.5 Relationship Between *Work Culture* and *TPM Implementation* (H_5)—Complex Model 2

It has been previously mentioned that it is convenient to have a *Work culture* before starting the *TPM implementation*, but that aspect is attributable only at the administrative level, and it is required that there is also a *Work culture* among the operators and general managers. In this section, the relationship between *Work culture* and *TPM implementation* focusing on operators is analyzed. In addition, the high and low scenarios for each variable and four scenarios are analyzed in a combined way.

Table 10.16 shows a summary of the probabilities of occurrence for the variables, and from this data analysis, the following can be stated:

- The optimistic scenario is presented when the *TPM implementation* and *Work culture* have high indexes. The probability of independent occurrence for the first variable is 0.193, while for the second is 0.160, but the probability of occurring in those scenarios together is only 0.098, a very low probability that represents an opportunity for improvement. Likewise, the probability of having a *TPM implementation* at its high level, since there is a high *Work culture,* is 0.61, so managers must focus on obtaining that *Work culture* in that scenario, which they can do along with the *Suppliers* and management support.
- The previous data is evaluated when analyzing the scenario where the *TPM implementation* is low, but *Work culture* is high, since together these two scenarios only have a probability of occurrence by 0.008; in other words, that scenario will almost never happen. However, the probability of the first variable occurring in this scenario, since the second exists, is 0.051. Similarly, when there are inverse levels, high *TPM implementation,* but low *Work culture,* they can occur simultaneously with a probability of 0.003, which indicates that the situation is almost nonexistent, but the probability of presenting the first in its

Table 10.16 Sensitivity analysis: *Work culture* and *TPM implementation*—Complex Model 2

Work culture	*TPM implementation*	
	High	Low
High	TPM implementation+ = 0.193 Work culture+= 0.160 TPM implementation+ & Work culture+ = 0.098 TPM implementation+ *if* Work culture+ = 0.610	TPM implementation− = 0.168 Work culture+ = 0.160 TPM implementation− & Work culture+ = 0.008 TPM implementation− *if* Work culture+ = 0.051
Low	TPM implementation+ = 0.193 Work culture− = 0.166 TPM implementation+ & Work culture− = 0.003 TPM implementation+ *if* Work culture− = 0.016	TPM implementation− = 0.168 Work culture− = 0.166 TPM implementation− & Work culture− = 0.073 TPM implementation− *if* Work culture− = 0.443

scenario, since the second variable has happened, is 0.16 and it does not represent a risk.

- Finally, the pessimistic scenario happens when the *TPM implementation* and *Work culture* are low. The probability of occurrence of the scenario from the first variable is 0.168, while for the second is 0.166, but the probability that they occur simultaneously is 0.073, and this scenario is a high risk. Finally, the probability of having a *TPM implementation* at its low level, because there is a low *Work culture,* is 0.443, a risk that the managers in charge of the implementation process should overlook.

10.2.5.6 Relationship Between *Technological Status* and *TPM Implementation* (H_6)—Complex Model 2

It is generally thought that, if the *Technological status* is low, there will be more issues for the *TPM implementation*; however, there is no statistical evidence for this. In this section, the relationships that these variables have at low and high levels independently, but also the probabilities of occurrence simultaneously, and conditionally, are analyzed. In addition, Table 10.17 evaluates the results of the relationships between these two variables and the following can be concluded:

- The optimistic scenario occurs when the *TPM implementation* and *Technological status* variables have high levels. Additionally, the scenario for the first variable is presented independently with a probability of 0.193 while the second is 0.163, but the probability that both variables are presented simultaneously in their high levels is only 0.103, so the management must work to increase that scenario. Also, reliably, the probability of having a *TPM*

Table 10.17 Sensitivity analysis: *Technological status* and *TPM implementation*—Complex Model 2

Technological status	*TPM implementation*	
	High	Low
High	TPM implementation+ = 0.193 Technological status+ = 0.163 TPM implementation+ & Technological status+ = 0.103 TPM implementation+ *if* Work culture+ = 0.633	TPM implementation− = 0.168 Technological status+ = 0.163 TPM implementation− & Technological status+ = 0.003 TPM implementation− *if* Technological status+ = 0.017
Low	TPM implementation+ = 0.193 Technological status− = 0.182 TPM implementation+ & Technological status− = 0.000 TPM implementation+ *if* Technological status− = 0.00	TPM implementation− = 0.168 Technological status− = 0.182 TPM implementation− & Technological status− = 0.114 TPM implementation− *if* Technological status− = 0.627

implementation at its high level, because there is a raised *Technological status,* is 0.633, which indicates that these two variables are closely related.

- The previous data is demonstrated when analyzing other scenarios of the variables, for example, that the *TPM implementation* is lower and that *Technological status* is high, which have a probability of being presented in a simultaneous way by 0.003, and the probability that the first variable is presented in its low scenario, because the second variable has occurred in its high scenario, it is 0.017, which indicates that almost always there is a *Technological status*, and the possibility of having a *TPM implementation* in a low scenario is almost zero.
- Finally, the pessimistic scenario occurs when there are low levels in the *TPM implementation* and in the *Technological status*. Also, the probability of occurrence for the first variable in this scenario is 0.168 while for the second is 0.182, which they can be considered high risks, but the probability that they occur simultaneously is 0.114. Consequently, the probability that a low *TPM implementation* is presented, because there is a low level in the *Technological status,* is 0.627, which represent high-risk values for the implementation process.

References

Abhishek J, Rajbir B, Harwinder S (2014) Total productive maintenance (TPM) implementation practice: a literature review and directions. Int J Lean Six Sigma 5(3):293–323. https://doi.org/10.1108/IJLSS-06-2013-0032

Ahuja IPS (2009) Total productive maintenance. In: Ben-Daya M, Duffuaa SO, Raouf A, Knezevic J, Ait-Kadi D (eds) Handbook of maintenance management and engineering. Springer London, London, pp 417–459. https://doi.org/10.1007/978-1-84882-472-0_17

Ahuja IPS, Khamba JS (2007) An evaluation of TPM implementation initiatives in an Indian manufacturing enterprise. J Qual Maintenance Eng 13(4):338–352

Ahuja IPS, Khamba JS (2008) Strategies and success factors for overcoming challenges in TPM implementation in Indian manufacturing industry. J Qual Maintenance Eng 14(2):123–147. https://doi.org/10.1108/13552510810877647

Attri R, Grover S, Dev N, Kumar D (2013) Analysis of barriers of total productive maintenance (TPM). Int J Syst Assur Eng Manag 4(4):365–377. https://doi.org/10.1007/s13198-012-0122-9

Bai C, Sarkis J (2017) Improving green flexibility through advanced manufacturing technology investment: modeling the decision process. Int J Prod Econ 188:86–104. https://doi.org/10.1016/j.ijpe.2017.03.013

Bon AT, Lim Ping P (2011) Implementation of total productive maintenance (TPM) in automotive industry. In: 2011 IEEE symposium on business, engineering and industrial applications (ISBEIA), 25–28 Sept 2011, pp 55–58. https://doi.org/10.1109/isbeia.2011.6088881

Bourke J, Roper S (2016) AMT adoption and innovation: an investigation of dynamic and complementary effects. Technovation 55–56:42–55. https://doi.org/10.1016/j.technovation.2016.05.003

Cardoso RdR, Pinheiro de Lima E, Gouvea da Costa SE (2012) Identifying organizational requirements for the implementation of advanced manufacturing technologies (AMT). J Manuf Syst 31(3):367–378. https://doi.org/10.1016/j.jmsy.2012.04.003

Collins H, Gordon C, Terra JC (2006) Chapter 9—The culture and change dimension. In: Collins H, Gordon C, Terra JC (eds) Winning at collaboration commerce. Butterworth-Heinemann, Boston, pp 157–181. https://doi.org/10.1016/B978-0-7506-7817-9.50013-9

Damiana C, Gianni G (2010) Maintenance management in Italian manufacturing firms: matters of size and matters of strategy. J Qual Maintenance Eng 16(2):156–180. https://doi.org/10.1108/13552511011048904

Eti MC, Ogaji SOT, Probert SD (2006) Impact of corporate culture on plant maintenance in the Nigerian electric-power industry. Appl Energy 83(4):299–310. https://doi.org/10.1016/j.apenergy.2005.03.002

Fulton M, Hon B (2010) Managing advanced manufacturing technology (AMT) implementation in manufacturing SMEs. Int J Prod Perform Manag 59(4):351–371. https://doi.org/10.1108/17410401011038900

Goyal S, Grover S (2012) Advanced manufacturing technology effectiveness: a review of literature and some issues. Front Mech Eng 7(3):256–267. https://doi.org/10.1007/s11465-012-0330-7

James J, Ikuma L, Nahmens I, Aghazadeh F (2014) The impact of Kaizen on safety in modular home manufacturing. Int J Adv Manuf Technol 70(1–4):725–734. https://doi.org/10.1007/s00170-013-5315-0

Singh RK, Garg SK, Deshmukh SG, Kumar M (2007) Modelling of critical success factors for implementation of AMTs. J Model Manag 2(3):232–250. https://doi.org/10.1108/17465660710834444

Krolikowski M, Yuan X (2017) Friend or foe: customer-supplier relationships and innovation. J Bus Res 78:53–68. https://doi.org/10.1016/j.jbusres.2017.04.023

Kutucuoglu KY, Hamali J, Irani Z, Sharp JM (2001) A framework for managing maintenance using performance measurement systems. Int J Oper Prod Manag 21(1/2):173–195. https://doi.org/10.1108/01443570110358521

Lewis MW, Boyer KK (2002) Factors impacting AMT implementation: an integrative and controlled study. J Eng Tech Manage 19(2):111–130. https://doi.org/10.1016/S0923-4748(02)00005-X

Li K (2018) Innovation externalities and the customer/supplier link. J Bank Finance 86:101–112. https://doi.org/10.1016/j.jbankfin.2017.09.003

Lodgaard E, Ingvaldsen JA, Gamme I, Aschehoug S (2016) Barriers to lean implementation: perceptions of top managers, middle managers and workers. Proc CIRP 57:595–600. https://doi.org/10.1016/j.procir.2016.11.103

Maggard BN, Rhyne DM (1992) Total productive maintenance: a timely integration of production and maintenance. Prod Inventory Manag J 33(4):6

Mandeep K, Kanwarpreet S, Inderpreet Singh A (2012) An evaluation of the synergic implementation of TQM and TPM paradigms on business performance. Int J Prod Perform Manag 62(1):66–84. https://doi.org/10.1108/17410401311285309

McDermott CM, Stock GN (1999) Organizational culture and advanced manufacturing technology implementation. J Oper Manag 17(5):521–533. https://doi.org/10.1016/S0272-6963(99)00008-X

McKone KE, Weiss EN (2000) Analysis of investments in autonomous maintenance activities. IIE Trans 32(9):849–859. https://doi.org/10.1023/a:1007647329259

Ng KC, Goh GGG, Eze UC (2011) Critical success factors of total productive maintenance implementation: a review. In: 2011 IEEE international conference on industrial engineering and engineering management, 6–9 Dec 2011, pp 269–273. https://doi.org/10.1109/ieem.2011.6117920

Okure MAE, Mukasa N, Elvenes BO (2006) Modelling the development of advanced manufacturing technologies (AMT) in developing countries. In: Mwakali JA, Taban-Wani G (eds) Proceedings from the international conference on advances in engineering and technology. Elsevier Science Ltd., Oxford, pp 488–494. https://doi.org/10.1016/B978-008045312-5/50053-4

Park J (2016) The impact of depreciation savings on investment: evidence from the corporate alternative minimum tax. J Publ Econ 135:87–104. https://doi.org/10.1016/j.jpubeco.2016.02.001

Park KS, Han SW (2001) TPM—total productive maintenance: impact on competitiveness and a framework for successful implementation. Hum Factors Ergon Manuf Serv Ind 11(4):321–338. https://doi.org/10.1002/hfm.1017

Pinjala SK, Pintelon L, Vereecke A (2006) An empirical investigation on the relationship between business and maintenance strategies. Int J Prod Econ 104(1):214–229. https://doi.org/10.1016/j.ijpe.2004.12.024

Riedel R, Neumann N, Franke M, Müller E (2010) The behavior of suppliers in supplier-customer relationships. In: Vallespir B, Alix T (eds) Advances in production management systems. New challenges, new approaches, Berlin, Heidelberg. Springer, Berlin Heidelberg, pp 259–266

Sade AB, Bojei BJ, Donaldson GW (2015) Customer service: managerial commitment and performance within industrial manufacturing firms. In: Sidin SMD, Manrai AK (eds) Proceedings of the 1997 world marketing congress, Cham. Springer International Publishing, pp 544–547

Sani SIA, Mohammed AH, Misnan MS, Awang M (2012) Determinant factors in development of maintenance culture in managing public asset and facilities. Proc Soc Behav Sci 65:827–832. https://doi.org/10.1016/j.sbspro.2012.11.206

Shamsuddin A, Masjuki Hj H, Zahari T (2005) TPM can go beyond maintenance: excerpt from a case implementation. J Qual Maintenance Eng 11(1):19–42. https://doi.org/10.1108/13552510510589352

Singh H, Kumar R (2013) Measuring the utilization index of advanced manufacturing technologies: a case study. IFAC Proc Volumes 46(9):899–904. https://doi.org/10.3182/20130619-3-RU-3018.00395

Singh R, Gohil AM, Shah DB, Desai S (2013) Total productive maintenance (TPM) implementation in a machine shop: a case study. Proc Eng 51:592–599. https://doi.org/10.1016/j.proeng.2013.01.084

Swink M, Nair A (2007) Capturing the competitive advantages of AMT: design–manufacturing integration as a complementary asset. J Oper Manag 25(3):736–754. https://doi.org/10.1016/j.jom.2006.07.001

Tong DYK, Rasiah D, Tong XF, Lai KP (2015) Leadership empowerment behaviour on safety officer and safety teamwork in manufacturing industry. Saf Sci 72:190–198. https://doi.org/10.1016/j.ssci.2014.09.009

Waldeck NE, Leffakis ZM (2007) HR perceptions and the provision of workforce training in an AMT environment: an empirical study. Omega 35(2):161–172. https://doi.org/10.1016/j.omega.2005.05.001

Yamada TT, Poltronieri CF, Gambi LdN, Gerolamo MC (2013) Why does the implementation of quality management practices fail? A qualitative study of barriers in Brazilian Companies. Proc Soc Behav Sci 81:366–370. https://doi.org/10.1016/j.sbspro.2013.06.444

Zammuto RF, O'Connor EJ (1992) Gaining advanced manufacturing technologies' benefits: the roles of organization design and culture. Acad Manag Rev 17(4):701–728. https://doi.org/10.2307/258805

Chapter 11
Structural Equation Models-Human Factor—Part II

11.1 Complex Model 3

In the two previous complex models, the benefits from the actions on the variables have not been integrated. However, this Complex Model 3 has the purpose of integrating two latent variables associated with the activities, but already related to two types of benefits in the analysis. In addition, the variables that comprehend the model are: *Managerial commitment*, *Suppliers*, *Benefits for the organization*, and *Productivity benefits*. In addition, the objective of the model is to analyze the effect that *Managerial commitment* has as a latent independent variable on *Suppliers*, which depends on in order to quantify their effect on *Benefits for the organization* and *Productivity benefits* as a latent dependent variable.

Thus, in this model, it is assumed that *Productivity benefits* are seen because of the activities that management performs as well as those that are developed during the TPM implementation. Finally, six hypotheses are proposed; they link the variables that integrate the model, which are justified and described in the section below.

11.1.1 Hypotheses

The manufacturing industry has experienced an unprecedented degree of changes in the past three decades, which implies drastic changes in management approaches, product and process technologies, customer expectations, *Suppliers* attitudes, and competitive behavior (Ahuja et al. 2006). In fact, nowadays the supplier is considered an essential part of a supply chain and this phenomenon cannot be an exception during the process of implementing techniques already installed in the production system such as TPM (Cardoso et al. 2012).

J. R. Díaz-Reza et al., *Impact Analysis of Total Productive Maintenance*,
https://doi.org/10.1007/978-3-030-01725-5_11

Currently, cooperation with *Suppliers* may influence the purchaser to be more efficient, therefore, it allows the goods to be purchased at lower prices, and also it may force the manufacturer to look for its core competence to remain more competitive (Chavhan et al. 2018). However, it must be remembered that the decision maker is the senior management, and it is the one who has a direct relationship with *Suppliers*, it determines the economic amounts to invest, and the ideal time to do so (Aravindan and Punniyamoorthy 2002).

Equally important, something that management must take into account is that the new technology *Suppliers* have many attributes, and that its cost is not the only aspect that must be considered in order to make a purchase decision (Chan et al. 2001). For instance, Efstathiades et al. (2002) indicate that this type of investment is a strategic decision, and the only ones who know about this type of plans are the senior management, therefore, through multidisciplinary decision groups, they make the investments (Fulton and Hon 2010). Consequently, it is considered that there is a relationship between *Managerial commitment* and *Suppliers* and the following hypothesis is proposed.

H_1: The *Managerial commitment* during the TPM implementation process has a direct and positive effect on machinery and equipment *Suppliers.*

Furthermore, management needs to have a strong commitment toward the TPM implementation program, and it must strive to develop multilevel communication mechanisms with all employees, explaining the importance and benefits of the program, spreading the benefits of TPM to the organization and employees, linking TPM to the strategy and the organization general objectives (Ahuja 2009).

In addition, Management's contributions toward successful implementations of TPM, among its activities may include: reviewing business plans to include goals, affect appropriate cultural transformations in the organizational culture, build success stories to promote motivation, communicate goals, providing adequate financial resources to affect business improvements, promote and foster the culture of multifunctional teams, provide training and upgrading skills for production as well as for workers in maintenance, develop appropriate incentive and rewards mechanisms to promote continuous improvement, ensuring total participation of employees, supporting changes and improvements in the workplace, eliminating barriers related to middle management, and improving interdepartmental energy (Ahuja 2009).

However, all the facilities that management offers when implementing TPM must be monitored, since it must expect a series of benefits from it (Willmott and McCarthy 2001b). For example, greater machines and equipment availability, lower repair costs, greater spare parts availability in the warehouse, lower administrative costs, a decrease in delays in the delivery of production orders, among others. In addition, it must be able to link or relate those activities and investments with the benefits, since only in this way the TPM implementation process is justified (Willmott and McCarthy 2001c). In effect, the following hypothesis can be proposed:

H_2: The *Managerial commitment* during the TPM implementation has a direct and positive effect on the *Benefits for the organization* that are obtained.

Moreover, as it has been already mentioned throughout the previous chapters, TPM provides an integral approach to the life cycle for the equipment management that minimizes equipment failures, production defects, and accidents, which involve all members from the organization to the external *Suppliers*, whose objective is to continuously improve the availability and avoid the equipment degradation (Shetty and Rodrigues 2010). Additionally, each company that manufactures products or provides services has its own set of equipment and tools *Suppliers* as well as consumables for the machines. Also, the set of *Suppliers* from an original equipment manufacturer (OEM) differs clearly according to the industry, where different materials, supplies, and even practices are displayed (Nagel 2006).

In order to take advantage of the benefits from the collaboration with an equipment supplier, it is important that the information exchanged is carefully adapted to the specific partner needs (Bruch and Bellgran 2013) and the instructions of use and handling that are in the manuals are followed, since otherwise, the profit that has been planned about the acquired equipment is diminished. Also, an inappropriate use in handling represents a breakdown risk, maintenance costs, low product quality, and low availability (Kumar and Soni 2015; Patrik and Magnus 1999).

In addition, even in situations where the equipment supplier plays an important role in the production equipment design, it seems necessary to maintain certain competencies also within the manufacturing company to ensure that this equipment work properly, since it will not be needed to have specialized equipment if it is not properly used (Bruch and Bellgran 2013). Therefore, senior management along with engineers who are related to the acquired equipment and machines must verify that the technical design specifications are achieved when they are operated in the production systems, as well as call out to the *Supplier* when there is a deficiency, since the company operational capacity depends on it (Saliba et al. 2017). According to the previous information, the following hypothesis is established:

H_3: Machinery and equipment *Suppliers* have a direct and positive effect on the *Productivity benefits* obtained from the TPM implementation.

Equally important, the senior management has the main responsibility for preparing an adequate and supportive environment before the official TPM launch within its organization, since only with the sustained commitment and enthusiasm from the Senior Direction and Management, also the changes in the corporation strategy can be designed and executed properly and TPM is not the exception (Attri et al. 2013). In addition, the senior management must strive to train and develop the employees skills by updating their abilities, knowledge, and attitude to allow greater productivity to reach the highest quality standards, eliminate product defects, equipment failures (breakdowns) and accidents, develop workforce with multiple skills and so, to create a pride sense and belonging among all employees (Ahuja and Khamba 2008).

However, the greatest senior *Managerial commitment* in a company to implement TPM, it is to guarantee a series of benefits that can be efficiency and productivity parameters in supply chains. In addition, it is the resources manager assigned to TPM, so it must be able to obtain measurements that are associated with the executed activities, because if the company does not fulfill to link its activities with the benefits obtained in the supply chain, it will surely have issues to justify the TPM implementation or other lean manufacturing techniques (Wyrwicka and Mrugalska 2017; Zhou 2016). From the previous data, the following hypothesis is settled:

H_4: The *Managerial commitment* during the TPM implementation has a direct and positive effect on the *Productivity benefits* obtained for the company.

By the same token, equipment *Suppliers* are important sources of innovations in the technology production process (Reichstein and Salter 2006). In addition, the supplier must know that, if it desires to continue in the market offering its machines, it must be able to provide a competitive advantage to the manufacturers, in such a way that their purchase is justified, because by having access to the new technology and installing them in their production processes, the manufacturing companies can benefit themselves from a faster introduction in the market, less development risks, and a better productivity, always when they are used in the same appropriate manner (Pisano 1997). Therefore, creating a win-win situation is not only expected in the relationship between the equipment supplier and the manufacturer, but essential in order to maintain competitiveness in the future (Thomas and Johan 2010).

Furthermore, *Suppliers* should always seek to exceed the production parameters of their equipment, establishing improvement systems focused on greater productive and operational efficiency, but also to occupy less space in supply chains, having a greater maintainability, precision, and quality, and in these times, a higher commitment toward the environment, where it should aim to be less noisy, emit less pollutants, and be more friendly with the environment (Cao and Wang 2017). Therefore, the following hypothesis can be proposed:

H_5: Machines and equipment *Suppliers* for production systems have a direct and positive impact on the *Productivity benefits* obtained for the company.

Moreover, in the increasingly competitive business environment of the globalized world, it is essential for any company to survive to be adaptable, competitive in terms of prices, receptive, and proactive, and get the ability to offer world-class products according to the different customer requirements with existing technology (Baysal et al. 2015) and TPM is a methodology that improves the potential to use machines and tools to achieve better productivity goals. Also, for each manufacturing company, the objective is to produce goods with a profit, and this can only be acquired by using an effective maintenance system that helps to maximize the equipment availability by minimizing machine downtime due to unwanted (Fore and Zuze 2010). In addition, what TPM does is to attack the hidden losses and guarantee the quality–price ratio for the direct manufacturing effort, this combined strategy will result in a dramatic benefit (Willmott and McCarthy 2001a).

Likewise, it has been reported that TPM eliminates the losses resulting from the equipment and employees unplanned downtime, reduces speed and quality (Duffuaa and Raouf 2015). In other words, administrative benefits are obtained for the organization, but these can be transformed into benefits related to productivity and efficiency indexes of the company, therefore, the following hypothesis can be presented:

H_6: The *Benefits for the organization* that are obtained when implementing TPM has a direct and positive impact on the acquired *Productivity benefits*.

The six hypotheses that have been proposed above can be expressed graphically according to Fig. 11.1. In addition, the latent variables that intervene in this model have been validated in an exclusive chapter, in that case, this activity is not described here, it is only convenient to mention that all the variables have fulfilled the validity indexes established in the methodology section.

11.1.2 Efficiency Indexes—Complex Model 3

Before interpreting the outcomes from Complex Model 3, the efficiency indexes were analyzed. In addition, in the list that is observed next, values obtained from its evaluation are illustrated and it is observed that they are suitable for the model. For example, the ARS and AARS indexes indicate that the model has enough predictive validity, since the associated p-value is under 0.05; however, according to the AVIF and AFVIF, there are no problems of collinearity, because they are values less than 3.3 and an adequate data adjustment to the model is also perceived. Finally, there

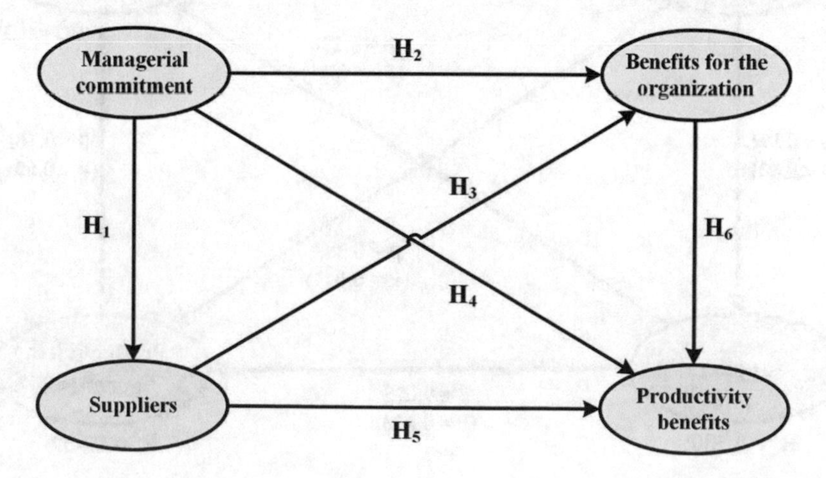

Fig. 11.1 Proposed hypotheses—Complex Model 3

are any problems in the direction where the hypotheses have been planned, consequently, the collected results are interpreted.

- Average path coefficient (APC) = 0.363, $P < 0.001$
- Average R-squared (ARS) = 0.468, $P < 0.001$
- Average adjusted R-squared (AARS) = 0.465, $P < 0.001$
- Average block VIF (AVIF) = 1.609, acceptable if ≤ 5, ideally ≤ 3.3
- Average full collinearity VIF (AFVIF) = 2.350, acceptable if ≤ 5, ideally ≤ 3.3
- Tenenhaus GoF (GoF) = 0.567, small ≥ 0.1, medium ≥ 0.25, large ≥ 0.36
- Sympson's paradox ratio (SPR) = 1.000, acceptable if ≥ 0.7, ideally = 1
- R-squared contribution ratio (RSCR) = 1.000, acceptable if ≥ 0.9, ideally = 1
- Statistical suppression ratio (SSR) = 1.000, acceptable if ≥ 0.7
- Nonlinear bivariate causality direction ratio (NLBCDR) = 1.000, acceptable if ≥ 0.7

11.1.3 Results—Complex Model 3

Figure 11.2 illustrates the parameters from the proposed relationships in the Complex Model 3. Also, as in all the previous models, the value of the β is presented for each of the relationships between the variables, the p-value associated with the statistical test of significance, as well as the R-squared values in the latent dependent variables.

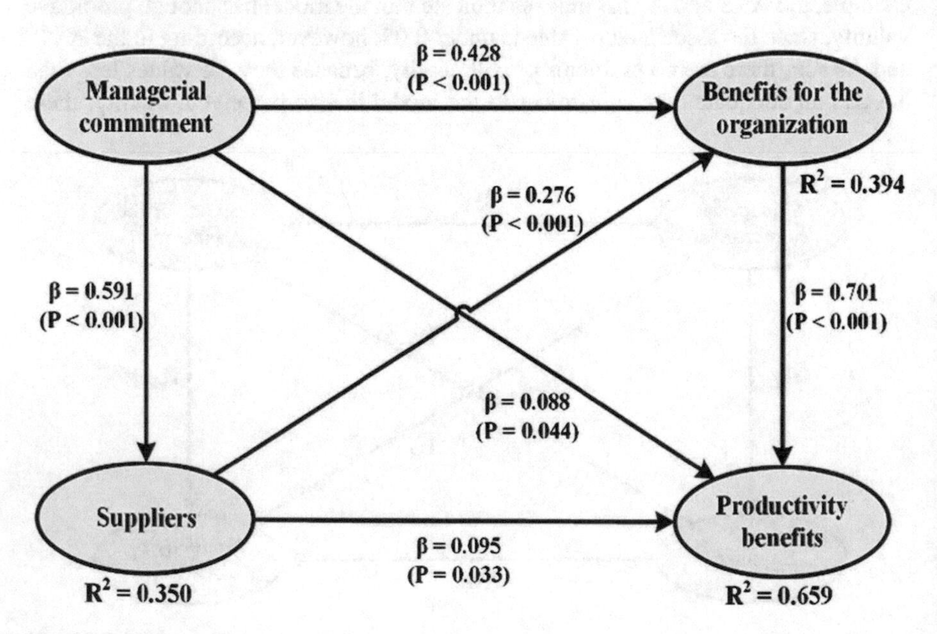

Fig. 11.2 Evaluated Complex Model 3

11.1.3.1 Direct Effects—Complex Model 3

The hypotheses that have been previously proposed can be validated through the direct effects, the value of the β, and the associated *p*-value, which is described next.

H_1: There is enough statistical evidence to declare that the *Managerial commitment* during the TPM implementation process has a direct and positive effect on machinery and equipment *Suppliers*, since when the first latent variable increases its standard deviation in one unit, the second does it in 0.591 units.
H_2: There is enough statistical evidence to declare that the *Managerial commitment* during the TPM implementation has a direct and positive effect on the *Benefits for the organization* that are obtained, since when the first latent variable increases its standard deviation in one unit, the second one does in 0.428 units.
H_3: There is enough statistical evidence to state that the Machinery and equipment *Suppliers* have a direct and positive effect on the *Productivity benefits* obtained from the TPM implementation, since when the first latent variable increases its standard deviation in one unit, the second goes up in 0.276 units.
H_4: There is enough statistical evidence to declare that the *Managerial commitment* during the TPM implementation has a direct and positive effect on the *Productivity benefits* obtained for the company, since when the first latent variable increases its standard deviation by one unit, the second does it in 0.088 units.
H_5: There is enough statistical evidence to declare that the Machines and equipment *Suppliers* for production systems have a direct and positive impact on the *Productivity benefits* obtained for the company, since when the first latent variable increases its standard deviation in one unit, it does it in 0.095 units.
H_6: There is enough statistical evidence to state that the *Benefits for the organization* that are obtained when implementing TPM has a direct and positive impact on the acquired *Productivity benefits*, since when the first latent variable increases its standard deviation in one unit, the second does it in 0.701 units.

11.1.3.2 Effects Size—Complex Model 3

This model is integrated with four latent variables, where three variables play the role of dependent variables, so they are related to an *R*-squared value. However, these dependent variables are affected by a dependent variable or more; in Table 11.1, the decomposition process is illustrated in order to identify those variables that are more significant or critical.

According to the information in Table 11.1, it can be concluded regarding the importance of independent variables in the dependent variables, as indicated below.

- The *Suppliers* variable is explained in 35% by the *Managerial commitment* variable, since they are who have a direct relationship in real life. In addition, the previous statement indicates that the management carried out by the senior

Table 11.1 R-squared contribution—Complex Model 3

Dependent variable	Independent variable			R^2
	Benefits of the organization	*Suppliers*	*Managerial commitment*	
Benefits of the organization		0.144	0.250	0.394
Suppliers			0.350	0.350
Productivity benefits	0.562	0.048	0.049	0.659

management must be directed to have better relationships with equipment and machinery *Suppliers* in order to be competitive.

- The *Benefits for the organization* variable are explained in 39.4% by two variables; *Suppliers* explain 14.4% while *Managerial commitment* explains 25%, which indicates that companies should focus on the second variable, since it has a greater explanatory power.
- The *Productivity benefits* variable is explained in 65.9% by three variables, which 4.9% is from *Managerial commitment*, 4.8% from *Suppliers*, and 56.2% from *Benefits for the organization*. According to this data, it is concluded that companies should focus their efforts on achieving *Benefits for the organization*, since these can quickly become *Productivity benefits*, because this variable is the one with the highest explanatory value.

11.1.3.3 Total Indirect Effects—Complex Model 3

When an independent variable has effects on a dependent variable through a mediating variable, indirect effects are generated with two or more segments. In the case of Complex Model 3, the total of these effects is shown in Table 11.2, where it is observed that each effect is statistically significant, and the following can be concluded:

- *Suppliers* have an indirect effect on *Productivity benefits*, which is because through *Benefits for the organization*, it has a value of 0.194, but the direct effect is only 0.095, which indicates that the indirect effect is twice as high as the

Table 11.2 Sum of indirect effects—Complex Model 3

Dependent variable	Independent variable	
	Suppliers	*Managerial commitment*
Benefits for the organization		0.163 ($P < 0.001$) ES = 0.096
Productivity benefits	0.194 ($P < 0.001$) ES = 0.099	0.470 ($P < 0.001$) ES = 0.263

direct effect. In addition, the previous statement indicates the importance of focusing on obtaining *Benefits for the organization* first, since they later *become Productivity benefits.*

- Another indirect effect that is relevant is the one that refers to the indirect effect that *Managerial commitment* has along with *Productivity benefits*, which is 0.470, since the direct effect was only 0.088; it was scarcely statistically significant. Additionally, the information from above indicates that management must focus on having an adequate relationship with its equipment and machinery *Suppliers* to develop *Benefits for the organization*, in order to these to be translated into *Productivity benefits*. In other words, *Suppliers* are an indispensable base for the management, since they allow it to create productive efficiency parameters, and it is through it that this relationship is possible.

11.1.3.4 Total Effects—Complex Model 3

The total direct and indirect effects outcomes are the result of the total effects, and in the Complex Model 3, in Table 11.3 these are illustrated, which are described together with the *p*-value to determine the statistical significance test as well as the effects size as a measure from the variance explained.

According to the data in Table 11.3, a series of conclusions can be acquired about the relationships between the variables, such as the following:

- When analyzing the magnitude of the values from the total effects, it is observed that the highest ratio is between the *Benefits for the organization* and the *Productivity benefits*, with a value of 0.702, which indicates that senior management should focus on obtaining a better employees moral toward TPM, improve their working conditions, assist them, and facilitate that the company is a means for personal improvement. In addition, it must always promote learning in its human resources and administration knowledge. Also, if the company focuses on these aspects, as a result, it will obtain better production rates.

Table 11.3 Total effects—Complex Model 3

Dependent variable	Independent variable		
	Benefits for the organization	*Suppliers*	*Managerial commitment*
Benefits for the organization		0.276 ($P < 0.001$) ES = 0.144	0.591 ($P < 0.001$) ES = 0.346
Suppliers			0.591 ($P < 0.001$) ES = 0.350
Productivity benefits	0.701 ($P < 0.001$) ES = 0.562	0.288 ($P < 0.001$) ES = 0.147	0.558 ($P < 0.001$) ES = 0.312

However, if the staff does not feel adapted to the work environment, they will probably have absences, low morale, and possibly workers will decide to leave the company.

- *Managerial commitment* is another variable that has high total effects on the subsequent variables, which indicates that, in order to obtain the *Productivity benefits* as well as *Benefits for the organization*, management is required to be a base that supports other departments and involved personnel.
- There are relationships that are directly small, such as the one between *Suppliers* and *Productivity benefits*, but the indirect effect is high, which shows that the total effect is also high. In addition, this indicates that the presence of *Benefits for the organization* as a mediating variable plays a very significant role, since it represents a focal point to the operation simplification and to the quality improvement.

11.1.4 Conclusions and Industrial Implications—Complex Model 3

This model integrates four variables and six hypotheses that have been established and validated statistically, therefore, a series of inferences can be presented regarding the found results and the industrial implications that they entail, such as the following:

- The company must strive to always have good relationships with the machines and equipment *Suppliers* that will be used in its supply chains, as well as those that provide consumables and supplies for them, since the first training sessions depend on them related to operational use and management. Also, if an operator does not learn to manage a team, far from being a competitive advantage, it will be a source of risk and hence the importance of the mutual education systems between the company and *Suppliers*. Surely a machine being operated by someone without experience, will be a source of poor quality, technical stoppages, among others; for instance, Spanos and Voudouris (2009) indicate that when the provider is not able to offer adequate training, that investment is a high risk.
- Between the management and *Suppliers*, they must reach an agreement on the quality parameters that the machines must offer when processing the raw material, which are in function of the customer's demands, so both must have the opportunity to listen to the client in order to be able to make improvements in the equipment that is already installed (Cardoso et al. 2012). However, aspects such as security must be integrated in the purchasing decisions not only productive aspects, since a very efficient team is useless if there is a risk for the operators and on the working environment (Waldeck and Leffakis 2007).

- Management should focus on obtaining *Benefits for the organization*, because it implies a better work environment quality, better the productive that are carried out, increase the employees morale by offering a safe and risk-free environment, encourage the operator to always read and take into account the operations manual of the team under his responsibility, and generate collaborative work networks between the maintenance department and other ones (Seth and Tripathi 2006). The previous data is because sooner or later those *Benefits for the organization* change into *Productivity benefits* that can be measured and monitored. It is important to notice that in the model shown in Fig. 11.2 that relationship is the highest.
- Managers must understand that *Productivity benefits* are significant, but that it is necessary to make investments in aspects that are not so easily measured, such as the increase in employee motivation and labor commitment. That is, many pillars are required to generate the productivity indexes, which is not fortuitous or just luck, it must be traversed in them and in all the aspects that precede it. In fact, Hana (2010) mentions that it is very important to maintain an adequate work environment in order to guarantee reliability not only in machines and equipment, but also in human resources. Recently, in a study conducted by Ahmadzadeh and Bengtsson (2017), the need for training and trainings is shown in order to avoid errors in the maintenance personnel, since otherwise the productivity parameters are lowered.
- There is something unique in the model, it is almost impossible to believe that *Managerial commitment* does not have a strong relationship with *Productivity benefits*, as it is shown in Fig. 11.2, but when it is used as variable for *Suppliers* and *Benefits for the organization*, the indirect effect becomes strong. In addition, the previous data indicates one more time that senior management must be focused on aspects associated with quality and facilitate its achievement, since at the end of the day, management is only a facilitator and director entity.

11.1.5 Sensitivity Analysis—Complex Model 3

This model integrates six hypotheses that link four latent variables. In this section, a Sensitivity analysis is carried out to determine the scenarios where these latent variables can happen.

11.1.5.1 Sensitivity Analysis: *Managerial Commitment* and *Suppliers* (H_1)—Complex Model 3

The relationship between *Managerial commitment* and machinery and equipment *Suppliers* is one of the most relevant in the study of TPM, since it is the first one that establishes commercial relationships with the second variable, in other words,

Table 11.4 Sensitivity analysis: *managerial commitment* and *suppliers* (H_1)—Complex Model 3

Managerial commitment	*Suppliers*	
	High	Low
High	Suppliers+ = 0.144	Suppliers− = 0.168
	Managerial commitment+ = 0.168	Managerial commitment+ = 0.168
	Suppliers+ and managerial commitment+ = 0.068	Suppliers− and managerial commitment+ = 0.000
	Suppliers+ *if* managerial commitment+ = 0.403	Suppliers− *if* managerial commitment+ =0.039
Low	Suppliers+ = 0.144	Suppliers− = 0.168
	Managerial commitment− = 0.158	Managerial commitment− = 0.158
	Suppliers+ and Managerial commitment− = 0.011	Suppliers− and managerial commitment− = 0.071
	Suppliers+ *if* Managerial commitment− = 0.069	Suppliers− *if* managerial commitment− = 0.448

there is no operators' intervention, and only senior managers have that decision power. In addition, Table 11.4 lists the high and low scenarios for these two variables, as well as the combinations of these, and based on that data, the following conclusions are presented:

- In the optimistic scenario, when *Managerial commitment* and *Suppliers* have high levels, the probability of occurrence independently for the first variable is 0.144 while for the second is 0.168, which can be considered as low values, but the probability that both variables are presented in those scenarios is 0.0 68. Finally, the probability of having high levels in the *Suppliers* variable if there is a high level in *Managerial commitment* is 0.403. The following data indicate that the senior management must make an effort to have good relationships with their machines and equipment *Suppliers*, since the commitment that is being held in that relationship will be the obtained results. As a result, whenever there are high levels of *Managerial commitment*, there will be high levels in the *Suppliers*.
- The information from above is clearly verified when analyzing the scenario where *Suppliers* have low levels, but *Managerial commitment* is very high, which has a probability of simultaneous occurrence in those scenarios in the variables of zero. That is, whenever *Managerial commitment* exists, there will be high levels for *Suppliers*. Similarly, the probability of having low levels in the *Suppliers*, because there is a senior *Managerial commitment*, is only 0.039, which is a very low value. Now, when the levels of the two variables are inverted and there are high levels for *Suppliers* and the levels for *Managerial commitment* are low, the probability of these scenarios happening simultaneously is only 0.011, and the probability that *Suppliers* will be presented with high levels, since they have a *Managerial commitment* under that only 0.069.

- Finally, in the most pessimistic scenario, when there are low levels in *Managerial commitment* and *Suppliers*, the probability that the first variable is presented in this scenario is 0.168 while for the second variable is 0.158. However, there is a high risk of 0.071 that both variables occur in these scenarios simultaneously. Likewise, there is a risk of 0.448 to have low levels in *Suppliers*, because there are low levels in *Managerial commitment*, where management must demonstrate a high commitment toward the TPM implementation and look forward to establishing good relationships with the machinery and equipment *Suppliers*.

11.1.5.2 Sensitivity Analysis de *Managerial Commitment* con *Benefits for the Organization* (H_2)—Complex Model 3

The senior management from the organization that should be concerned to generate benefits for it in a quantify tangible way. In this section, the sensitivity analysis is performed for the scenarios that *Managerial commitment* can have as well as the *Benefits for the organization*, where the high and low levels for each variable are analyzed, as well as the four combinations. In addition, Table 11.5 illustrates a summary of the results of these scenarios, and based on this information, the following can be concluded:

- In the most optimistic scenario, *Managerial commitment* and *Benefits for the organization* have high values and the probability that the first variable is

Table 11.5 Sensitivity analysis: *managerial commitment* and *benefits for the organization* (H_2)—Complex Model 3

Managerial commitment	*Benefits for the organization*	
	High	Low
High	Benefits for the organization+ = 0.163	Benefits for the organization− = 0.166
	Managerial commitment+ = 0.168	Managerial commitment+ = 0.168
	Benefits for the organization+ *y* managerial commitment+ = 0.065	Benefits for the organization− and managerial commitment+ = 0.000
	Benefits for the organization+ *if* managerial commitment+ = 0.387	Benefits for the organization− *if* managerial commitment+ = 0.000
Low	Benefits for the organization+ = 0.163	Benefits for the organization− = 0.166
	Managerial commitment− = 0.158	Managerial commitment− = 0.158
	Benefits for the organization+ and managerial commitment− = 0.024	Benefits for the organization− and managerial commitment− = 0.076
	Benefits for the organization+ *if* managerial commitment− = 0.155	Benefits for the organization− *if* managerial commitment− = 0.483

presented in this scenario independently is 0.163 while the probability that the second is presented is 0.168, which can be considered as low values, since one of the TPM objectives is to generate greater benefits associated with the machines quality and availability. In addition, it is observed that the probability that both variables are presented simultaneously in these scenarios, which would be desired, it is only 0.065; a very low value. However, the importance of *Managerial commitment* is presented when analyzing the probability of obtaining *Benefits for the organization* and its high levels, because the first variable also has high levels by 0.387.

- The question that should be asked now is, what is the probability that there are high levels on *Managerial commitment* and low *Benefits for the organization* are acquired, fortunately, that probability is zero. Also, the probability that *Benefits for the organization* will be obtained at its low levels because there are high levels on *Managerial commitment*, which is zero. The information from above indicates that whenever there are high levels on *Managerial commitment*, there will always be a response towards *Benefits for the organization*. Moreover, if the inverted role of the variables is observed, where *Benefits for the organization* are acquired from its high levels and *Managerial commitment* is low, the probability of occurrence of that scenario is 0.024. Also, the probability that the *Benefits for the organization* are high because there are low levels on *Managerial commitment* is 0.155. The previous information indicates that these *Benefits for the organization* can be obtained by more than one source, and not only because of the *Managerial commitment*.
- However, the pessimistic scenario for the relationship between these two variables has some troubles when *Benefits for the organization* and *Managerial commitment* have low levels. In addition, the probability of occurrence independently for the first variable is 0.166 whereas for the second is 0.158. Also, the worst scenario that could arise is when both variables have low levels and there is a probability of 0.076 that this may happen. Finally, the probability of having low levels on *Benefits for the organization*, because there is a low *Managerial commitment* is 0.483, as a result, managers in charge of the TPM implementation should always look to have the backup and management support before starting this process, because if there is no support, the failure chances are wide, and a company will not get the expected benefits.

11.1.5.3 Sensitivity Analysis: *Suppliers* and *Benefits for the Organization* (H_3)—Complex Model 3

Machinery and equipment *Suppliers* that are installed in a production system, without a doubt that help to achieve *Benefits for the organization*. In this section, the Sensitivity analysis is performed for these two variables considering their high and low scenarios, and a summary of the results is portrayed in Table 11.6, where the following conclusions can be obtained:

Table 11.6 Sensitivity analysis: *suppliers* and *benefits for the organization* (H_3)—Complex Model 3

Suppliers	*Benefits for the organization*	
	High	Low
High	Benefits for the organization+ = 0.163	Benefits for the organization− = 0.166
	Suppliers+ = 0.144	Suppliers+ = 0.144
	Benefits for the organization+ and suppliers+ = 0.057	Benefits for the organization− and suppliers+ = 0.011
	Benefits for the organization+ *if* suppliers+ = 0.396	Benefits for the organization− *if* suppliers+ = 0.075
Low	Benefits for the organization+ = 0.163	Benefits for the organization− = 0.166
	Suppliers− = 0.168	Suppliers− = 0.168
	Benefits for the organization+ and suppliers− = 0.014	Benefits for the organization− and suppliers− = 0.073
	Benefits for the organization+ *if* suppliers− = 0.081	Benefits for the organization− *if* suppliers− = 0.435

- In a totally optimistic scenario, when *Benefits for the organization* and *Suppliers* have high levels, there is a probability of 0.163 that the first variable is presented independently in that scenario while a probability of 0.144 that the second one is presented. However, the probability that both variables are presented simultaneously in their high levels is only 0.057, this probability can be considered low. In addition, the probability of holding *Benefits for the organization* at its high level because *Suppliers* have their high participation level by 0.396, and that value would represent an ideal to be achieved. Thus, senior management must worry about having good relationships with machinery and equipment *Suppliers* that it has, since it is a guarantee to obtain the *Benefits for the organization*.
- The previous conclusion is easily verified when reviewing the scenario where *Benefits for the organization* have low levels and *Suppliers* high levels, since the probability of simultaneous occurrence for those variables and their scenarios is only 0.011. In addition, the probability of presenting the first variable in its scenario, because the second one has been presented, is only 0.075; a very low probability that indicates a strong relationship between these two variables. Similarly, when reviewing the scenario where the variables have an inverted role, *Benefits for the organization* in its high levels and *Suppliers* in its lowest levels, it is observed that the probability of occurrence for these scenarios happening simultaneously is only 0.014, and the probability that the second one is presented at first is 0.081.
- The most pessimistic scenario occurs when *Benefits for the organization* and *Suppliers* have low levels, where the probability of occurrence independently for the first variable is 0.166 while the probability of occurrence for the second variable is 0.168, but the probability of presenting these two variables in their scenario simultaneously is 0.073; a very high probability that management must worn on to prevent it from happening, since the probability that the first variable

is presented in that scenario because the second one has been presented is 0.435, a high-risk value that forces us to suggest that the senior management must work closely with its professionals in machinery and equipment.

11.1.5.4 Sensitivity Analysis: *Managerial Commitment* and *Productivity Benefits* (H_4)—Complex Model 3

As it was mentioned above, any maintenance manager is entrusted to increase the productivity indexes in a company; however, the commitment that it observes in the management is fundamental before starting in this adventure. In this section, the Sensitivity analysis for the relationship between *Managerial commitment* and *Productivity benefits* is carried out, where the high and low scenarios for each of the variables are analyzed, as well as a combination of these.

Table 11.7 illustrates a summary of the results obtained, and based on this the following inferences and conclusions can be said:

- The most optimistic scenario can appear when there is a *Managerial commitment* and *Productivity benefits* at its high levels, where the probability of occurrence for the first variable independently in that scenario is 0.198 while for the second is 0.168; however, the best scenario happens when both variables are presented simultaneously, which happens with a probability of 0.071, and it represents a low probability, because it should always look for these two variables to be present during the TPM implementation process. Finally, the probability of having high *Productivity benefits* because there is a high *Managerial commitment* from 0.419, which motivates the maintenance manager

Table 11.7 Sensitivity analysis: *managerial commitment* and *productivity benefits* (H_4)—Complex Model 3

Managerial commitment	*Productivity benefits*	
	High	Low
High	Productivity benefits+ = 0.198	Productivity benefits− = 0.177
	Managerial commitment+ = 0.168	Managerial commitment+ = 0.168
	Productivity benefits+ and managerial commitment+ = 0.071	Productivity benefits− and managerial commitment+ = 0.000
	Productivity benefits+ *if* managerial commitment+ = 0.419	Productivity benefits− *if* managerial commitment+ = 0.000
Low	Productivity benefits+ = 0.198	Productivity benefits− = 0.177
	Managerial commitment− = 0.158	Managerial commitment− = 0.158
	Productivity benefits+ and managerial commitment− = 0.030	Productivity benefits− and managerial commitment− = 0.065
	Productivity benefits+ *if* managerial commitment− = 0.190	Productivity benefits− *if* managerial commitment− = 0.414

to always seek a commitment from management, which indicates a high ratio between these two variables.

- The question that needs to be asked at this time is what would happen if there is a high *Managerial commitment* even when low *Productivity benefits* are obtained. Fortunately, the probability that these two variables are presented together in this scenario is zero while the same happens with the second variable, because the first is present. The previous data indicates that there is always *Managerial commitment* in the TPM implementation process, a series of *Productivity benefits* will be answered, and hence the importance of always guaranteeing this commitment. However, it is important to mark that when the role and scenario of the variables are inverted, with high *Productivity benefits* and low *Managerial commitment*, the probability that these two variables are presented in that scenario simultaneously is the same; 0.030, but the probability that the first variable is presented in this scenario because the second one has happened is 0.190, which indicates that its *Productivity benefits* may have another source and not only from *Managerial commitment*.
- The worst scenario can happen when there are low *Managerial commitment* and *Productivity benefits*, where the probability of occurrence for the first variable independently is 0.177 while the second is 0.158, but the probability of simultaneous occurrence for two variables together in this scenario is 0.065, which represents a risk for the person in charge of implementing TPM. Finally, the probability of having low *Productivity benefits*, because there is also a low *Managerial commitment* is 0.414, and it represents the greatest risk to face.

11.1.5.5 Sensitivity Analysis: *Suppliers* and *Productivity Benefits* (H_5)—Complex Model 3

One of the criteria selection for *Suppliers* is the way they may help to increase the *Productivity benefits* that can be obtained in the supply chains, and the ease that the maintenance can be carried out to the machines and tools that they provide. Additionally, Table 11.8 illustrates a summary of the Sensitivity analysis that is performed on these variables when they are at their high and low levels independently as well as combined. Based on its data, the following conclusions and inferences can be stated:

- The most optimistic scenario is when the *Productivity benefits* and *Suppliers* variables have high levels. Also, the probability of presenting the first variable in this scenario is 0.198 while for the second is 0.144. However, efforts must be made to increase the probability of occurrence on these two variables in their optimistic scenarios, since it is currently only 0.065, and the probability of having high *Productivity benefits* because *Suppliers* are at high levels is 0.453, therefore, maintenance management should seek to increase those probabilities, since it indicates that whenever machinery and equipment *Suppliers* are

Table 11.8 Sensitivity analysis: *suppliers* and *productivity benefits* (H_5)—Complex Model 3

Suppliers	*Productivity benefits*	
	High	Low
High	Productivity benefits+ = 0.198	Productivity benefits− = 0.177
	Suppliers+ = 0.144	Suppliers+ = 0.144
	Productivity benefits+ and suppliers+ = 0.065	Productivity benefits− and suppliers+ = 0.005
	Productivity benefits+ *if* suppliers+ = 0.453	Productivity benefits− *if* suppliers+ = 0.038
Low	Productivity benefits+ = 0.198	Productivity benefits− = 0.177
	Suppliers− = 0.168	Suppliers− = 0.168
	Productivity benefits+ and suppliers− = 0.014	Productivity benefits− and suppliers− = 0.073
	Productivity benefits+ *if* suppliers − = 0.081	Productivity benefits− *if* suppliers− = 0.435

committed to the process of installation, training, and trainings, high levels are obtained in *Productivity benefits*.

- This favorable relationship between the two variables is easily demonstrated when *Productivity benefits* are observed at their low level, but *Suppliers* at their high levels, since the probability of occurrence for those two variables at the same time is 0.005, which indicates that this situation will almost never be presented. In addition, the probability of presenting the first variable because the second variable has been presented in their respective scenarios is only 0.038, which indicates that there is always a high *Suppliers* involvement in the TPM implementation process and *Productivity benefits* will be obtained. Also, this conclusion can be developed when obtaining the results with the levels invested in the variables, high *Productivity benefits* and low *Suppliers*, since the probability of occurrence for those scenarios happening together and simultaneously is only 0.014, therefore, the probability of the first variable occurring in its scenario, since the second variable has already occurred, it is 0.081, which means that *Productivity benefits* can be obtained from other sources than *Suppliers*, such as the machines operators.
- Finally, the most pessimistic scenario for the maintenance manager would be to have *Productivity benefits* and *Suppliers* at their low levels, where the probability of occurrence for the first variable in their scenario independently is 0.177 while for the second is 0.168, but the probability of occurrence for both variables occurring simultaneously in their low scenarios is only 0.073; a low probability that represents a risk for the manager. In the same way, the probability of having low *Productivity benefits* because *Suppliers* have been presented at their low levels is 0.435, an undesirable situation for the maintenance manager, which represents a risk, since it does not have a direct relationship with the supplier and should look forward to integrating it.

11.1.5.6 Sensitivity Analysis: *Benefits for the Organization* and *Productivity Benefits* (H_6)—Complex Model 3

As a matter of fact, it has been mentioned that the *Benefits for the organization* create *Productivity benefits,* which was demonstrated through the hypothesis H_6, but it demonstrates what would happen if there are different scenarios for those two variables, which is expected that they are always in their high levels. In addition, Table 11.9 displays the results from the Sensitivity analysis regarding these two variables when there are high and low levels for both. Also, the conclusions and inferences that can be collected from these results are the following:

- The most optimistic scenario occurs when the *Productivity benefits* and *Benefits for the organization* are at their high levels, where the probability of occurrence for the first variable in their scenario independently is 0.198 while for the second is 0.163, which can be considered as low values, and management should focus on increasing them. However, the probability that both variables are presented simultaneously with their high levels is 0.114, which could also be considered as a low probability for the maintenance manager. Finally, the probability of having high *Productivity benefits* because the *Benefits for the organization* has been obtained is 0.700, which indicates that senior management must focus on obtaining *Benefits for the organization* because these will later become into *Productivity benefits.*
- The conclusion from the above is easily demonstrated when analyzing the scenario where the *Productivity benefits* are low and the *Benefits for the*

Table 11.9 Sensitivity analysis: *benefits for the organization* and *productivity benefits* (H_6)—Complex Model 3

Benefits for the organization	*Productivity benefits*	
	High	Low
High	Productivity benefits+ = 0.198	Productivity benefits− = 0.177
	Benefits for the organization+ = 0.163	Benefits for the organization+ = 0.163
	Productivity benefits+ and benefits for the organization+ = 0.114	Productivity benefits− and benefits for the organization+ = 0.000
	Productivity benefits+ *if* benefits for the organization+ = 0.700	Productivity benefits− *if* benefits for the organization+ = 0.000
Low	Productivity benefits+ = 0.198	Productivity benefits− = 0.177
	Benefits for the organization− = 0.166	Benefits for the organization− = 0.166
	Productivity benefits+ and benefits for the organization− = 0.005	Productivity benefits− and benefits for the organization− = 0.122
	Productivity benefits+ *if* benefits for the organization− = 0.033	Productivity benefits− *if* benefits for the organization− = 0.738

organization are high, since the probability of simultaneous occurrence for both variables is zero and, the probability that the first variable is presented in its scenario since the second variable has been presented in its scenario is zero as well, which shows the high relation between both variables. In addition, when analyzing the invested roles in the variable levels, where the *Productivity benefits* are high and the *Benefits for the organization* are low, the probability that the variables are presented together in that scenario is 0.005, but the probability of presenting the first variable because the second one has been presented is 0.033, which indicates that these *Productivity benefits* can be acquired from another source and not only from the *Benefits for the organization*.

- The most pessimistic scenario in the relationship between these variables occurs when the *Productivity benefits* and *Benefits for the organization* are at their low levels. In addition, the probability of occurrence for the first variable in this scenario independently is 0.177 while for the second is 0.166, but there is a 0.122 probability that both variables are presented simultaneously, which represents a high risk for the manager in charge of TPM implementation. Likewise, the probability of having low *Productivity benefits*, because the *Benefits for the organization* are also low, is 0.738. Therefore, it can be concluded that managers should focus on obtaining *Benefits for the organization*, since it guarantees that *Productivity benefits* will be achieved.

11.2 Complex Model 4

The Complex Model 4 integrates four latent variables and six hypotheses that relate them. In addition, the variables to analyze are the following: *Suppliers*, *Layout*, *Benefits for the organization,* and *Security benefits*. In this model, it is assumed that there is no control over the machinery and equipment *Suppliers*, because they are external entities to the company, although they are easily influenced, and they can be turned into *Layout* or plant distribution influence in the production system machines. Also, these two variables are linked to two benefits; *Benefits for the organization*, as a dependent variable for all the others, and *Security benefits*, which the company management must be focused on as well.

In other words, the objective of this model is to determine the effect that machinery and equipment *Suppliers* have as an external entity toward the company, in the *Security benefits* that the company may acquire when implementing TPM, having as support the *Layout* and the *Benefits for the organization*. The hypotheses that relate these four variables are discussed in the section below.

11.2.1 Hypotheses—Complex Model 4

First, when considering the *Layout* from an installation, the neighboring facilities spatial distribution should also be considered to avoid any conflict between them (Seyed-Mahmoud et al. 2011). The amount of space used in an installation is positively related to the installation operational cost, so the administration and the machines and tools producers must consider the space location availability in the supply chains and the supplies storage, spare parts, and consumables that are required (Liang and Chao 2008). Also, it is applicable when considering the departmental offices location within an installation, since managers have a specific space to perform administrative aspects; the questions that should be asked by the production engineers to *Suppliers* is how big the equipment is, how it can be organized to obtain a maximum efficiency, which departments from the disposal facility that should be located at close range, how it will be affect product flows with the design decisions, what spatial conditions the equipment requires, among others (Seyed-Mahmoud et al. 2011).

Therefore, machines and tools *Suppliers* have a high impact on the plant distribution that is established in the company, where they should follow their recommendations to ensure the operation. Also, it is important to remember that the machine manufacturer is the person who has more knowledge about its capabilities and strengths, that's why it is convenient to always listen to them. According to the information above, the following hypothesis is established:

H_1: Machines and tools *Suppliers* have a direct and positive effect on the *Layout* in supply chains.

Moreover, competitive forces pressure companies to improve quality, delivery performance, and capacity response, while reducing costs, as a result, they are increasingly exploring ways to take advantage of their supply chains and particularly assess the *Suppliers* role in their activities (Vijay and Keah Choon 2006), therefore, an effort must be included to integrate them into their productive system. In addition, at the operational level, the benefit for a manufacturer by developing close relationships with *Suppliers* is presented in a better quality or delivery service, reduced cost, or some combination at the strategic level; it should lead to sustainable improvements in the quality and product innovation from greater competitiveness and participation in the market (Vijay and Keah Choon 2006), which is translated into *Benefits for the organization*.

In addition, manufacturers can reduce costs and products cycle times, as well as improve product quality when working closely to their *Suppliers* (Ragatz et al. 2002), since they depend directly on the machines capacities that they have installed and bought. Also, the successful management of these relationships contributes to the company performance (Keah-Choon et al. 1999), similarly, companies obtain competitive advantages through the resources usage, skills, and *Suppliers* capabilities, especially their design vision (Chang 2017) and knowledge, since they are the ones that initially provide training and coaching.

Nowadays, there is a proposal that seeks to relate to maintenance by process control graphs to recognize the effect and productive benefits between them (Shrivastava et al. 2016). There are also reported cases where TPM is associated to the time of delays due to breakdowns as *Benefits for the organization* and the negative impact that machines have on productivity indexes (Borkowski et al. 2014), For this reason, proposals have also been presented where TPM auxiliary programs, such as SMED, help to reduce these technical stoppages along with *Suppliers* (Cakmakci 2009). Therefore, the following hypothesis can be established:

H_2: Machines and tools *Suppliers* have a direct and positive effect on *Benefits for the organization* when TPM is implemented.

Furthermore, the facilities *Layout* planning depends on many factors, including the facilities adjacency, the facilities resources, the distance between the facilities, and the current facilities location (Seyed-Mahmoud et al. 2011). In addition, the efficiency within the installation takes into account the employees and products traveled distances, the installation resources distance, and the frequency that trips are made by the personnel in the installation, which allows to choose the best machines supplier (Seyed-Mahmoud et al. 2011).

Also, facilities within a company are constantly facing increasing competition and must find ways to maximize the production or service efficiency in order to remain competitive in the market, and in that aspect, *Suppliers* have a lot of commitment toward manufacturing managers (Seyed-Mahmoud et al. 2011), because together, they must be able to increase the production systems reliability and make them safer for operators and the environment. In fact, it is currently recommended that these *Suppliers* get an interference in the plant or *Layout* distribution in order to ensure that safety standards are always achieved (Camuffo et al. 2017).

Likewise, currently the products production by foreign manufacturers creates even greater pressure on domestic manufacturers in their attempt to produce quality products with competitive prices and low labor costs, consequently, manufacturers must use the *Layout* design in facilities to increase their efficiency (Seyed-Mahmoud et al. 2011), since the installation *Layout* has always affected the efficiency and benefits for a company (Tompkins et al. 2003). It can be considered that greater efficiency leads to better profits, better products, and services quality, as well as many other benefits (Seyed-Mahmoud et al. 2011). Therefore, the following hypothesis can be established:

H_3: The *Layout* in the company productive system has a direct and positive effect on the *Benefits for the organization* when implementing TPM.

Equally important, a machine must always be safe when it is operated by the personnel, and it must never be a source of risk. Also, machines and equipment *Suppliers* must provide security from the moment they are designed. In fact, there are regulations for it, as it reported by Porras-Vázquez and Romero-Pérez (2018), who present a methodology, as well as Sadeghi et al. (2015) who point out the need to create indicators and safety parameters from the design. However, once the machine has been implemented in the productive system, it is the senior management and *Suppliers* responsibility to generate a training program for the equipment safe, not only they should focus on operational and productive aspects, but also on security; the aspects that must be ensured are related to fire prevention (Gehandler 2017), a system or plant distribution that is safe for operators and environment (Vitayasak and Pongcharoen 2018), among others. For the above, it is considered that *Suppliers* are related to the *Security benefits* that a company can obtain, and the following hypothesis is proposed.

H_4: Machines and tools *Suppliers* have a direct and positive impact on the *Security benefits* obtained by a company when TPM is implemented.

Additionally, the installation *Layout* design has an essential impact on the performance of the entire manufacturing system, and it is always considered as a key for manufacturing systems to improve their productivity currently (Yang et al. 2013), it is an essential subject in the industrial engineers training and those who are focused on the industrial facilities design. Also, the *Layout* aims to design the workplace in such a way that it can be carried out efficiently and without defects, protect employees from accidents and long-term damage due to the nature of the tasks, eliminate defects causes in the work environment, reduce the effort of time required to complete a task or a series of tasks (Lunau et al. 2013).

In addition, the link between plant distribution and safety is not new, since the 80s, Ham (1987) indicates that there was a need to follow all the regulations from a *Layout* to ensure safety inside the factories, but above all, in the warehouses where the pieces move with excessive dimensions is required. Nowadays, models have been presented that help to propose plant distribution models, where risk is a variable to be integrated, as is the case of Medina-Herrera et al. (2014) and Yi et al. (2018). Therefore, the following hypothesis is presented:

H_5: The *Layout* that a company has in its productive system has a direct and positive effect on the *Security benefits* when implementing TPM.

Moreover, for each manufacturing company, the objective is to produce goods with a benefit, and it can only be achieved by using an effective maintenance system that helps to maximize the equipment availability by minimizing the inactivity time due to unwanted stoppages like security (Fore and Zuze 2010) and hygiene (Nakajima 1988). Similarly, Wakjira and Singh (2012) suggest that after the TPM implementation, strategic improvements can be made to the achievement of important manufacturing performance improvements, since this tool is a gradual strategy that combines the best preventive and productive maintenance features with full employees participation, so efforts should be made to fulfill a better work environment that is safe for the operator, improve operations control, and safely increase the traders morale because having accidents and even, generate employees of the month in terms of security (Tsarouhas 2013).

In order for the company to obtain *Security benefits*, it must ensure a system for learning the critical points, create a health and safety commission, and actively participate with local governmental and civilian security institutions. Likewise, it must create work networks with other departments to exchange information regarding the equipment safety and operation. Therefore, the following hypothesis can be proposed

H_6: The *Benefits for the organization* that are obtained when implementing TPM, have a direct and positive impact on the *Security benefits*.

A graphic representation from the relationships that have been previously presented is indicated graphically in Fig. 11.3. Also, it is significant to recall that there is a special paragraph dedicated to the validation of the variables in the model, so it is assumed that all the required indexes have been analyzed in the final model.

11.2.2 Efficiency Indexes—Complex Model 4

Since all the latent variables have particularly exceeded the efficiency indexes, they are integrated into the model that when it is evaluated, it allows obtaining the following efficiency indexes.

- Average path coefficient (APC) = 0.353, $P < 0.001$
- Average R-squared (ARS) = 0.445, $P < 0.001$
- Average adjusted R-squared (AARS) = 0.443, $P < 0.001$
- Average block VIF (AVIF) = 1.490, acceptable if ≤ 5, ideally ≤ 3.3
- Average full collinearity VIF (AFVIF) = 2.373, acceptable if ≤ 5, ideally ≤ 3.3
- Tenenhaus GoF (GoF) = 0.547, small ≥ 0.1, medium ≥ 0.25, large ≥ 0.36

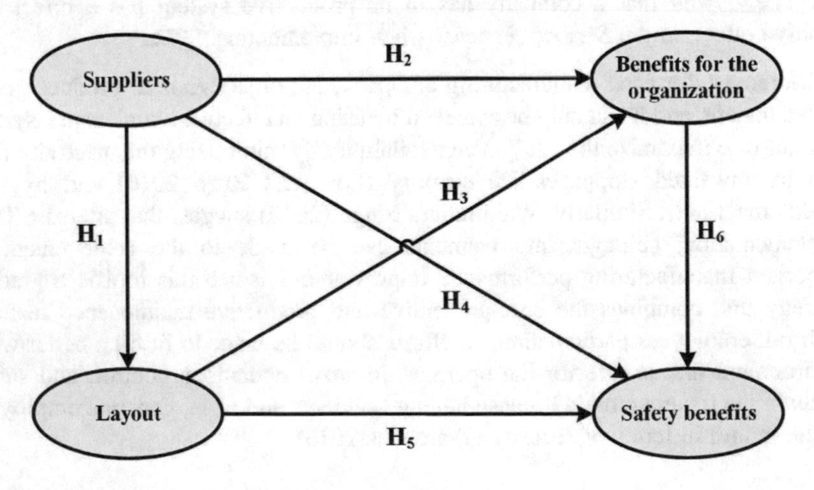

Fig. 11.3 Proposed hypotheses—Complex Model 4

- Sympson's paradox ratio (SPR) = 1.000, acceptable if ≥ 0.7, ideally = 1
- *R*-squared contribution ratio (RSCR) = 1.000, acceptable if ≥ 0.9, ideally = 1
- Statistical suppression ratio (SSR) = 1.000, acceptable if ≥ 0.7
- Nonlinear bivariate causality direction ratio (NLBCDR) = 1.000, acceptable if ≥ 0.7

From the efficiency indexes, the following conclusions are developed:

- The regression indexes are valid and statistically significant, indicating adequate relationships between the analyzed latent variables.
- There is enough predictive validity in the model, since the averages of ARS and AARS are high, and the associated *P*-value is under 0.05.
- There are no collinearity problems among the latent variables, since the AVIF and AFVIF indexes are less than 3.3; the maximum admitted value.
- The data is suitably adjusted to the model, since the GoF index is over 0.36.
- There are no problems with the hypotheses directionality.
- Therefore, considering that the model complies with the required efficiency indexes, it is proceeding to interpret it.

11.2.3 Results—Complex Model 4

Figure 11.4 illustrates the evaluated Complex Model 4, where the values of the β, the associated *p*-value for the statistical test of significance are indicated as well as

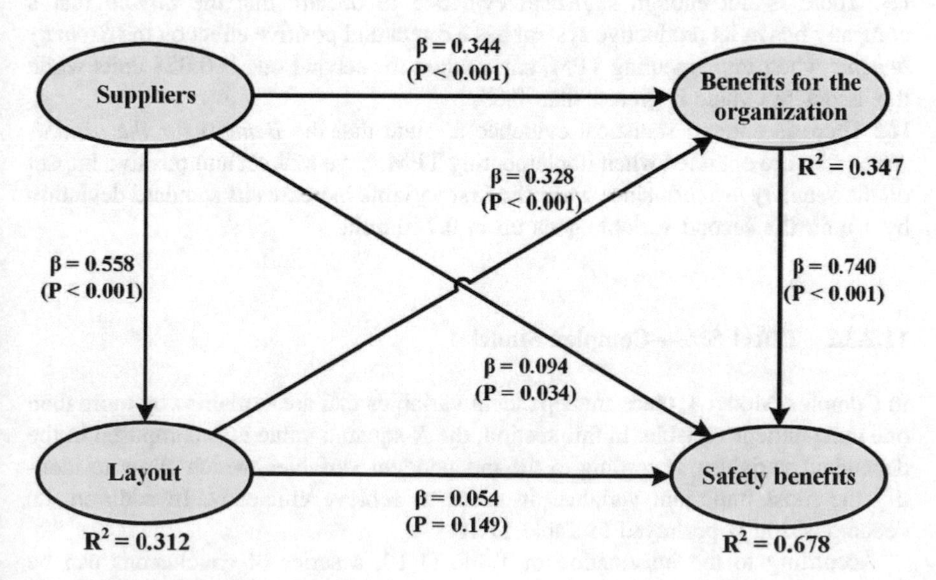

Fig. 11.4 Evaluated Complex Model 4

in the dependent variables, an *R*-squared value is indicated to measure the variance explained in the dependent variables by the independent variables.

11.2.3.1 Direct Effects—Complex Model 4

The direct effects allow the validation of the hypotheses that have been initially proposed, which is done in function of the β values and the *p*-value associated that is indicated in Fig. 11.4. The conclusions of these hypotheses are the following:

H_1: There is enough statistical evidence to declare that machines and tools *Suppliers* have a direct and positive effect on the *Layout* in supply chains, because when the first variable increases its standard deviation by one unit, the second variable does it in 0.558 units.
H_2: There is enough statistical evidence to declare that machines and tools *Suppliers* have a direct and positive effect on *Benefits for the organization* when TPM is implemented, since when the first variable increases its standard deviation in one unit, the second variable increases in 0.344 units.
H_3: There is enough statistical evidence to state that the *Layout* in the company productive system has a direct and positive effect on the *Benefits for the organization* when implementing TPM, since when the first variable increases its standard deviation in one unit, the second one goes up in 0.328 units.
H_4: There is enough statistical evidence to declare that machines and tools *Suppliers* have a direct and positive impact on the *Security benefits* obtained by a company when TPM is implemented, since when the first variable increases its standard deviation by one unit, the second variable does it in 0.094 units.
H_5: There is not enough statistical evidence to declare that the *Layout* that a company has in its productive system has a direct and positive effect on the *Security benefits* when implementing TPM, since when the second one is 0.054 units while the associated value is greater than 0.05.
H_6: There is enough statistical evidence to state that the *Benefits for the organization* that are obtained when implementing TPM, have a direct and positive impact on the *Security benefits*, since when the first variable increases its standard deviation by 1 unit, the second variable goes up in 0.740 units.

11.2.3.2 Effect Size—Complex Model 4

In Complex Model 4, there are dependent variables that are explained by more than one independent variable. In this section, the *R*-squared value is decomposed in the dependent variables, according to the independent variables, which allow to identify the most important variables in order to achieve efficiency. In addition, its decomposition is portrayed in Table 11.10.

According to the information in Table 11.10, a series of conclusions can be made, such as the following:

Table 11.10 R-squared contribution—Complex Model 4

Dependent variable	Independent variable			R^2
	Benefits for the organization	*Suppliers*	*Layout*	
Benefits for the organization		0.179	0.168	0.347
Layout		0.312		0.312
Security benefits	0.604	0.048	0.026	0.678

- The *Layout* variable is explained only by the *Suppliers* in this model in 31.2%, which indicates that the technical specifications and machines need that sell, determine the distribution that is given to them in the supply chains in the company.
- The *Benefits for the organization* variable are explained in 34.7% by two independent variables. In this case, *Suppliers* can explain 17.9% while *Layout* 16.8%; very similar values are observed. It is concluded that, in order to achieve a better quality in work, better operations control, and increase the morale in people, *Suppliers* are simultaneously required to be able to facilitate these aspects, but also, they require an efficient *Layout* that allows the creation of an accountability culture towards TPM and the machines preservation.
- The *Security benefits* variable is explained by three independent variables, although one of them did not have a statistically significant direct effect (*Layout*). In this case, *Layout* can only explain 2.6%, *Suppliers* 4.8% and finally, the *Benefits for the organization* explain 60.4%, which indicate that the last one is the most significant variable in this model.

11.2.3.3 Total Indirect Effects—Complex Model 4

In the Complex Model 4, there is an independent variable (*Suppliers)* and another dependent variable (*Security benefits*) clearly distinguished, but there are two variables that are mediating, and their presence generates indirect effects between them. In this section, in Table 11.11 the total indirect effects are illustrated, which

Table 11.11 Total indirect effects—Complex Model 4

Dependent variable	Independent variable	
	Suppliers	*Layout*
Benefits for the organization	0.183 ($P < 0.001$) ES = 0.095	
Security benefits	0.420 ($P < 0.001$) ES = 0.214	0.243 ($P < 0.001$) ES = 0.115

are presented in the model, where the p-value is also indicated for the statistical significance of the model as well as the effect size or variance that is explained.

The outcomes in Table 11.11 are interesting, since they show an important aspect of the mediating variables and allow to conclude the following:

- In this case, it is relevant to remember that the hypothesis represented by H_5 was statistically not significant and it refers to the relationship between *Layout* and *Security benefits*. However, indirectly it is observed that these two variables have an effect of 0.243, which occurs in the presence of the *Benefits for the organization* variable, which again indicates that managers must focus on that aspect and play a mediating role.
- The direct effect between *Suppliers* and *Security benefits* was only 0.094, slightly statistically significant. However, the indirect effect in the presence of *Benefits for the organization* and *Layout* is 0.420, it is over 400% larger, which again indicates the importance that management should pay attention on such aspects, since they are imputable for them.
- Finally, it is observed that the relationship between *Suppliers* and *Benefits for the organization* is appropriate with an indirect effect of 0.183 when *Layout* is present as a mediating variable. In other words, *Suppliers* have a direct effect on the plant distribution, which can be translated into *Security benefits*.

11.2.3.4 Total Effects—Complex Model 4

The total direct and indirect effects results in the total effects, which often give a better perspective on the relationship between the variables, since many times the direct effects are small size. Table 11.12 presents the total effects from the Complex Model 4, where the β, value the p-value for the statistical significance test, and the effect size as a measure of the explained variability are included.

Table 11.12 Total effects—Complex Model 4

Dependent variable	Independent variables		
	Benefits of the organization	*Suppliers*	*Layout*
Benefits for the organization		0.527 ($P < 0.001$) ES = 0.274	0.328 ($P < 0.001$) ES = 0.168
Suppliers			
Layout		0.558 ($P < 0.001$) ES = 0.312	
Security benefits	0.740 ($P < 0.001$) ES = 0.604	0.515 ($P < 0.001$) ES = 0.262	0.297 ($P < 0.001$) ES = 0.141

From the results obtained in Table 11.12, a series of results and conclusions are acquired such as the following:

- Machines and tools *Suppliers* play a vital role in several internal aspects of the company during the TPM implementation process, since these are considered external. Among the most important aspects are the plant distribution or *Layout*, the *Benefits for the organization* and the *Security benefits*, where the senior management must have an adequate relationship with those *Suppliers*.
- The total effects that the *Suppliers* variable exerts on the other subsequent variables have values over 0.5, and they are statistically significant, which indicates that it can explain a lot on their variability.
- It is observed that the relationship between *Benefits for the organization and security Benefits* is the highest in value and that it is a direct effect, since there are no indirect effects.
- The direct effect between *Suppliers* and *Security benefits* was scarcely significant, with a value of 0.094 and an associated *p*-value of 0.034, even lower than 0.05. However, when the indirect effect is observed, it is shown that it is 0.420, which gives a total effect of 0.515 (rounded to three figures).
- The direct effect between *Layout* and *Security benefits* was statistically not significant, but the indirect effect and the total effect were, which shows the importance of focusing on obtaining *Benefits for the organization*.

11.2.4 Conclusions and Industrial Implications—Complex Model 4

Based on the direct, indirect, and total effects, a series of conclusions and recommendations can be made due to the industrial implications that this may have, such as the following:

- Senior management should focus on improving the work environment where operations are carried out, increase the traders morale at all levels, create a work culture and manage the knowledge generated from manufacturing and maintenance practices, since all that becomes *Security benefits*; because in this way the culture of errors and accidents prevention is increased, traders are able to identify risk points, and understand the norms and legislations associated with safety (Gosavi 2006). From the previous data, it is concluded that the direct effect has been barely significant, but the total effect it is, since it considers aspects linked to the *Layout* and *Benefits for the organization*.
- Although the direct relationship between *Layout* and *Security benefits* was not statistically significant, it is observed that the indirect effect through the *Benefits for the organization* has been, therefore, the total effects are equally important. Also, it demonstrates the importance of management in focusing on plant distribution to obtain *Benefits for the organization*, which are reflected in *Security*

benefits. In addition, to achieve these benefits, the company must strive to achieve a *Layout* where machines are organized by families or technology groups to reduce raw materials transportation costs in the production process, it must always make a distribution that facilitates the maintenance and not only focuses on productive aspects, sacrificing safety and maintenance programs efficiency.

- *Suppliers* should be able to facilitate a better plant distribution or *Layout* about the machines and tools sold in the supply chains as well as be able to guarantee *Benefits for the organization*, where operations are simplified, and costs are reduced. It is the senior management responsibility to establish the contracts with *Suppliers* in order to guarantee that process.

11.2.5 Sensitivity Analysis—Complex Model 4

The Complex Model 4 presents a total of four latent variables, two of them refer to activities associated with TPM while the other two explain the benefits obtained as a consequence of performing them. Also, these variables are linked by means of six hypotheses, and in this section, Sensitivity analysis is portrayed to show the low and high states that these variables can have independently, as well as their combination

11.2.5.1 Sensitivity Analysis: *Suppliers* and *Layout* (H_1)—Complex Model 4

Machines and tools *Suppliers* are the ones who know their technical specifications as well as the requirements to be installed in the supply chains. In this section, Sensitivity analysis of the *Suppliers* variable and *Layout* is presented, in their high and low levels, as well as their combination. Table 11.13 shows a summary of the results, conclusions, and inferences that are obtained; they are interpreted below:

Table 11.13 Sensitivity analysis: *suppliers* and *layout* (H_1)—Complex Model 4

Suppliers	*Layout*	
	High	Low
High	Layout+ = 0.163	Layout− = 0.177
	Suppliers+ = 0.144	Suppliers+ = 0.144
	Layout+ and suppliers+ = 0.06	Layout− and suppliers+ = 0.008
	Layout+ *if* suppliers+ = 0.415	Layout− *if* suppliers+ = 0.057
Low	Layout+ = 0.163	Layout− = 0.177
	Suppliers− = 0.168	Suppliers− = 0.168
	Layout+ and suppliers− = 0.003	Layout− and suppliers− = 0.082
	Layout+ *if* suppliers− = 0.016	Layout− *if* suppliers− = 0.484

- The most optimistic scenario occurs when the *Layout* and *Suppliers* variables have their high levels, where the first variable probability of occurrence in this scenario is 0.163 independently, while for the second is 0.144, but the probability that they are presented simultaneously is only 0.06; a very low probability of concurrence, which indicates that the senior management must work to integrate *Suppliers*, their production, and maintenance managers to increase that probability. Finally, the probability that *Layout* has high levels, because *Suppliers* have high participation levels is 0.415; It has been previously indicated that *Suppliers* have a strong influence on the plant distribution because of the machines in the production system.
- The high relationship between *Layout* and *Suppliers* is demonstrated when analyzing the low levels for the first variable and the high levels for the second, since the probability that these two variables are presented simultaneously is 0.008; a very low probability. However, the probability of having a *Layout* at its low level, because *Suppliers* have a high level, is only 0.057. In other words, as long as machinery and equipment *Suppliers* promote training for the employees, they will also have high compliance levels in the *Layout*. The previous statement is clearly demonstrated when analyzing the invested roles in the variables as well, where the *Layout* has high values, but *Suppliers* low values, since the probability of occurrence is only 0.003; very close to zero. But, the probability of having high levels in the *Layout* since there have been low levels in the *Suppliers*, is only 0.016.
- Finally, the worst scenario that can be presented to the manager in charge of implementing TPM is that *Layout* and *Suppliers* have low levels, where the probability that the first variable is presented in this scenario in an independent manner from 0.177 while the second is 0.168. In addition, the probability that both variables are presented simultaneously is 0.082, and the probability of having a *Layout* at its low level since *Suppliers* also have a low level, is 0.484, which represents a risk to the maintenance manager.

11.2.5.2 Sensitivity Analysis: *Suppliers* and *Benefits for the Organization* (H_2)—Complex Model 4

Suppliers are a variable that must be linked to the benefits that can be offered to the company, these are variables that are considered before taking the choice to acquire any modern technology that is going to be implemented in the production process. In this section, the Sensitivity analysis that relates *Suppliers* with *Benefits for the organization* that can be obtained when implementing TPM is presented. Table 11.14 summarizes the high and low scenarios for both variables, as well as their combination. The conclusions that can be seen are the following:

- The most optimistic scenario in the analysis of that relationship is when the *Benefits for the organization* and *Suppliers* have high levels, where the first variable has a probability of 0.163 to be presented in that scenario in an independent manner, while the second has a probability of 0.144. However,

Table 11.14 Sensitivity analysis: *suppliers* with *benefits for the organization* (H_2)—Complex Model 4

Suppliers	*Benefits for the organization*	
	High	Low
High	Benefits for the organization+ = 0.163	Benefits for the organization− = 0.166
	Suppliers+ = 0.144	Suppliers+ = 0.144
	Benefits for the organization+ and suppliers+ = 0.057	Benefits for the organization− and suppliers+ = 0.011
	Benefits for the organization+ *if* suppliers+ = 0.396	Benefits for the organization− *if* suppliers+ = 0.075
Low	Benefits for the organization+ = 0.163	Benefits for the organization− = 0.166
	Suppliers− = 0.168	Suppliers− = 0.168
	Benefits for the organization+ and suppliers− = 0.014	Benefits for the organization− and suppliers− = 0.073
	Benefits for the organization+ *if* suppliers− = 0.081	Layout− *if* suppliers− = 0.435

management must strive to increase the probability of having these two at high variability levels, since the probability of occurrence simultaneously is only 0.057, but both variables are strongly related, because the probability of having *Benefits for the organization* at its high level, since *Suppliers* are also at their high level, is 0.396. This information indicates that *Suppliers* who are committed to the fulfillment of educational processes and training for workers; *Benefits for the organization* will be obtained.

- The previous assertion is confirmed when analyzing the scenario where the *Benefits for the organization* are low and *Suppliers* have high levels, since the probability of occurrence that these two scenarios happen simultaneously is only 0.011; a very low probability. In the same way, the probability of holding low *Benefits for the organization* because *Suppliers* are high is 0.075. However, if the roles of the analyzed variables are inverted and the *Benefits for the organization* have high levels and *Suppliers* low levels, the probability that these two scenarios are presented simultaneously is only 0.014, which again strengthens the conclusion. Likewise, the probability of having *Benefits for the organization* at its high levels, because *Suppliers* have a low level, is 0.081, which indicates that these *Benefits for the organization* can be obtained from another source and not only from *Suppliers.*
- The most pessimistic scenario that can be found in the relationship *between Benefits for the organization* and *Suppliers* is when both have low levels, and in this case, the probability of occurrence for the first variable in a nondependent manner in that scenario is 0.166 while for *Suppliers* is 0.168, but the probability that both variables are present in their low levels simultaneously is 0.073, which represents a high organizational risk. Finally, it is observed that the probability of having low *Benefits for the organization*, because there are low levels with *Suppliers* is 0.435, which represents a high risk.

11.2.5.3 Sensitivity Analysis: *Layout* and *Benefits for the Organization* (H_3)—Complex Model 4

The machines *Layout* along with a productive system is one of the most used techniques to generate *Benefits for the organization*, in this section the Sensitivity analysis is presented where these two variables are related in their high and low levels. Table 11.15 shows a summary of the results obtained from the analysis, the conclusions, and the inferences that can be stated about the collected data are the following:

- The optimistic scenario for these variables occurs when the *Benefits for the organization* and *Layout* have high levels. The probability that the first variable is presented independently in this scenario is 0.166 while the probability for the second is 0.163, but the probability that both variables are presented simultaneously in this scenario is 0.063; a relatively low probability. Finally, the probability that the *Benefits for the organization* are obtained at its high level since there is a high *Layout* is 0.330, which indicates that these two variables are strongly related, and the presence of a good *Layout* guarantees the *Benefits for the organization.*
- The previous information is demonstrated by analyzing the scenario, where the *Benefits for the organization* have low levels and *Layout* has high levels, where the probability of occurrence that this scenario for the two variables happens simultaneously is only 0.005. In addition, the probability that these *Benefits for the organization* are presented at their low levels since there are high levels in the *Layout* is 0.033, as a result, it is concluded that the high levels for the second variable imply high levels in the first. In the same way, if those roles are inverted in the latent variables, where the *Benefits for the organization* have high levels and *Layout* has low values, the probability of occurrence that this scenario occurs simultaneously for the variables is only 0.011 and the probability that the

Table 11.15 Sensitivity analysis: *layout* and *benefits for the organization* (H_3)—Complex Model 4

Layout	*Benefits for the organization*	
	High	Low
High	Benefits for the organization+ = 0.166	Benefits for the organization− = 0.166
	Layout+ = 0.163	Layout+ = 0.163
	Benefits for the organization+ and layout+ = 0.063	Benefits for the organization− and layout+ = 0.005
	Benefits for the organization+ *if* layout+ = 0.33	Benefits for the organization− *if* layout+ = 0.033
Low	Benefits for the organization+ = 0.163	Benefits for the organization− = 0.166
	Layout− = 0.177	Layout− = 0.177
	Benefits for the organization+ and layout− = 0.011	Benefits for the organization− and layout− = 0.087
	Benefits for the organization+ *if* layout − = 0.062	Benefits for the organization− *if* layout − = 0.492

first variable is presented in its scenario, because the second one has occurred, is only 0.062.
The most pessimistic scenario related to the *Benefits for the organization* and *Layout* occurs when there are low levels in both variables, where the first variable has a probability of occurrence in that scenario independently, which is 0.166 while for the second is 0.177, which can be considered high risks. However, the probability that these two variables are presented simultaneously in the TPM implementation process is 0.087, and the probability that the *Benefits for the organization* will be at its low levels if there are low levels in *Layout* is 0.492; a high risk level that the maintenance manager faces and for this reason, it is recommended that a lot of emphasis is placed on the plant distribution in the production system.

11.2.5.4 Sensitivity Analysis: *Suppliers* and *Security Benefits* (H_4)—Complex Model 4

When *Suppliers* install a machine in a supply chain, it is important to consider that the safety and operation manuals must be delivered, as well as to establish an education and training program that is aimed when generating *Security benefits*, which becomes necessary to know the pact and sensitivity in the relationship that these two variables have. Table 11.16 presents a summary of four possible scenarios that the combination of high and low levels from that this relationship may have. Based on this, the following conclusions and inferences are proposed:

- The best scenario is when the *Security benefits* and *Suppliers* have high levels and, in this case, the probability of having the first variable in its scenario independently is 0.220 while the second is 0.144, but the probability that both variables occur simultaneously and together is only 0.084; a low probability of occurrence. As a result, the maintenance manager should focus on increasing it, since this scenario is desired; however, the probability of having *Security benefits* at its high level, because there are high levels in *Suppliers* is 0.585. The previous data indicates that, if *Suppliers* develop an acceptable role by offering adequate operating manuals and courses focused on machine maintenance and operator integrity, *Security benefits* will always be obtained.
- The previous statement is easily demonstrated when analyzing the scenario where the *Security benefits* have low levels and *Suppliers* high levels, since the probability that these variables are presented simultaneously and together is 0.011, and the probability that the first variable appears in that scenario when the second variable has appeared is only 0.075, which again indicates that these variables are strongly related. In addition, a similar conclusion is reached when analyzing the variables with their levels invested, and the probability of having high *Security benefits* when *Suppliers* have low levels simultaneously, is only 0.011. In addition, the probability of having the first variable in this scenario, since you have the second is only 0.065.

Table 11.16 Sensitivity analysis: *suppliers* and *security benefits* (H_5)—Complex Model 4

Suppliers	*Security benefits*	
	High	Low
High	Security benefits+ = 0.220	Security benefits− = 0.166
	Suppliers+ = 0.144	Suppliers+ = 0.144
	Security benefits+ and suppliers+ = 0.084	Security benefits− and suppliers+ = 0.011
	Security benefits+ *if* suppliers+ = 0.585	Security benefits− *if* suppliers+ = 0.075
Low	Security benefits+ = 0.220	Security benefits− = 0.166
	Suppliers− = 0.168	Suppliers− = 0.168
	Security benefits+ and suppliers− = 0.011	Security benefits− and suppliers− = 0.065
	Security benefits+ *if* suppliers− = 0.065	Security benefits− *if* suppliers− = 0.387

- The most pessimistic scenario in the relationship between *Security benefits* and *Suppliers* is found when they have low levels, where the probability of occurrence for the first variable independently in this scenario is 0.166 while for the second is 0.168, but the greatest risk is when these variables can be presented simultaneously in their low levels, which has a probability of 0.065 that could seem low. Finally, the probability of having *Security benefits* at its low level because *Suppliers* having low levels is also 0.387, and it represents a high risk for the manager in charge of the TPM implementation.

11.2.5.5 Sensitivity Analysis: *Layout* and *Security Benefits* (H_5)—Complex Model 4

As a matter of fact, the senior management commitment in a company or machines *Suppliers* are not the only factors that may have an impact on security, because the physical distribution in the productive system is an aspect that may affect security, which is in charge of maintenance and production managers. Table 11.17 illustrates the Sensitivity analysis where *Security benefits* are related to the *Layout*; four possible scenarios for the high and low levels where both variables can be presented. Also, it is important to recall that the direct effect regarding these two variables was statistically not significant, but the direct effect that relies on the *Benefits for the organization* was. Based on this data, the following conclusions and inferences are proposed:

- The optimistic scenario for the relationship between *Security benefits* and *Layout* is when both variables have high levels, where the probability of occurrence for the first variable independently is 0.220 while for the second variable is 0.163, which must be considered satisfactory, although an effort must be made to increase them; however, the probability that both variables are presented in these scenarios simultaneously is 0.071, which can be considered

Table 11.17 Sensitivity analysis: *layout* and *security benefits* (H_5)—Complex Model 4

Layout	*Security benefits*	
	High	Low
High	Security benefits+ = 0.220	Security benefits− = 0.166
	Layout+ = 0.163	Layout+ = 0.163
	Security benefits+ and layout+ = 0.071	Security benefits− and layout+ = 0.008
	Security benefits+ *if* layout+ = 0.433	Security benefits− *if* layout+ = 0.050
Low	Security benefits+ = 0.220	Security benefits− = 0.166
	Layout− = 0.177	Layout− = 0.177
	Security benefits+ and layout− = 0.016	Security benefits− and layout− = 0.079
	Security benefits+ *if* layout− = 0.092	Security benefits− *if* layout− = 0.446

low, since this would be the desired scenario. Finally, the probability of having high *Security benefits* because there are high levels in the *Layou*t is 0.433, which indicates that the two variables are strongly related (although not directly).

- The previous information is easily verified by reviewing the scenario where the *Security benefits* have low levels and the *Layout* high levels, since the probability of simultaneous occurrence about these two variables is only 0.008. Also, if the levels are inverted in the variables, where the *Security benefits* have high levels and *Layout* low levels, the probability that those two variables happen simultaneously is only 0.016, which again demonstrates the high level in this relationship.
- The most pessimistic scenario in the study of the relationship between *Security benefits* and *Layout* occurs when both variables have their low levels, where the probability of occurrence for the first variable in that scenario independently is 0.166, while for the second is 0.177. Unfortunately, the probability that these two variables are presented simultaneously in this scenario is only 0.079; which represents a risk. Finally, the probability of having *Security benefits* at its low level because there is a low *Layout*, is 0.446, which represents one of the biggest risks to consider by the maintenance and production manager.

11.2.5.6 Sensitivity Analysis: *Benefits for the Organization* and *Security Benefits* (H_6)—Complex Model 4

Benefits for the organization are focused on improving quality, operations control, and employee morale, which undoubtedly improves traders' motivation, and therefore, there will be greater *Security benefits*. In Table 11.18, the Sensitivity analysis is performed, which details the high and low levels results as well as their combinations. From the data analysis, the following conclusions and statistical inferences are established:

- The most optimistic scenario where the relationship between *Benefits for the organization* and *Security benefits* happens, it is when both variables have high

Table 11.18 Sensitivity analysis: *benefits for the organization* and *security benefits* (H_6)—Complex Model 4

Benefits for the organization	*Security benefits*	
	High	Low
High	Security benefits+ = 0.220	Security benefits− = 0.166
	Benefits for the organization+ = 0.163	Benefits for the organization+ = 0.163
	Security benefits+ and benefits for the organization+ = 0.125	Security benefits− and benefits for the organization+ = 0.003
	Security benefits+ *if* benefits for the organization+ = 0.763	Security benefits− *if* benefits for the organization+ = 0.017
Low	Security benefits+ = 0.220	Security benefits− = 0.166
	Benefits for the organization− = 0.166	Benefits for the organization− = 0.166
	Security benefits+ and benefits for the organization− = 0.005	Security benefits− and benefits for the organization− = 0.111
	Security benefits+ *if* benefits for the organization− = 0.033	Security benefits− *if* benefits for the organization− = 0.679

levels, in the case of the first variable, the probability of being present in that scenario independently is 0.220, while the probability for the second is 0.163, but the probability of occurrence for both variables in those high levels is only 0.125; which can be considered as a low probability. Finally, the probability of having high *Security benefits*, because *Benefits for the organization* has been presented at its high levels is 0.763. The previous data demonstrate that managers must strive to obtain *Benefits for the organization*, since this will sooner or later become *Security benefits*.

- The data is easily demonstrated when analyzing the scenario where *Security benefits* have low levels and the *Benefits for the organization* high levels, since the probability of occurrence for these two variables simultaneously is only 0.003 whereas the probability that the first variable happens, because the second one has happened in this scenario is 0.017; a low probability. In the same way, the probability that there will be high *Security benefits* as well as *Benefits for the organization* at the same time is 0.005, which indicates that it will almost never happen.
- Finally, the most pessimistic scenario in the relationship between *Security benefits* and *Benefits for the organization* when they are at their lowest levels, and the probability of occurrence for the first variable happening independently in this scenario is 0.166 while for the second is 0.166. However, the true risk for a manager is when these variables are presented simultaneously, which has a probability of 0.111, which can be considered a high risk. Also, when analyzing the conditional probability, it is observed that the probability of having low *Security benefits*, because *Benefits for the organization* are low is of 0.679, which indicates that the maintenance manager along with other managements must always look to acquire those *Benefits for the organization* at high levels.

References

Ahmadzadeh F, Bengtsson M (2017) Using evidential reasoning approach for prioritization of maintenance-related waste caused by human factors—a case study. Int J Adv Manuf Technol 90(9):2761–2775. https://doi.org/10.1007/s00170-016-9377-7

Ahuja IPS (2009) Total productive maintenance. In: Ben-Daya M, Duffuaa SO, Raouf A, Knezevic J, Ait-Kadi D (eds) Handbook of maintenance management and engineering. Springer, London, pp 417–459. https://doi.org/10.1007/978-1-84882-472-0_17

Ahuja IPS, Khamba JS (2008) Strategies and success factors for overcoming challenges in TPM implementation in Indian manufacturing industry. J Qual Maint Eng 14(2):123–147. https://doi.org/10.1108/13552510810877647

Ahuja IS, Khamba JS, Choudhary R (2006) Improved organizational behavior through strategic total productive maintenance implementation. (47748):91–98. https://doi.org/10.1115/imece2006-15783

Aravindan P, Punniyamoorthy M (2002) Justification of advanced manufacturing technologies (AMT). Int J Adv Manuf Technol 19(2):151–156. https://doi.org/10.1007/s001700200008

Attri R, Grover S, Dev N, Kumar D (2013) An ISM approach for modelling the enablers in the implementation of total productive maintenance (TPM). Int J Syst Assur Eng Manag 4(4):313–326. https://doi.org/10.1007/s13198-012-0088-7

Baysal ME, Sümbül MO, Ekicioğlu E (2015) A total productive maintenance implementation in a manufacturing company operating in insulation sector in Turkey. In: 2015 6th international conference on modeling, simulation, and applied optimization (ICMSAO), 27–29 May 2015. pp 1–7. https://doi.org/10.1109/icmsao.2015.7152205

Borkowski S, Czajkowska A, Stasiak-Betlejewska R, Borade AB (2014) Application of TPM indicators for analyzing work time of machines used in the pressure die casting. J Ind Eng Int 10(2):55. https://doi.org/10.1007/s40092-014-0055-9

Bruch J, Bellgran M (2013) Critical factors for successful user-supplier integration in the production system design process. In: Emmanouilidis C, Taisch M, Kiritsis D (eds) Advances in production management systems. Competitive manufacturing for innovative products and services. Berlin, Heidelberg, pp 421–428

Cakmakci M (2009) Process improvement: performance analysis of the setup time reduction-SMED in the automobile industry. Int J Adv Manuf Technol 41(1):168–179. https://doi.org/10.1007/s00170-008-1434-4

Camuffo A, De Stefano F, Paolino C (2017) Safety reloaded: lean operations and high involvement work practices for sustainable workplaces. J Bus Ethics 143(2):245–259. https://doi.org/10.1007/s10551-015-2590-8

Cao B, Wang S (2017) Opening up, international trade, and green technology progress. J Clean Prod 142:1002–1012. https://doi.org/10.1016/j.jclepro.2016.08.145

Cardoso RdR, Pinheiro de Lima E, Gouvea da Costa SE (2012) Identifying organizational requirements for the implementation of advanced manufacturing technologies (AMT). J Manuf Syst 31(3):367–378. https://doi.org/10.1016/j.jmsy.2012.04.003

Chan FTS, Chan MH, Lau H, Ip RWL (2001) Investment appraisal techniques for advanced manufacturing technology (AMT): a literature review. Integr Manuf Syst 12(1):35–47. https://doi.org/10.1108/09576060110361528

Chang J (2017) The effects of buyer-supplier's collaboration on knowledge and product innovation. Ind Mark Manag 65:129–143. https://doi.org/10.1016/j.indmarman.2017.04.003

Chavhan R, Mahajan DSK, Joshi Sarang P (2018) Supplier development success factors in Indian manufacturing practices. Materials today: proceedings 5(2, Part 1):4078–4096. https://doi.org/10.1016/j.matpr.2017.11.669

Duffuaa SO, Raouf A (2015) Total productive maintenance. In: Duffuaa SO, Raouf A (eds) Planning and control of maintenance systems: modelling and analysis. Springer International Publishing, Cham, pp 261–270. https://doi.org/10.1007/978-3-319-19803-3_12

Efstathiades A, Tassou S, Antoniou A (2002) Strategic planning, transfer and implementation of advanced manufacturing technologies (AMT). Development of an integrated process plan. Technovation 22(4):201–212. http://dx.doi.org/10.1016/S0166-4972(01)00024-4

Fore S, Zuze L (2010) Improvement of overall equipment effectiveness through total productive maintenance. World Acad Sci Eng Technol 37:402–410

Fulton M, Hon B (2010) Managing advanced manufacturing technology (AMT) implementation in manufacturing SMEs. Int J Prod Perform Manage 59(4):351–371. https://doi.org/10.1108/17410401011038900

Gehandler J (2017) The theoretical framework of fire safety design: reflections and alternatives. Fire Saf J 91:973–981. https://doi.org/10.1016/j.firesaf.2017.03.034

Gosavi A (2006) A risk-sensitive approach to total productive maintenance. Automatica 42 (8):1321–1330. https://doi.org/10.1016/j.automatica.2006.02.006

Ham R (1987) 7—seating layout and safety regulations. In: Ham R (ed) Theatres. Architectural Press, pp 45–49. https://doi.org/10.1016/B978-0-442-20497-6.50012-8

Hana P (2010) Human reliability in maintenance task. Front Mech Eng China 5(2):184–188. https://doi.org/10.1007/s11465-010-0002-4

Keah-Choon T, Vijay RK, Robert BH, Soumen G (1999) Supply chain management: an empirical study of its impact on performance. Int J Oper Prod Manage 19(10):1034–1052. https://doi.org/10.1108/01443579910287064

Kumar J, Soni VK (2015) An exploratory study of OEE implementation in Indian manufacturing companies. J Inst Eng (India): Series C, 96(2):205–214. https://doi.org/10.1007/s40032-014-0153-x

Liang LY, Chao WC (2008) The strategies of tabu search technique for facility layout optimization. Autom Constr 17(6):657–669. https://doi.org/10.1016/j.autcon.2008.01.001

Lunau S, Meran R, John A, Roenpage O, Staudter C (2013) Improve. In: Meran R, John A, Roenpage O, Staudter C, Lunau S (eds) Six Sigma+Lean toolset: mindset for successful implementation of improvement projects. Springer, Berlin, pp 265–344. https://doi.org/10.1007/978-3-642-35882-1_5

Medina-Herrera N, Jiménez-Gutiérrez A, Grossmann IE (2014) A mathematical programming model for optimal layout considering quantitative risk analysis. Comput Chem Eng 68:165–181. https://doi.org/10.1016/j.compchemeng.2014.05.019

Nagel M (2006) Environmental quality in the supply chain of an original equipment manufacturer: what does it mean? In: Sarkis J (ed) Greening the supply chain. Springer, London, pp 325–340. https://doi.org/10.1007/1-84628-299-3_18

Nakajima S (1988) Introduction to TPM: total productive maintenance. Productivity Press, Cambridge, MA

Patrik J, Magnus L (1999) Evaluation and improvement of manufacturing performance measurement systems—the role of OEE. Int J Oper Prod Manage 19(1):55–78. https://doi.org/10.1108/01443579910244223

Pisano GP (1997) The development factory: unlocking the potential of process innovation. Harvard Business Press

Porras-Vázquez A, Romero-Pérez J-A (2018) A new methodology for facilitating the design of safety-related parts of control systems in machines according to ISO 13849:2006 standard. Reliab Eng Syst Saf 174:60–70. https://doi.org/10.1016/j.ress.2018.02.018

Ragatz GL, Handfield RB, Petersen KJ (2002) Benefits associated with supplier integration into new product development under conditions of technology uncertainty. J Bus Res 55(5):389–400. https://doi.org/10.1016/S0148-2963(00)00158-2

Reichstein T, Salter A (2006) Investigating the sources of process innovation among UK manufacturing firms. Ind Corp Change 15(4):653–682

Sadeghi L, Mathieu L, Tricot N, Al Bassit L (2015) Developing a safety indicator to measure the safety level during design for safety. Saf Sci 80:252–263. https://doi.org/10.1016/j.ssci.2015.08.006

Saliba MA, Zammit D, Azzopardi S (2017) A study on the use of advanced manufacturing technologies by manufacturing firms in a small, geographically isolated, developed economy:

the case of MHigh. Int J Adv Manuf Technol 89(9):3691–3707. https://doi.org/10.1007/s00170-016-9294-9
Seth D, Tripathi D (2006) A critical study of TQM and TPM approaches on business performance of Indian manufacturing industry. Total Qual Manag Bus Excell 17(7):811–824. https://doi.org/10.1080/14783360600595203
Seyed-Mahmoud A, Saeedreza H, Lotfollah N, Ziaul H (2011) The influence of work-cells and facility layout on the manufacturing efficiency. J Facil Manage 9(3):213–224. https://doi.org/10.1108/14725961111148117
Shetty PK, Rodrigues LLR (2010) Total productive maintenance of a diesel power generating unit of a institution campus. In: 2010 international conference on mechanical and electrical technology, 10–12 Sept 2010. pp 68–71. https://doi.org/10.1109/icmet.2010.5598493
Shrivastava D, Kulkarni MS, Vrat P (2016) Integrated design of preventive maintenance and quality control policy parameters with CUSUM chart. Int J Adv Manuf Technol 82(9):2101–2112. https://doi.org/10.1007/s00170-015-7502-7
Spanos YE, Voudouris I (2009) Antecedents and trajectories of AMT adoption: the case of Greek manufacturing SMEs. Res Policy 38(1):144–155. https://doi.org/10.1016/j.respol.2008.09.006
Thomas L, Johan F (2010) Equipment supplier/user collaboration in the process industries: In search of enhanced operating performance. J Manuf Technol Manage 21(6):698–720. https://doi.org/10.1108/17410381011064003
Tompkins JA, White JA, Bozer YA, Frazelle EH, Tanchoco JMA, Trevino J (2003) Facility planning. Wiley, New York, NY
Tsarouhas PH (2013) Evaluation of overall equipment effectiveness in the beverage industry: a case study. Int J Prod Res 51(2):515–523. https://doi.org/10.1080/00207543.2011.653014
Vijay RK, Keah Choon T (2006) Buyer-supplier relationships: the impact of supplier selection and buyer-supplier engagement on relationship and firm performance. Int J Phys Distrib Logistics Manage 36(10):755–775. https://doi.org/10.1108/09600030610714580
Vitayasak S, Pongcharoen P (2018) Performance improvement of teaching-learning-based optimisation for robust machine layout design. Exp Syst Appl 98:129–152. https://doi.org/10.1016/j.eswa.2018.01.005
Wakjira MW, Singh AP (2012) Total productive maintenance: a case study in manufacturing industry. Global J Res Eng 12 (1-G)
Waldeck NE, Leffakis ZM (2007) HR perceptions and the provision of workforce training in an AMT environment: an empirical study. Omega 35(2):161–172. https://doi.org/10.1016/j.omega.2005.05.001
Willmott P, McCarthy D (2001a) 1—putting TPM into perspective from total productive maintenance to total productive manufacturing. In: Total productivity maintenance, 2nd edn. Butterworth-Heinemann, Oxford, pp 1–16. https://doi.org/10.1016/B978-075064447-1/50004-7
Willmott P, McCarthy D (2001b) 2—Assessing the true costs and benefits of TPM. In: Total productivity maintenance, 2 edn. Butterworth-Heinemann, Oxford, pp 17–22. http://dx.doi.org/10.1016/B978-075064447-1/50005-9
Willmott P, McCarthy D (2001c) 3—The top-down and bottom-up realities of TPM. In: Total productivity maintenance, 2 edn. Butterworth-Heinemann, Oxford, pp 23–61. https://doi.org/10.1016/B978-075064447-1/50006-0
Wyrwicka MK, Mrugalska B (2017) Mirages of lean manufacturing in practice. Procedia Eng 182:780–785. https://doi.org/10.1016/j.proeng.2017.03.200
Yang L, Deuse J, Jiang P (2013) Multiple-attribute decision-making approach for an energy-efficient facility layout design. Int J Adv Manuf Technol 66(5):795–807. https://doi.org/10.1007/s00170-012-4367-x
Yi W, Chi H-L, Wang S (2018) Mathematical programming models for construction site layout problems. Autom Constr 85:241–248. https://doi.org/10.1016/j.autcon.2017.10.031
Zhou B (2016) Lean principles, practices, and impacts: a study on small and medium-sized enterprises (SMEs). Ann Oper Res 241(1):457–474. https://doi.org/10.1007/s10479-012-1177-3

Chapter 12
Structural Equation Models-Technical Factors

Abstract In this chapter two structural equation models are presented, which are related to the technical factors associated with the total preventive maintenance programs implementation such as the *PM Implementation*, *TPM implementation*, *Technological status*, *Layout* and *Warehouse management*. In addition, these technical factors are linked to the *Benefits for the organization*, productivity, and worker and environment safety, which generate a series of hypotheses. Also, least squares algorithms are implemented to evaluate the models, and the hypotheses are validated by estimating the direct effects, although indirect and total effects are also acquired as well as the effect size. Finally, for each relationship between the variables, a sensitivity analysis is performed.

12.1 Complex Model 5

This model integrates four latent variables into the operative factors category, which are *Technological status*, *Layout*, *PM implementation*, and *TPM implementation* that are related to generate six research hypotheses. Additionally, the model objective is to demonstrate that depending on the company *Technological status* in its production system, with the support of an adequate machines or *Layout* distribution as well as a *PM implementation* plan, an adequate *TPM implementation* can be generated. In other words, it is assumed that the *Technological status* is the independent variable and the *TPM implementation* is the dependent variable. Also, the hypotheses from this model are illustrated below.

12.1.1 Hypotheses

When starting the *TPM implementation* process, an analysis from the equipment setting that is installed in the production systems should always be performed, for that reason, Bourke and Roper (2016) consider that these complementary effects should be

J. R. Díaz-Reza et al., *Impact Analysis of Total Productive Maintenance*,
https://doi.org/10.1007/978-3-030-01725-5_12

taken into account. Currently, machines are usually small and perform multiple functions with fast changes in the programming, however, those that are older usually occupy more space in the plant and reduce the production processes flexibility (Lewis and Boyer 2002). Similarly, it is common that technological advanced machines require a space with special weather conditions such as thermal and vibrations isolation or they require a connection to special wire (Cardoso et al. 2012).

In addition, due to these special maintenance needs that new machines may require, it is recommended that some aspects are considered when evaluating any new equipment for its purchasing, since very often special adaptations are necessary, because there are physical installations already established on machines, therefore it is crucial to acknowledge them in advance (Fulton and Hon 2010). Considering that the technological level that machines and equipment have affects their physical distribution in the productive system, the following hypothesis is proposed:

H_1: The *Technological status* that a company has in its supply chains has a direct and positive effect on the plant *Layout* or distribution.

Moreover, usually machines and equipment with high technological level are very expensive, as a result, managers must ensure that they have optimal use conditions to justify their investment, and it can only be achieved through a *PM implementation* plan (Eti et al. 2006). In fact, one of the purchase and selection attributes must be the maintainability that is possible to provide to the machines as well as their cost, since frequently equipment has a low cost, but their maintenance represents a strong economic outlay for the company (Choe 2004). Also, a maintenance assessment cost and its *Technological status* may help to take a better purchase decision.

In addition, it is the management responsibility to guarantee that equipment warranties are shown when they have a high technological level due to the high economic risk that they represent and, often preventive maintenance plans can be contracted for the machines with the supplier, therefore, it guarantees that they are the experts who carry out those activities. Also, this type of agreements are usually implemented only in the first years of the machines life and as time goes by those plans become more expensive (Jin et al. 2017) besides, operators and maintenance managers learn to predict the faults that machines could have. According to the previous information and considering the machines modern level and equipment; they have an influence on preventive maintenance and the following hypothesis is proposed:

H_2: The *Technological status* that a company has in its supply chains has a direct and positive effect on the *PM implementation*.

Furthermore, sometimes the productive systems are frequently changed in their physical distribution in order to group activities and generate demanded products (Yi et al. 2018). However, this equipment in its new physical distribution still have the maintenance and adjustments need, especially when going from one product to another, then, calibrations are essential before starting the production of a new batch (Mohsen and Hassan 2002).

Thus, preventive maintenance plans must consider the physical equipment distribution, since it depends on the tools that are going to be used in this activity because if the *Layout* does not allow it, it is not possible to get in with forklifts and elevated cranes. For this reason, proper maintenance planning must be taken into account in the *Layout* restrictions that there may be, although this distribution should always be looked for as an advantage and not as a disadvantage to keep the equipment in an appropriate operational status (Medina-Herrera et al. 2014). Considering the previous information, the following hypothesis is proposed:

H_3: The plant *Layout* or distribution that a company has in its production system has a direct and positive effect on the *PM implementation*.

In addition, *Technological status* should not only facilitate the maintenance planning but should have a more holistic concept and focus on integrating each element in the company, therefore, it should also facilitate the *TPM implementation* (Pang et al. 2016), where all resources converge in favor of the machines and equipment preservation as a wealth mean.

Also, the equipment *Technological status* must be considered when generating a TPM plan, since those that are very old or old are critical and can easily have technical stoppages due to breakdowns, and it may be reflected in late deliveries with high production costs (Arnaiz et al. 2013; Eti et al. 2006). In addition, if the equipment is newer, it may require less maintenance attention and should only focus on monitoring sensors and interpreting their signs. Therefore, considering that the equipment *Technological status* in the supply chains affects the *TPM implementation* process, the following hypothesis is proposed:

H_4: The *Technological status* that a company has in its supply chains has a direct and positive impact on the *TPM implementation* process.

There are many factors that affect the *TPM implementation* process and one of the most important is the machines physical distribution or *Layout* in the production system (Vilarinho et al. 2018). For instance, Chand and Shirvani (2000) indicate that the machines arrangement in manufacturing cells facilitates the maintenance process, since the personnel that operates them is highly trained and specialized, and they are a great support in the preservation equipment tasks.

Similarly, Singh et al. (2013) have found that the *TPM implementation* process is streamlined when the machines have a distribution oriented to facilitate this work and report a case study. Also, Mwanza and Mbohwa (2015) have reported a case study where a TPM plan is designed in a pharmaceutical company considering the equipment physical distribution in the production system and in their multifunctionality. According to the previous information, considering that the *Layout* influences the *TPM implementation*, the following hypothesis is proposed:

H_5: The plant *Layout* or distribution that a company has in its supply chains has a direct and positive effect on the *TPM implementation* process.

Undoubtedly, the beginning of a *TPM implementation* program starts with a work culture focused on the equipment preservation or with a preventive maintenance plan that considers the useful life of the elements that integrate them (Gouiaa-Mtibaa et al. 2018), since keeping a record of the components changes and replacements avoids the unexpected technical stoppages that are reflected in productive system delays (Fumagalli et al. 2017).

Nowadays, there are studies where the price of an extended preventive maintenance guarantee along with the supplier and the period of time that the contract must be performed is analyzed, in such a way that the benefits for the company are maximized (Anand et al. 2018). In addition, other studies relate the *PM implementation* with other techniques such as the product quality output, which discusses the need to have a *TPM implementation* that is comprehensive and related to other lean techniques that are already implemented, and not only focus on preventing but in a more holistic maintenance (Shrivastava et al. 2016). Therefore, considering that the *PM implementation* is the basis of a more comprehensive program that affects the *TPM implementation*, the following hypothesis is proposed:

H_6: The *PM implementation* process in a production system has a direct and positive effect on the *TPM implementation* process.

Figure 12.1 illustrates the relationships between the latent variables that are integrated into the Complex Model 5. It is relevant to recall that these latent variables have been validated in an earlier chapter and that all have achieved the corresponding indexes, consequently, the model evaluation is performed.

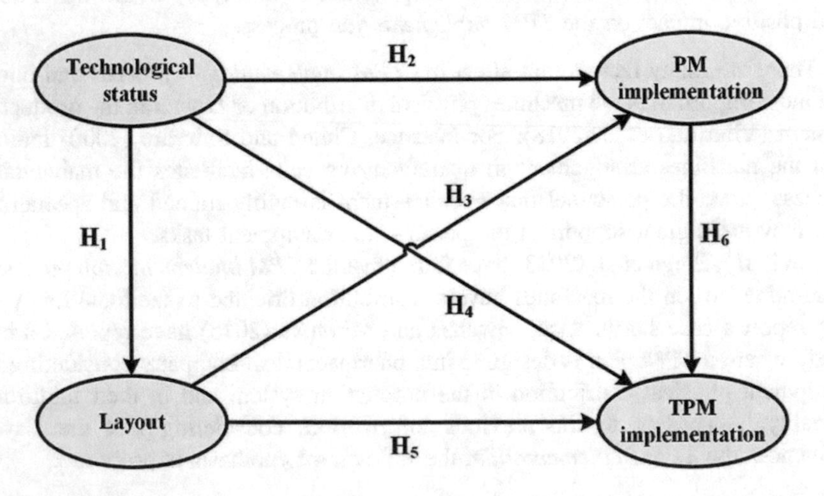

Fig. 12.1 Proposed hypotheses from the Complex Model 5

12.1.2 Efficiency Indexes—Complex Model 5

After executing the Complex Model 5 according to the methodology described above, the following efficiency indexes are acquired.

- Average path coefficient (APC) = 0.380, $P < 0.001$
- Average R-squared (ARS) = 0.493, $P < 0.001$
- Average adjusted R-squared (AARS) = 0.490, $P < 0.001$
- Average block VIF (AVIF) = 1.764, acceptable if $\leq$ 5, ideally $\leq$ 3.3
- Average full collinearity VIF (AFVIF) = 2.342, acceptable if $\leq$ 5, ideally $\leq$ 3.3
- Tenenhaus GoF (GoF) = 0.555, small $\geq$ 0.1, medium $\geq$ 0.25, large $\geq$ 0.36
- Sympson's paradox ratio (SPR) = 1.000, acceptable if $\geq$ 0.7, ideally = 1
- R-squared contribution ratio (RSCR) = 1.000, acceptable if $\geq$ 0.9, ideally = 1
- Statistical suppression ratio (SSR) = 1.000, acceptable if $\geq$ 0.7
- Nonlinear bivariate causality direction ratio (NLBCDR) = 1.000, acceptable if $\geq$ 0.7

From the analysis of the previous indexes list, the following is concluded, therefore, the model is interpreted:

- On average, all regression or β indexes are statistically significant, since the associated p-value is under 0.05.
- There is enough predictive validity, since the ARS and AARS indexes are over 0.02 and the associated p-value is less than 0.05.
- There are no collinearity problems between the variables analyzed, since the AVIF and AFVIF indexes are under 3.3; the maximum admitted value.
- The data is adjusted appropriately to the model, since the GoF index is over 0.36.
- Finally, it is observed that there are no problems in the variables dependence direction.

Due to the previous conclusions, the model is interpreted in the following sections.

12.1.3 Results—Complex Model 5

Since the model efficiency indexes indicate its suitability, it proceeds to interpret it, which is presented in Fig. 12.2, where the β value appears for each of the relationships between the variables, their p-value associated to the statistical validation as well as the R-squared value for each dependent latent variable.

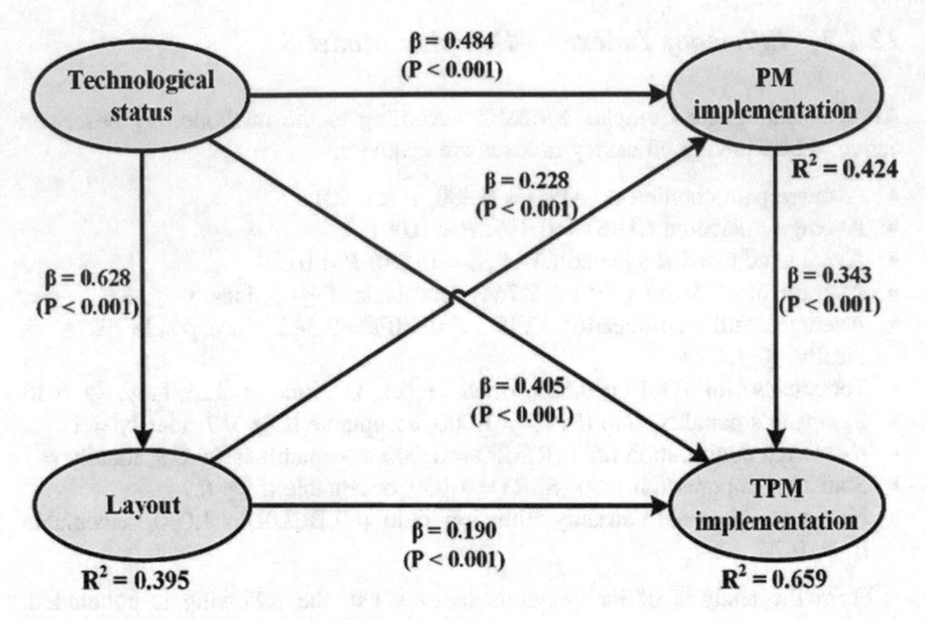

Fig. 12.2 Evaluated Complex Model 5

12.1.3.1 Direct Effects—Complex Model 5

According to the direct effects from Fig. 12.2 and the associated *p*-value, the following are concluded regarding the initially stated hypotheses:

H_1: There is enough statistical evidence to declare that *Technological status* that a company has in its supply chains has a direct and positive effect on the plant *Layout* or distribution, since when the first variable increases its standard deviation in a unit, the second one does it in 0.628 units.
H_2: There is enough statistical evidence to declare that the *Technological status* that a company has in its supply chains has a direct and positive effect on the *PM implementation*, since when the first variable increases its standard deviation by one unit, the second one increases in 0.484 units.
H_3: There is enough statistical evidence to declare that the plant *Layout* or distribution that a company has in its production system has a direct and positive effect on the *PM implementation*, since when the first variable increases its standard deviation in one unit, the second one increases in 0.228 units.
H_4: There is enough statistical evidence to declare that the *Technological status* that a company has in its supply chains has a direct and positive impact on the *TPM implementation* process, since when the first variable increases its standard deviation by one unit, the second one does it in 0.405 units.
H_5: There is enough statistical evidence to state that the plant *Layout* or distribution that a company has in its supply chains has a direct and positive effect on the *TPM*

implementation process, since when the first variable increases its standard deviation in one unit, the second one increases in 0.190 units.

H_6: There is enough statistical evidence to state that the *PM implementation* process in a production system has a direct and positive effect on the *TPM implementation* process, since when the first variable increases its standard deviation by one unit, the second variable increases in 0.343 units.

12.1.3.2 Effect Size—Complex Model 5

Figure 12.2 shows that dependent variables can be explained by more than one independent variable, as a result, Table 12.1 illustrates the *R*-squared decomposition to determine which of these variables is more significant in order to make recommendations to the maintenance management.

According to the information in Table 12.1, the following is concluded:

- The *PM implementation* variable is explained in 42.4% by two variables: The *Technological status* in 30.3% and the *Layout* or plant distribution in 12.1%. From the previous information, it is concluded that in order to achieve an adequate *PM implementation*, it is important that management pay attention to the *Technological status* that their machines have, since currently those that are more modern are facilitated by sensors including man–machine interfaces, a greater operative status of communication, and a quick intervention process as well as decision making.
- The *TPM implementation* process is explained by three variables that are before it at 65.9%, and represents the highest *R*-squared value in the model, but 24.0% is due to the *PM implementation*, 30.0% is from the *Technological status*, finally, 11.9% is from the *Layout*. According to the variables explanatory power, it is concluded that the most relevant aspect in the *TPM implementation* is to consider the *Technological status*.
- When observing the contributions values that the *Technological status* has along with the other variables, it is observed that it contributes the most, which allows to conclude that this is a variable that must always be monitored by the senior management, since the TPM success depends a lot on it.

Table 12.1 *R*-squared contribution—Complex Model 5

Dependent variable	Independent variable			
	PM implementation	*Technological status*	*Layout*	R^2
PM implementation		0.303	0.121	0.424
TPM implementation	0.240	0.300	0.119	0.659
Layout		0.395		0.395

Table 12.2 Total indirect effects—Complex Model 5

Dependent variable	Independent variable	
	Technological status	*Layout*
PM implementation	0.143 ($P < 0.001$) ES = 0.090	
TPM implementation	0.334 ($P < 0.001$) ES = 0.247	0.078 ($P = 0.016$) ES = 0.049

12.1.3.3 Total Indirect Effects—Complex Model 5

As it was mentioned above, the model objective is to determine the impact that the *Technological status* has on the *TPM implementation*, having as mediating variables the *Layout* and the *PM implementation*. According to the indirect effects analysis, the role of these mediating variables can be determined, therefore, Table 12.2 indicates the total indirect effects from the Complex Model 5.

From the data analysis in Table 12.2, the following conclusions can be obtained:

- All indirect effects are statistically significant, since the associated value is under 0.05.
- The *Technological status* has an indirect effect on the *PM implementation* by 0.143 while the *Layout* is a mediating variable. In other words, the technological level can affect the *PM implementation* process when there is an adequate machine distribution in plant on the supply chains.
- Although there is a direct effect between the *Layout* and the *TPM implementation,* there is also an indirect effect of 0.078 when there is a *PM implementation* as a mediating variable. This aspect is significant because it indicates the importance to give a preventive focus to the machinery maintenance and distribution that the must be considered.
- Finally, there is a direct effect between the *Technological status* and the *TPM implementation* process by 0.405, but an indirect effect of 0.334 is observed, a very similar value, which is by using the *Layout* and the *PM implementation* as mediating variables, this demonstrates the importance of the machines *Layout* in the plant as well as the preventive approach to maintenance, since these modern and sophisticated machines are useless if they do not have a good distribution in the production systems, therefore, efforts to be preserved are performed.

12.1.3.4 Total Effects—Complex Model 5

Table 12.3 presents the total effects from the Complex Model 5, which represents the total direct and indirect effects. In each total effect, the β value is shown, the p-value for the statistical significance test, and the effect size as a measure of the

Table 12.3 Total effects—Complex Model 5

Dependent variable	Independent variable		
	PM implementation	*Technological status*	*Layout*
PM implementation		0.627 ($P < 0.001$) ES = 0.392	0.228 ($P < 0.001$) ES = 0.121
TPM implementation	0.343 ($P < 0.001$) ES = 0.240	0.740 ($P < 0.001$) ES = 0.548	0.268 ($P < 0.001$) ES = 0.168
Layout		0.628 ($P < 0.001$) ES = 0.395	

variance explained. According to the total effects, the following conclusions can be obtained:

- All the total effects are statistically significant, since the associated *p*-value is under 0.05.
- The *Technological status* variable is one of the most important to guarantee the *TPM implementation* success, since it has the highest total effect among the six relationships from the Complex Model 5. However, the same importance is observed in the *Layout* and the *PM implementation*, with values over 0.6.
- The plant or *Layout* distribution is a crucial variable in the TPM success, however, they are lower than those from the machines technological level.

12.1.4 Conclusions and Industrial Implications—Complex Model 5

The Complex Model 5 integrates four variables and six hypotheses, after analyzing it and acknowledges the several types of effects, the following conclusions, and industrial implications are proposed:

- Business managers must include a special emphasis on the machines and tools *Technological status* that they acquire for their production systems, since the total effects that this variable has on the adequate plant's distribution success are high (Cardoso et al. 2012).
- Due to the *Technological status* importance, it is recommended that managers hold frequent meetings in order to identify operational needs that can be added to the equipment already installed, and to determine if they have the ability to perform them, otherwise new machines designs or improved versions must be performed (Spanos and Voudouris 2009).
- It is crucial to consider that the acquired machines and tools have maintainability as a purchase criterion (Szőke et al. 2017), otherwise, there will be problems when the equipment is installed or there will be a lot of dependence on suppliers, which increases operating costs. Another reason, it is that sometimes,

due to the difficulty to perform the machines maintenance, it is frequently delayed and the products generation without the quality standards is incurred (Gouiaa-Mtibaa et al. 2018).

- Similarly, it is observed that the *Technological status* has a strong impact on the *TPM implementation*, as a result, management must focus on acquiring increasingly modern technologies that have an effective man–machine interface that allows to know in real time the operating conditions instead of waiting for the equipment to fail (Gao et al. 2016).
- In order to ensure the *TPM implementation* success, it is important that through an appropriate work culture, the *PM implementation* be achieved, first to focus on the equipment prevention and preservation (Fumagalli et al. 2017).
- It is significant that the maintenance managers keep a working log about all the equipment, which is written along with operators, since in this way it becomes more reliable because during the work shift they are who observed the operational parameters and are familiar with their trends.
- The company managers must demonstrate commitment and leadership in the *TPM implementation*, but, they must be more dedicated towards the *PM implementation*, since in this way, it supports the preservation work culture (Aga et al. 2016; Davila and Elvira 2012).

12.1.5 *Sensitivity Analysis—Complex Model 5*

In this section, a Sensitivity analysis is presented, which the variables low and high scenarios for each hypothesis are analyzed, as well as the combinations of these levels.

12.1.5.1 Sensitivity Analysis: *Technological Status* and *Layout* (H_1)—Complex Model 5

In this section, the relationship between the *Technological status* and the *Layout* in its high and low levels are analyzed, as well as their combinations; the results are shown in Table 12.4. Also, it is observed that there is a probability of 0.163 that the *Layout* is presented at its high level and with 0.177 at its low level; in the same way, the *Technological status* has a probability of 0.163 at its high level and 0.182 of presenting itself at a low level. The combined scenarios are the following:

- In the optimistic scenario, the probability that the *Layout* and the *Technological status* will be presented simultaneously at their high level is only 0.076, which is a low probability that requires immediate attention from the manager. However, the probability of having a *Layout* at its high level, because it has a high *Technological status* is 0.467, which indicates that efforts must be done to have adequate technological levels in the equipment, since this guarantees a better

Table 12.4 Sensitivity analysis: *Technological status* and *Layout* (H_1)—Complex Model 5

Technological status	*Layout*	
	High	Low
High	Layout+ = 0.163 Technological status+ = 0.163 Layout+ & Technological status + = 0.076 Layout+ *if* Technological status+ = 0.467	Layout− = 0.177 Technological status+ = 0.163 Layout− & Technological status+ = 0.000 Layout− *if* Technological status+ = 0.000
Low	Layout+ = 0.163 Technological status− = 0.182 Layout+ & Technological status− = 0.008 Layout+ *if* Technological status− = 0.045	Layout− = 0.177 Technological status− = 0.182 Layout− & Technological status− = 0.082 Layout− *if* Technological status− = 0.448

plant distribution, therefore, it indicates a strong relationship between the variables.

- The above is easily demonstrated when analyzing the situation in where there are *Layout* low levels and high *Technological status*, which probability of occurrence is zero; likewise, the probability that the scenario from the first variable will occur, because the second one has occurred, is zero. The previous information indicates that when there are high *Technological status* levels, there will always be adequate *Layout* levels.
- In the particular case, where the *Layout* is at its high level whereas the *Technological status* is at its low level, the probability of occurrence for those variables happening simultaneously is only 0.008, while the probability of occurrence for the first variable in its high scenario, because the second variable is at its low level is 0.045. The information from the above indicates that it is possible to have high *Layout* levels, which may be due to other situations than *Technological status*, although those values are small.
- The most pessimistic scenario in the relationship between the *Layout* and the *Technological status* is when both have low levels. The probability that both variables are presented in this scenario is 0.082, and the probability of the first variable in its scenario, because the second and its scenario are presented is 0.0448; it indicates that managers must pay special attention to the machines and tools technological levels that they purchase.

12.1.5.2 Sensitivity Analysis: *Technological Status* and *PM Implementation* (H_2)—Complex Model 5

The second hypothesis (H_2) relates to the *Technological status* and the *PM implementation* in their different presentation scenarios. It is observed that the probability of occurrence independently for the first latent variable in its high level is 0.163 while in its low level is 0.182. However, the probability of occurrence for the second variable at its high level is 0.179, whereas at its low level is 0.190. The conclusions from these levels combined in the variables are the following:

- In its optimistic scenario, when the *PM implementation* and *Technological status* have a high level simultaneously, the probability of occurrence for that scenario is 0.079; a very low value that requires the manager intervention. In the same way, the probability of having a *PM implementation* at its high level, because it has a high *Technological status* is 0.483, and this indicates a strong relationship between these variables and, therefore, it is recommended that high technological levels always be sought to have equipment, since that facilitates the *PM implementation*.
- The information from above is checked when analyzing the scenario where the *PM implementation* has low levels and the *Technological status* high levels, since the probability of occurrence simultaneously is 0.003; a very low probability; In addition, when analyzing the probability scenario when the *PM implementation* is low, because the *Technological status* is high, it is only 0.017, which is a sign that it will almost never happen.
- Otherwise, where the *PM implementation* has high levels, but the *Technological status* low levels, the probability of simultaneous occurrence for these variables is 0.008. Similarly, the probability of the first variable occurring in its scenario, because the second one has happened is 0.045, which confirms that these variables have a strong relationship.
- Finally, in the pessimistic scenario, where the *PM implementation* and *Technological status* have low levels simultaneously, it has a probability of occurrence of 0.101; a relatively high probability and even more critical is the probability of occurrence for the first variable in its low scenario, since the second has also happened in its low scenario, which is 0.552. For this reason, managers are recommended to look forward to increasing the machines and tools technological level to avoid having low *PM implementation* levels (Table 12.5).

12.1.5.3 Sensitivity Analysis: *Layout* and *PM Implementation* (H_3)—Complex Model 5

Hypothesis H_3 relates the *Layout* with the *PM implementation*, and Table 12.6 illustrates the outcomes of analyzing these variables high and low scenarios, as well as their combinations. It is shown that the probability that a *Layout* is found at its high level is 0.163 while at its low level is 0.177; while the probability of having a *PM implementation* at its high level is 0.179 and at its low is 0.190. Some conclusions about these variables different combined scenarios are the following:

Table 12.5 Sensitivity analysis: *Technological status* and *PM implementation* (H_2)—Complex Model 5

Technological status	*PM implementation*	
	High	Low
High	PM implementation+ = 0.179 Technological status+ = 0.163 PM implementation+ *&* Technological status+ = 0.079 PM implementation+ *if* Technological status+ = 0.483	PM implementation− = 0.190 Technological status+ = 0.163 PM implementation− & Technological status+ = 0.003 PM implementation− *if* Technological status+ = 0.017
Low	PM implementation+ = 0.179 Technological status− = 0.182 PM implementation+ *&* Technological status− = 0.008 PM implementation+ *if* Technological status− = 0.045	PM implementation− = 0.1900 Technological status− = 0.182 PM implementation− & Technological status− = 0.101 PM implementation− *if* Technological status− = 0.552

Table 12.6 Sensitivity analysis: *Layout* and *PM implementation* (H_3)—Complex Model 5

Layout	*PM implementation*	
	High	Low
High	PM implementation+ = 0.179 Layout+ = 0.163 PM implementation+ *&* Layout+ = 0.073 PM implementation+ *if* Layout+ = 0.450	PM implementation− = 0.190 Layout+ = 0.163 PM implementation− *&* Layout+ = 0.014 PM implementation− *if* Layout+ = 0.083
Low	PM implementation+ = 0.179 Layout− = 0.177 PM implementation+ *&* Layout− = 0.005 PM implementation+ *if* Layout− = 0.031	PM implementation− = 0.190 Layout− = 0.177 PM implementation− *&* Layout− = 0.084 PM implementation− *if* Layout− = 0.477

- In the most optimistic scenario, where the *PM implementation* and the *Layout* are presented with high occurrence scenarios, it has a probability of 0.073, which is a low value, since it is intended to be a high value. Also, it is observed that, if there is an adequate distribution, then the probability of having a *PM implementation* at its high level is 0.450. The previous statements indicate that these two variables are always associated and that managers must focus their attention on having a good plant distribution in order to adequately implement preventive maintenance.
- The previous information is easily demonstrated when analyzing the scenario where the *PM implementation* has low levels and the *Layout* high levels, since the probability of simultaneous occurrence for the variables in that scenario is 0.014; a low probability. Likewise, the probability for the first variable

occurring in its scenario, because the second one has occurred, is 0.083; a probability that must be analyzed carefully, since this indicates that the *PM implementation* low levels may be due to other factors and not only because of those associated with the *Layout*.

- In the opposite case, where the *PM implementation* has high levels and the *Layout* low levels, the probability that these scenarios are presented simultaneously is 0.005; a very low probability that indicates that it will almost never happen. In the same way, the probability that the first variable is presented in its scenario, because the second one has been presented is 0.031, which indicates that whenever there are low levels in the *Layout*, there will be low levels in the *PM implementation*.
- Finally, in the most pessimistic scenario, when the *PM implementation* and the *Layout* are presented simultaneously in their low scenarios, the probability of occurrence is 0.084, which is a probability that can generate managerial risk concerns, this is confirmed when analyzing the probability of occurrence for the first variable in its scenario, because the second one has occurred, which is 0.477 and it indicates that if the managers do not make an effort to have an adequate plant distribution, then the appropriate *PM implementation* cannot be guaranteed.

12.1.5.4 Sensitivity Analysis: *Technological Status* and *TPM Implementation* (H_4)—Complex Model 5

In the fourth hypothesis (H_4), the *Technological status* is related to the *TPM implementation* as a more holistic and integrating program. Table 12.7 portrays the Sensitivity analysis for different scenarios in the variables levels, where it can be

Table 12.7 Sensitivity analysis: *Technological status* and *TPM implementation* (H_4)—Complex Model 5

Technological status	*TPM implementation*	
	High	Low
High	TPM implementation+ = 0.193 Technological status+ = 0.163 TPM implementation+ & Technological status+ = 0.103 TPM implementation+ *if* Technological status+ = 0.633	TPM implementation− = 0.168 Technological status+ = 0.163 TPM implementation− & Technological status+ = 0.003 TPM implementation− *if* Technological status+ = 0.017
Low	TPM implementation+ = 0.193 Technological status− = 0.182 TPM implementation+ & Technological status− = 0.000 TPM implementation+ *if* Technological status− = 0.000	TPM implementation− = 0.168 Technological status− = 0.182 TPM implementation− & Technological status− = 0.114 TPM implementation− *if* Technological status− = 0.627

observed that the probability that the *Technological status* is presented in its high scenario is 0.163 and for its low scenario is 0.182. Likewise, it is observed that the probability of having a *TPM implementation* at its high level is 0.193 while at its low level is 0.163.

The results and conclusions about the combined status for these variables are the following:

- In the most optimistic scenario, when the *TPM implementation* and the *Technological status* have high occurrence scenarios, the probability of simultaneous occurrence for these variables is 0.103, which can be considered as a low probability. However, the probability of having high levels in the *TPM implementation*, because the *Technological status* is high is 0.633. The information from the above indicates what is necessary to have a high *Technological status* level to be able to guarantee an adequate *TPM implementation*.
- The previous assertion is clearly demonstrated when there is a scenario where the *TPM implementation* has low levels and the *Technological status* has high levels, where the probability of simultaneous occurrence for those two variables in its status is only 0.003. Also, the probability of presenting the first variable in its scenario, because the second variable has been presented in its comma scenario is 0.017, which indicates the strong relationship between them.
- Also, when analyzing the scenario where the *TPM implementation* has high levels and *Technological status* low levels, the probability of occurrence for these two variables simultaneously is 0. In addition, the probability of having the first variable in its scenario, because the second variable has occurred in its scenario, is 0. It shows that it is practically impossible to achieve an adequate *TPM implementation* when the supply chains have machinery and equipment with a low *Technological status*.
- Finally, the worst scenario that can be presented to a maintenance manager is that there are a *TPM implementation* and a *Technological status* at their low levels, where the probability of occurrence is 0.114, which can be considered as a high risk. Likewise, it is observed that the probability of occurrence for the first variable in its scenario, because the second one has occurred in its scenario is 0.627, which is considered a high risk for the maintenance manager and indicates that they must always focus on having machinery and equipment with a high technological level.

12.1.5.5 Sensitivity Analysis: *Layout* and *TPM Implementation* (H_5)—Complex Model 5

In this section, the Sensitivity analysis is developed between the *Layout* and the *TPM implementation* variables, as these are undoubtedly related. In addition, Table 12.8 indicates the results obtained, where the probability of occurrence for the *TPM implementation* at its high level is 0.193, but the probability of occurrence

Table 12.8 Sensitivity analysis: *Layout* and *TPM implementation* (H_5)—Complex Model 5

Layout	*TPM implementation*	
	High	Low
High	TPM implementation+ = 0.193 Layout+ = 0.163 TPM implementation+ & Layout+ = 0.103 TPM implementation+ *if* Layout+ = 0.633	TPM implementation− = 0.168 Layout+ = 0.163 TPM implementation− & Layout+ = 0.003 TPM implementation− *if* Layout+ = 0.017
Low	TPM implementation+ = 0.193 Layout− = 0.177 TPM implementation+ & Layout− = 0.000 TPM implementation+ *if* Layout− = 0.000	TPM implementation− = 0.168 Layout− = 0.177 TPM implementation− & Layout− = 0.082 TPM implementation− *if* Layout− = 0.432

at its low level is 0.168. Similarly, the probability of having the *Layout* at its low level is 0.163, while at its high level is 0.177.

However, the following scenarios are obtained when there is a combination of the two variables at diverse levels:

- The most optimistic scenario for a maintenance manager occurs when the *TPM implementation* and the *Layout* have high levels. The probability of occurrence for these variables occurring in this scenario simultaneously is 0.103, which can be considered as a low probability, since the ideal scenario would be that these two variables had high levels. However, the probability of presenting the first variable in its scenario, because the second variable has been presented is 0.633. The previous information indicates that the maintenance manager should always seek to have an adequate machines distribution facilitating the *TPM implementation* process.
- The previous statement is clearly demonstrated when the *TPM implementation* is analyzed at its low level and the *Layout* at its high level, where the probability of simultaneous occurrence for these two variables is only 0.003, which indicates that it occurs what is an unlikely event to happen. Likewise, the probability that the first variable is found in its scenario, because the second variable has occurred is 0.017; a low probability of occurrence.
- Also, if the opposite case is analyzed, where the *TPM implementation* has high levels and the *Layout* low levels, the probability of occurrence for these two variables simultaneously is 0. In addition, it is the same probability that the first variable occurs in its scenario, since the second one has happened, because is 0 as well. It indicates that it is not possible to have high *TPM implementation* levels when there is a poor machinery and equipment distribution.
- The most pessimistic scenario for the maintenance manager is when the *TPM implementation* and the *Layout* have low levels, which has a probability of simultaneous occurrence of 0.082, and can be considered as a moderate risk.

Similarly, the probability that the first variable is presented in its scenario, because the second one has been presented is 0.432. Finally, this probability is considered a high risk, and indicates that maintenance managers should focus on having a good machinery and equipment physical distribution in the production systems to ensure an adequate *TPM implementation*.

12.1.5.6 Sensitivity Analysis: *PM Implementation* and *TPM Implementation* (H_6)—Complex Model 5

Undoubtedly, the *TPM implementation* in its most holistic state requires the equipment preservation culture, therefore, in this section, the Sensitivity analysis for that variable is described along with the *PM implementation*. In addition, Table 12.9 shows the results from this analysis, where it can be observed that the *TPM implementation* has a probability of occurrence of 0.193 at its high level, and a probability of 0.168 at its low level, while the *PM implementation* has a probability of 0.179 at its high level, and a probability of 0.190 at its low level.

According to the simultaneously or conditionally probabilities of occurrence analysis from the two variables, the following results and industrial implications are obtained:

- The best scenario for the maintenance engineer occurs when the *TPM implementation* and the *PM implementation* are at their high levels. The probability of such variables occurring simultaneously in such scenarios is 0.101, which may be considered a moderate probability of occurrence, although a manager would like that probability to be higher. Also, the probability that the first variable is in its high scenario, because the second one is also in its high scenario is 0.561. The previous information indicates that the maintenance manager should focus

Table 12.9 Sensitivity analysis: *PM implementation* and *TPM implementation* (H_6)—Complex Model 5

PM implementation	*TPM implementation*	
	High	Low
High	TPM implementation+ = 0.193 PM implementation+ = 0.179 TPM implementation+ & PM implementation+ = 0.101 TPM implementation+ *if* PM implementation+ = 0.561	TPM implementation− = 0.168 PM implementation+ = 0.179 TPM implementation− & PM implementation+ = 0.005 TPM implementation− *if* PM implementation+ = 0.030
Low	TPM implementation+ = 0.193 PM implementation− = 0.190 TPM implementation+ & PM implementation− = 0.003 TPM implementation+ *if* PM implementation− = 0.014	TPM implementation− = 0.168 PM implementation− = 0.190 TPM implementation− & PM implementation− = 0.098 TPM implementation− *if* PM implementation− = 0.514

their efforts in general on a *PM implementation* paying attention to the equipment maintenance, before generating a more holistic *TPM implementation*.

- The previous statement is clearly exposed when analyzing the scenario where the *TPM implementation* has low levels and the *PM implementation* high levels, where the probability of simultaneous occurrence for these two variables in those scenarios is 0.005, which can be cataloged as an event that will almost never happen. Also, the probability that the first variable will be found in its low scenario, because the second variable is in its high scenario, is 0.030. This shows that low *TPM implementation* levels are not associated with high *PM implementation* levels.
- When analyzing the opposite variables levels, where the *TPM implementation* has high levels and the *PM implementation* has low values, it is observed that the probability of simultaneous occurrence for these two variables in their scenarios is only 0.003, an almost non-existing probability that this event will happen. Likewise, the probability that the first variable is presented in its high scenario and the second one in its low scenario is 0.014. The previous information indicates that the *TPM implementation* at its high levels is not associated with low *PM implementation* levels.
- Finally, the worst scenario for the maintenance manager is when the *TPM implementation* and the *PM implementation* have low levels, and the probability of simultaneous occurrence for these two variables is 0.098, which can be considered as a moderate risk. Also, the probability of having the first variable at its low level, because the second variable is already at its low level is 0.514, which is a high risk. In addition, it indicates that low *PM implementation* levels are also associated with low *TPM implementation* levels, as a result, managers should focus on generating a preservation culture towards the installed machinery and equipment in the production process to guarantee the TPM holistic application and benefits.

12.2 Complex Model 6

Complex Model 5 was about only the operative factors among themselves, in order to determine the impact on the benefits that can be obtained from these factors, now in the Complex Model 6, two operative factors and two benefits are integrated; the variables to analyze are the following: *PM implementation*, *Warehouse management*, *Benefits for the organization*, and *Productivity benefits* that are acquired with TPM.

In this model, it is assumed that the independent variable is the *PM implementation*, where the others depend on and it is placed at the top of the model. In addition, the dependent variable is *Productivity benefits*, having as media variables the spare parts and consumables *Warehouse management* and the *Benefits for the organization*. Finally, six hypotheses have been established to relate these variables, which are discussed in the section below.

12.2.1 Hypotheses—Complex Model 6

The production performance is affected by certain types of uncertainties, such as machine failures, processing times, customer demands, and quality failures. As a matter of fact, it can be said that machine failures may disturb the production process, which may result in losses or quality failures, such as, slow production, production losses, and production costs increment (Kang and Subramaniam 2018), in order to ensure an appropriate production, machine maintenance is crucial, as well as having the necessary items to perform it on time. In addition, maintenance is a set of activities (machine defects inspection, repair, replacement, etc.) that help to cope with the deterioration, restore systems or machines status where they can perform the required functions (Colledani and Tolio 2012), however, when machines have failures, it is mandatory to have the required items such as spare parts.

An important part from the maintenance program is to establish that the correct elements and parts are obtained and stored correctly (Sutton 2017). A philosophy and a spare parts program must be developed for each installation, and it is essential that there are enough spare parts available for critical safety functions; In addition, it is relevant that there is an effective program to manage and distribute the parts once they have been purchased (Sutton 2017). Therefore, the following hypothesis can be proposed:

H_1: The *PM implementation* in a productive system has a direct and positive effect on the spare parts and consumables *Warehouse management*.

Furthermore, the *PM implementation* describes a wide range of designed activities that are carried out to improve the reliability and general availability from a system, which is based on the inspection, cleaning, lubrication, adjustment, alignment, and/or worn-out systems replacement (Kamran and John 2010). The *PM implementation* must be done to reduce the corrective maintenance costs each time it reduces the failure probability (Vilarinho et al. 2017). In addition, the equipment status can be effectively improved by a daily PM program (Pan et al. 2010).

In addition, it is mentioned that *PM implementation* can reduce the probability of costly and corrective repairs, as well as avoid excessive maintenance, which significantly reduces maintenance costs (Yang et al. 2017). In order to achieve excellence in maintenance, the balance between performance, risks and maintenance costs must be considered to acquire good quality solutions (Campbell et al. 2016). According to the previous information, the next hypothesis can be established:

H_2: The *PM implementation* in a productive system has a direct and positive effect on the *Benefits for the organization* that can be obtained.

Moreover, spare parts are necessary to ensure the critical equipment operation in many companies because it plays a central role in their operations, both manufacturers and service suppliers need to have spare parts in their stock to minimize the financial and commercial downtime costs (Turrini and Meissner 2017).

Spare parts are kept in stock to reduce the downtime consequences in the equipment, playing a significant role to achieve the desired availability from the same at a minimum economic cost (Hu et al. 2018). The considerable spare parts availability when repairs are required will lead to an enormous economic loss, especially for those industries with sophisticated technologies, which objective is to have a massive and continuous production. Therefore, the spare parts management plays a relevant role to complete the desired equipment availability at a minimum cost (Hu et al. 2018).

Also, to make a competent spare parts management, nowadays, there are different types of labels such as bar codes, 2D data codes, and radio frequency identification that facilitates automatic reading, and operators can carry handheld scanners all over the warehouse to accelerate tasks and deliveries; In addition, barcode readers and radio frequency scanners can be placed at the warehouse entrance and exit to keep the database updated with all products arriving and departing, offering a continuous stock in real time and eliminating theoretically the need for a regular or manual inventory (Christine 2008). According to the previous information, the following hypothesis can be presented:

H_3: The spare parts and consumables *Warehouse management* has a direct and positive effect on the *Benefits for the organization* that can be obtained.

Moreover, manufacturers need to plan and carry out the maintenance of critical resources to avoid production in poor performance conditions, while ensuring the desired resources availability (Das et al. 2007). Therefore, effective maintenance planning is one of the essential activities to achieve high productivity and a good cost–benefit ratio in advanced manufacturing systems (Ni and Jin 2012) and the *PM implementation* is to maintain a specific equipment status according to the equipment failure data and its deterioration (Zhang and Chen 2018).

Thus, PM plans are an effective means to avoid equipment failures (Amik and Deshmukh 2006). In addition, a significant maintenance objective is to ensure the production equipment safety and all assets in general in order to avoid accidents and risks for operators (Pintelon and Muchiri 2009). According to the information from above, the following hypothesis can be proposed:

H_4: The *PM implementation* in a productive system has a direct and positive effect on the *Security benefits* that the company can obtain.

In order to manage a warehouse efficiently, it is needed to know what exactly is on it, and where each item is stored, which is essential for the efficient orders selection, as a result, the simple physical labels give a unique address to each shelf and rack in the warehouse, as well as the databases record the address about each item (Christine 2008). Additionally, there are several specific software programs to support *Warehouse management* operations; one of these is known as the *Warehouse management* system and its main advantages rely in the storage space reduction in the warehouse, as well as greater precision in stock information, greater speed, and operational quality as well as an increasement in the warehouse staff and team productivity (Smith 1998).

The industrial plant reliability and availability represent a critical aspect in many modern manufacturing and service organizations, where increasing their efficiency requires minimizing machine downtime (Ilgin and Tunali 2007). Along with the widespread use of advanced manufacturing technologies, many modern companies are paying more attention to the maintenance management systems development (Tu et al. 2001), therefore, the heavy-duty equipment can be used efficiently. It must be considered that the spare parts availability and their prompt adhesion are key to the maintenance management system success (Ilgin and Tunali 2007), consequently, the next hypothesis can be proposed:

H_5: The spare parts and consumables *Warehouse management* has a direct and positive effect on the *Productivity benefits* that the company can obtain.

The hypothesis H_6 has been justified from the Complex Model 6, which has been defined in another model, in that case, it is only described below:

H_6: The *Benefits for the organization* when implementing TPM in its supply chains have a direct and positive effect on the *Productivity benefits* obtained by the company.

In addition, Fig. 12.3 graphically illustrate the hypotheses that have been presented between the variables from the Complex Model 6. Also, it is important to recall that the variables involved in that model have been validated previously, and their evaluations, as well as their efficiency indexes, are analyzed.

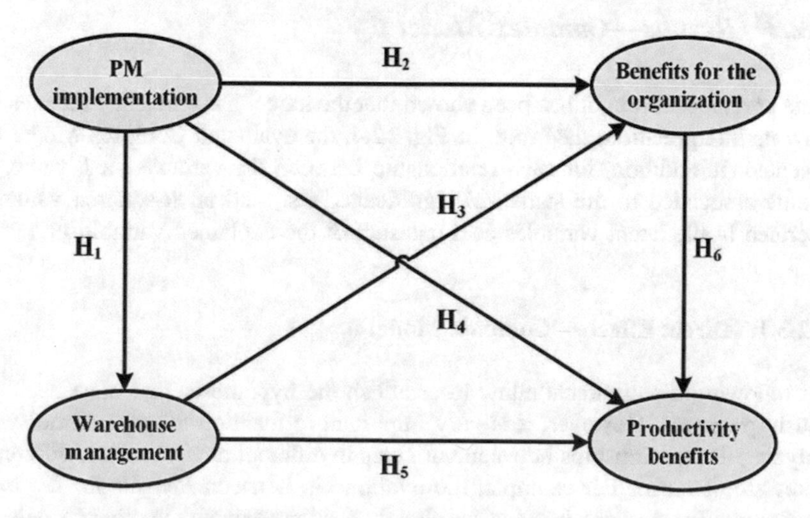

Fig. 12.3 Proposed hypotheses from Complex Model 6

12.2.2 Efficiency Indexes—Complex Model 6

As a matter of fact, the efficiency indexes from the Complex Model 6 are listed in order to verify if the established values are achieved.

- Average path coefficient (APC) = 0.356, $P < 0.001$
- Average R-squared (ARS) = 0.442, $P < 0.001$
- Average adjusted R-squared (AARS) = 0.439, $P < 0.001$
- Average block VIF (AVIF) = 1.491, acceptable if $\leq$ 5, ideally $\leq$ 3.3
- Average full collinearity VIF (AFVIF) = 2.337, acceptable if $\leq$ 5, ideally $\leq$ 3.3
- Tenenhaus GoF (GoF) = 0.542, small $\geq$ 0.1, medium $\geq$ 0.25, large $\geq$ 0.36
- Simpson's paradox ratio (SPR) = 0.833, acceptable if $\geq$ 0.7, ideally = 1
- R-squared contribution ratio (RSCR) = 0.987, acceptable if $\geq$ 0.9, ideally = 1
- Statistical suppression ratio (SSR) = 1.000, acceptable if $\geq$ 0.7
- Nonlinear bivariate causality direction ratio (NLBCDR) = 1.000, acceptable if $\geq$ 0.7

From the previous information, it can be concluded that the model has the required predictive validity, because the ARS and AARS indexes are over 0.02, and the p-values associated with these indexes are under 0.05. Similarly, it is observed that there are no collinearity problems, since the AVIF and AFVIF indexes are less than 3.3; acceptable values. Also, it is noticed that the data obtained from the industry have an adequate adjustment to the model, since the GoF index has a value greater than 0.36, finally, there are no problems are observed in the relationships between the variables in the model.

12.2.3 Results—Complex Model 6

In the previous section, it has been shown that the model has efficiency indexes that allow its interpretation, therefore, in Fig. 12.4, the evaluated Complex Model 6 is presented. In addition, for each relationship between the variables, a β value, the p-value associated to the statistical significance test, and an R-squared value are described in the latent variables as a measure of the explained variability.

12.2.3.1 Direct Effects—Complex Model 6

The following direct effects allow to establish the hypotheses that have been previously proposed. However, it is very important to mention that sometimes when analyzing the relationships between variables in different environments, different β values are obtained. For example, the relationship between *Benefits for the organization* and *Productivity benefits* has already been investigated in other models, but

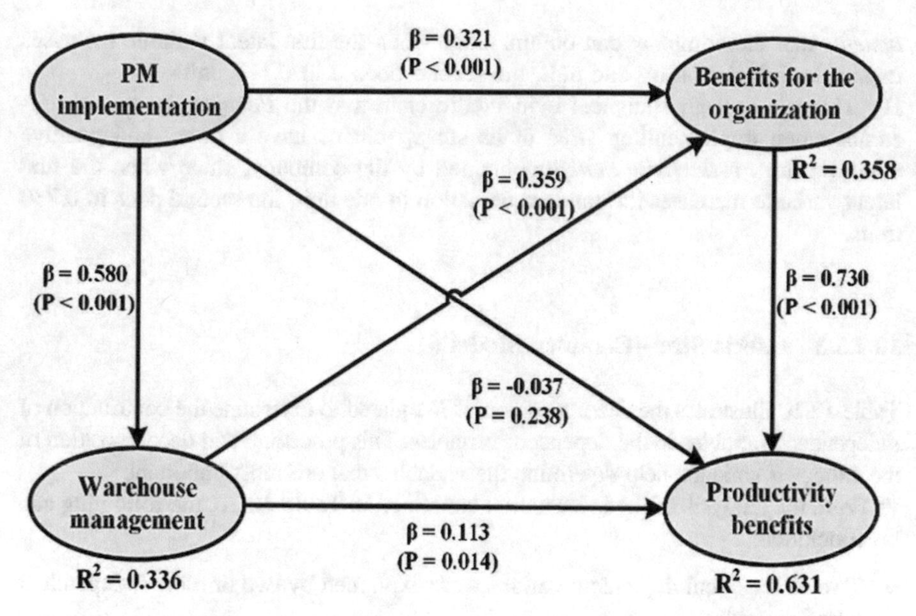

Fig. 12.4 Evaluated Complex Model 6

in a different environment and with other integrated variables, in this case, that relationship can retrieve different results.

The conclusions obtained from the direct effects analysis, the β value and the associated value p, are the following:

H_1: There is enough statistical evidence to state that the *PM implementation* in a productive system has a direct and positive effect on the spare parts and consumables *Warehouse management*, since when the first latent variable increases its standard deviation in one unit, the second increases 0.580 units.

H_2: There is enough statistical evidence to state that the *PM implementation* in a productive system has a direct and positive effect on the *Benefits for the organization* that can be obtained, since when the first latent variable increases its standard deviation in one unit, the second one increases 0.321 units.

H_3: There is enough statistical evidence to state that the spare parts and consumables *Warehouse management* has a direct and positive effect on the *Benefits for the organization* that can be obtained, since when the first latent variable increases its standard deviation in a unit, the second one goes up in 0.359 units.

H_4: There is not enough statistical evidence to state that the *PM implementation* in a productive system has a direct and positive effect on the *Security benefits* that the company can obtain, since the p-value associated with the β in this relationship is over 0.05, therefore, it is not statistically significant.

H_5: There is enough statistical evidence to state that the spare parts and consumables *Warehouse management* has a direct and positive effect on the *Productivity benefits* that the company can obtain, since when the first latent variable increases its standard deviation by one unit, the second does it in 0.113 units.

H_6: There is enough statistical evidence to state that the *Benefits for the organization* when implementing TPM in its supply chains have a direct and positive effect on the *Productivity benefits* obtained by the company, since when the first latent variable increases its standard deviation in one unit, the second does in 0.730 units.

12.2.3.2 Effects Size—Complex Model 6

Table 12.10 illustrates the decomposition of *R*-squared to determine the contribution of independent variables in the dependent variables. This procedure and decomposition of the extracted variance help determine the variables that are most important.

From the analysis of the information contained in Table 12.10, the following can be concluded:

- Two of the latent dependent variables are explained by two or more independent latent variables.
- The *Benefits for the organization* variable are explained in 35.8% by two independent variables; 19.2% is from the *Warehouse management* while 16.6% is from the *PM implementation*. Based on the effect size, it shows that the *Warehouse management* variable is the most important for achieving *Benefits for the organization*, although that does not mean that *PM implementation* is not, since they have very similar values.
- The *Warehouse management* variable is explained only by the *PM implementation* in 33.6% and a decomposition of the variance is not required.
- *Productivity benefits* are explained by three independent latent variables with 63.1%; 59.0% is from the *Benefits for the organization*, 5.8% from the *Warehouse management*, and there is a 1.7% negative contribution from the *PM implementation*, therefore, based on the effect size, it is concluded that the

Table 12.10 *R*-squared decomposition—Complex Model 6

Dependent variable	Independent variable			R^2
	Benefits for the organization	*Warehouse management*	*PM implementation*	
Benefits for the organization		0.192	0.166	0.358
Warehouse management			0.336	0.336
Productivity benefits	0.590	0.058	−0.017	0.631

company must focus on obtaining *Benefits for the organization* in order to get Productivity *benefits*, since the contribution that it has regarding *R*-squared is the highest.

12.2.3.3 Total Indirect Effects—Complex Model 6

Table 12.11 illustrates the total indirect effects that the independent variables have on the dependent variables through the mediating variables, which helps to understand the phenomena and relationships that are often not easily identified.

From the data analysis in Table 12.11, the following is concluded:

- It is observed that the *Warehouse management* has an indirect effect on the *Productivity benefits*, which is because of the mediating variable; *Benefits for the organization* that has a value of 0.262, which represent a high value around 73% from the direct effect, which is only 0.359 that demonstrates the importance of focusing on obtaining associated benefits, creating a better work environment and maintaining workers motivated with easily controllable operations.
- The *PM implementation* has an indirect effect on the *Benefits for the organization* through the *Warehouse management*, which has a value of 0.208, but the direct effect is only 0.321, as a result, this indirect effect represents 64.79% from the direct effect. In addition, the previous information indicates that management should focus on creating preventive maintenance plans that are adequate, but that should include appropriate planning about the supplies and spare parts required to obtain *Benefits for the organization*.
- Finally, it is convenient to observe that the direct effect between the *PM implementation* and *Productivity benefits* was statistically not significant, which does not make sense in the industrial praxis. Fortunately, in this model there is an indirect effect between these two variables with a value of 0.451 that is the highest that appears in the model, which is because of the moderating variables; *Warehouse management* and the *Benefits for the organization*. The previous statement indicates that the *PM implementation* should be focused on obtaining *Benefits for the organization*, and that it should have a support from the *Warehouse management* to strengthen the established plans.

Table 12.11 Total indirect effects—Complex Model 6

Dependent variable	Independent variable	
	Warehouse management	*PM implementation*
Benefits for the organization		0.208 ($P < 0.001$) ES = 0.108
Productivity benefits	0.262 ($P < 0.001$) ES = 0.135	0.451 ($P < 0.001$) ES = 0.210

Table 12.12 Total effects—Complex Model 6

Dependent variable	Independent variable		
	Benefits of the organization	*Warehouse management*	*PM implementation*
Benefits for the organization		0.359 ($P < 0.001$) ES = 0.192	0.529 ($P < 0.001$) ES = 0.274
Warehouse management			0.580 ($P < 0.001$) ES = 0.336
Productivity benefits	0.730 ($P < 0.001$) ES = 0.590	0.375 ($P < 0.001$) ES = 0.194	0.414 ($P < 0.001$) ES = 0.193

12.2.3.4 Total Effects—Complex Model 6

Table 12.12 illustrates the total indirect and direct effects, which is label as total effects, and the following can be concluded based on those results:

- This model confirms what had already been demonstrated before by the *Benefits for the organization* and *Productivity benefits* relationship, since it is the largest direct effect observed in Fig. 12.4 and Table 12.6.
- To obtain *Benefits for the organization*, it is necessary to have a proper *Warehouse management*, because the spare parts and consumables stock is guaranteed, therefore, machines are always in optimal working conditions.
- The *PM implementation* has strong effects on the *Warehouse management*, which indicates that there must be coordination between the raw materials stock and the spare parts and components useful life for the machines, since that ratio is 0.580 units.
- The *PM implementation* has a strong overall effect on *Benefits for the organization*, which is due to a direct and indirect effect that is obtained through the *Warehouse management*. The previous data indicates that, if the mediating variable is not considered, the effect would be only 0.321, but its integration generates a total effect of 0.529 units. In other words, the *PM implementation* should focus on obtaining *Benefits for the organization* but considering the *Warehouse management*.

12.2.4 Conclusions and Industrial Implications—Complex Model 6

The objective of this model was to relate the *PM implementation* with the *Productivity benefits* when they have as mediating variables the spare parts and consumables *Warehouse management* and the *Benefits for the company*. From the direct, indirect, and total effects analysis, the following relationships between these variables can be stated:

- It has been found that the *PM implementation* does not directly affect the *Productivity benefits*, which does not make sense. However, when indirect effects are analyzed, it is found that there is such a relationship, but that the presence of an adequate *Warehouse management* is required focusing on obtaining *Benefits for the organization*. In other words, the *PM implementation* should be focused on having a better work environment, better operation control, increasing employee morale through compliance with maintenance plans and programs, and creating a responsibility culture in the workers who manage the production means. The previous information is clearly defined when analyzing the existing relationship between *Benefits for the organization* and *Productivity benefits*, which is the highest direct and total effect.
- In order to achieve the *Benefits for the organization* through the *PM implementation*, plans and programs are not enough, because they must be coordinated with the *Warehouse management*, since it would be disappointing if the traders are doing every effort to keep machines in good condition and constant materials flows, but when a component is required, it would not exist in the warehouse and there would be delivery delays.
- The *Warehouse management* and stock that is required depend a lot on the senior management, consequently, it is convenient to define the optimal stocks levels that there should be and let managers know about them. In addition, these levels should be estimated based on the parts and components useful life, although fortuitous or unplanned events should be considered as well as the components replacement.
- Operators must have the authority to suggest the machine components purchase, and for this purpose, each spare part or component must be clearly identified with a universal code to avoid confusion in their organization and while ordering them.
- The management must avoid technical stoppages by high replacement materials or spare parts, since that directly affects the fulfillment of the production orders that the client has established. In the same way, it can be deceptive for the operators, since the supply of components is the senior management responsibility.
- When there is an adequate preventive maintenance plan, this can be reflected indirectly in a reliability improvement in the machines and production process, since there is higher machines availability levels, in consequence, costs for this item are reduced.
- Equipment that has adequate preventive maintenance will always be calibrated and in better operating conditions, therefore, a better quality can be acquired.

12.2.5 Sensitivity Analysis—Complex Model 6

In this section, the Sensitivity analysis from the Complex Model 6 is presented, where the different scenarios about the variables that intervene in the model at their high and low levels, as well as their combination are analyzed, therefore, a subsection to describe the analysis for each of the initial hypotheses is displayed.

12.2.5.1 Sensitivity Analysis: *PM Implementation* and *Warehouse Management* (H_1)—Complex Model 6

Undoubtedly, the *PM implementation* requires an adequate *Warehouse management* in which replacements and spare parts for the equipment are safeguarded. In addition, Table 12.13 presents the Sensitivity analysis for the relationship between these two variables, where it is observed that the *PM implementation* has a probability of occurrence of 0.179 that is in its high independently level while a probability of 0.190 being in its low level. On the other hand, *Warehouse management* has a probability at its high level of 0.188 whereas at its low level of 0.141.

According to the simultaneously or conditionally probabilities of occurrence, the following can be concluded:

- The best scenario for the maintenance engineer is the *Warehouse management* and the *PM implementation* is at their low levels. The probability that both variables are presented simultaneously in this scenario is 0.092, which is a low probability, since it is desirable to have high values. In the same way, the probability of having an inventory administration at its high level, because there is a *PM implementation* at its high level is 0.515, which indicates that managers

Table 12.13 Sensitivity analysis: *PM implementation* and *Warehouse management* (H_1)—Complex Model 6

PM implementation	*Warehouse management*	
	High	Low
High	Warehouse management+ = 0.188 PM implementation+ = 0.179 Warehouse management+ & PM implementation+ = 0.092 Warehouse management+ *if* PM implementation+ = 0.515	Warehouse management− = 0.141 PM implementation + = 0.179 Warehouse management− & PM implementation+ = 0.005 Warehouse management− *if* PM implementation+ = 0.030
Low	Warehouse management+ = 0.188 PM implementation− = 0.190 Warehouse management+ & PM implementation− = 0.000 Warehouse management+ *if* PM implementation− = 0.000	Warehouse management− = 0.141 PM implementation− = 0.190 Warehouse management− & PM implementation− = 0.060 Warehouse management− *if* PM implementation− = 0.314

are required focus on obtaining the second variable, since this guarantees its occurrence.

- The previous information is easily demonstrated when analyzing the scenario where the *Warehouse management* has low levels and the *PM implementation* high levels, since it is observed that the probability of occurrence for these variables simultaneously in those scenarios is only 0.005; a probability very close to zero. Likewise, the probability of presenting the first variable in its scenario, because the second has occurred, is 0.030, which indicates that high *PM implementation* levels are not associated with low *Warehouse management* levels.
- In the same way, when analyzing the variables with the levels exchanged, when the *Warehouse management* has high levels and the *TPM implementation* has low levels, the probability of occurrence for those variables simultaneously is zero, but the probability of the first variable occurring in its scenario, because the second has occurred, is also zero, which indicates that low *PM implementation* levels cannot generate high *Warehouse management* levels.
- Finally, the worst scenario for the maintenance manager occurs when there are low *Warehouse management* and *PM implementation* levels simultaneously, which has a probability of occurrence of 0.060. However, the probability of the first one occurring in its scenario, because the second one has occurred is 0.314. The previous data indicates that low *PM implementation* levels will result in low levels in the *Warehouse management*.

12.2.5.2 Sensitivity Analysis: *PM Implementation* and Benefits for the Company (H_2)—Complex Model 6

The *PM implementation* is carried out to obtain some type of *Benefits for the organization*. In this subsection, the Sensitivity analysis for the relationship between these two variables regarding their low and high levels, as well as their the combinations is performed. In addition, Table 12.14 illustrates the results obtained, where it is observed that the probability of obtaining *Benefits for the organization* at its high levels is 0.163, but the probability that it is at low levels is 0.166. Similarly, the probability of having high levels in the *PM implementation* is 0.179, but if there are low levels, the probability is 0.190.

From the analysis about the high and low scenarios, as well as the conditional probabilities, the following results and conclusions are obtained:

- The best scenario in the relationship for these variables occurs when the *Benefits for the organization* and the *PM implementation* have high levels simultaneously, which has a probability of occurrence of 0.063 and is considered a low probability. However, the probability of presenting the first variable in its scenario, because the second has occurred is 0.348. This indicates that high *PM implementation* levels result in high *Benefits for the organization* levels, therefore, managers must focus on better PM planning processes.

Table 12.14 Sensitivity analysis: *PM implementation* and Benefits for the Company (H_2)—Complex Model 6

PM implementation	*Benefits for the company*	
	High	Low
High	Benefits for the organization + = 0.163 PM implementation+ = 0.179 Benefits for the organization+ *&* PM implementation+ = 0.063 Benefits for the organization+ *if* PM implementation+ = 0.348	Benefits for the organization− = 0.166 PM implementation+ = 0.179 Benefits for the organization− *&* PM implementation+ = 0.000 Benefits for the organization− *if* PM implementation+ = 0.000
Low	Benefits for the organization+ = 0.163 PM implementation− = 0.190 Benefits for the organization+ *&* PM implementation− = 0.022 Benefits for the organization+ *if* PM implementation− = 0.114	Benefits for the organization− = 0.166 PM implementation− = 0.190 Benefits for the organization− *&* PM implementation− = 0.065 Benefits for the organization− *if* PM implementation− = 0.343

- The previous statement is easily demonstrated when analyzing the scenario where the *Benefits for the organization* have low levels and the *PM implementation* high levels, which has a probability of zero occurrences. In the same way, the probability of the first variable occurring in its scenario, because the second has happened, is also zero, which indicates that the *PM implementation* is not associated with low *Benefits for the organization* levels.
- The probability of having *Benefits for the organization* at its high levels and low *PM implementation* levels simultaneously is 0.022; a low value. However, when the probability for the first variable occurring in its scenario, because the second one has occurred is 0.114, which indicates that a company is likely to have *Benefits for the organization* even when there are low *PM implementation* levels, because these benefits may have other sources.
- Finally, the worse scenario that can be presented is that there are *Benefits for the organization* and *PM implementation* at their low level in a simultaneously, which has a probability of occurrence of 0.065, and can be considered as a moderate risk. Likewise, the probability of presenting the first variable in its scenario, because the second one has happened is 0.343, and this indicates that low *PM implementation* levels also imply low levels in the *Benefits for the organization*.

12.2.5.3 Sensitivity Analysis: *Warehouse Management* and Benefits for the Company (H_3)—Complex Model 6

The proper replacements and spare parts *Warehouse management* is intended to avoid technical delays and stoppages in the repairs, therefore, the relationship with

Table 12.15 Sensitivity analysis: *Warehouse management* and Benefits for the company (H_3)—Complex Model 6

Warehouse management	*Benefits for the organization*	
	High	Low
High	Benefits for the organization+ = 0.163 Warehouse management+ = 0.188 Benefits for the organization+ & Warehouse management+ = 0.087 Benefits for the organization+ if Warehouse management+ = 0.464	Benefits for the organization− = 0.166 Warehouse management+ = 0.188 Benefits for the organization− & Warehouse management+ = 0.005 Benefits for the organization− *if* Warehouse management+ = 0.029
Low	Benefits for the organization+ = 0.163 Warehouse management− = 0.141 Benefits for the organization+ & Warehouse management− = 0.008 Benefits for the organization+ if Warehouse management− = 0.0058	Benefits for the organization− = 0.166 Warehouse management− = 0.141 Benefits for the organization− & Warehouse management− = 0.071 Benefits for the organization− *if* Warehouse management− = 0.500

the *Benefits for the organization* should be analyzed. In addition, Table 12.15 shows a summary about the results obtained and presents that the *Benefits for the organization* have a probability of occurrence independently at its high level of 0.163, but at its low level is 0.166. In the same way, the *Warehouse management* has a probability of occurrence in its high level of 0.188 and in its low level of 0.141.

According to the analysis of the high and low scenarios combinations that the variables can have, the following results and conclusions are obtained:

- The best scenario occurs when the *Benefits for the organization* and the *Warehouse management* have high levels simultaneously, which has a probability of occurrence of 0.087, and can be considered a low value, since ideally, they are presented at a higher value. However, it is observed that the probability of occurrence for the first variable in its scenario, because the second has occurred is 0.464, which indicates that the *Warehouse management* in its high levels has therefore on the *Benefits for the organization* at its high levels too.
- The previous information is verified when analyzing the scenario where *Benefits for the organization* is at its low level and the *Warehouse management* at its high level, since the probability of occurrence for both variables simultaneously in this scenario is 0.005; which is a low value. Similarly, the probability that the first variable is presented in its scenario, because the second variable is presented is 0.029. In addition, it indicates that the high levels in the *Warehouse management* are not associated with low levels in the *Benefits for the organization*.
- The previous assertion is verified when reviewing the scenario where the *Benefits for the organization* have high levels, but the *Warehouse management* has low levels, which has a probability of occurrence of only 0.008; an almost

invalid value. Likewise, the probability that the first variable is presented in its scenario, because the second one has been presented is 0.005. In addition, it indicates that the low *Warehouse management* levels are not associated with high levels in the *Benefits for the organization.*

- Finally, the worst scenario occurs when the *Benefits for the organization* and the *Warehouse management* are at their low levels simultaneously, which has a probability of occurrence of 0.071, and can be considered as a moderate risk. In addition, the probability of occurrence for the first variable in its scenario, since the second variable has occurred is 0.500, which indicates that low *Warehouse management* levels are associated with low levels in the *Benefits for the organization.*

12.2.5.4 Sensitivity Analysis: *PM Implementation* and Productivity Benefits (H_4)—Complex Model 6

The *PM implementation* also generates *Productivity benefits* and not only related to the company. In addition, Table 12.16 illustrates the Sensitivity analysis results for the relationship between these two latent variables in their high and low scenarios where they may occur, as well as their combination. Also, it is observed that the *Productivity benefits* have a probability of occurrence of 0.190 in their high level, and a probability of 0.155 in their low level; in the same way, the *PM implementation* at its high level has a probability of occurrence of 0.179, while at its low level, it has a probability of 0.190.

According to the data analysis in Table 12.16 regarding the dominated scenarios, the following conclusions can be obtained:

Table 12.16 Sensitivity analysis: *PM implementation* and Productivity benefits (H_4)—Complex Model 6

PM implementation	*Productivity benefits*	
	High	Low
High	Productivity benefits+ = 0.190 PM implementation+ = 0.179 Productivity benefits+ & PM implementation+ = 0.063 Productivity benefits+ *if* PM implementation+ = 0.348	Productivity benefits− = 0.155 PM implementation+ = 0.179 Productivity benefits− & PM implementation+ = 0.003 Productivity benefits− *if* PM implementation+ = 0.015
Low	Productivity benefits+ = 0.190 PM implementation− = 0.190 Productivity benefits+ & PM implementation− = 0.024 Productivity benefits+ *if* PM implementation− = 0.129	Productivity benefits− = 0.155 PM implementation− = 0.190 Productivity benefits− & PM implementation− = 0.065 Productivity benefits− *if* PM implementation− = 0.343

- The most optimistic scenario occurs when the *Productivity benefits* and the *PM implementation* have high levels simultaneously, which has a probability of occurrence of 0.063, and can be considered as a low value, since ideally for the maintenance manager is that those two variables have high values. However, the probability that the first variable is presented in its scenario, because the second one has been presented is 0.348, which indicates that the high *PM implementation* levels are related to high *Productivity benefits* levels, therefore, managers must have an adequate planning for the first variable.
- The previous information is clearly demonstrated when analyzing the scenario where the *Productivity benefits* have low levels and the *PM implementation* high values, where the probability of simultaneous occurrence for these two variables in this scenario is 0.003, while the probability for the first variable occurring in its scenario because the second has occurred is 0.015, which indicates that high *PM implementation* levels are not associated with low levels in the *Productivity benefits*.
- Similarly, the previous conclusion is reached when analyzing the scenario where the *Productivity benefits* have high levels and the *PM implementation* low levels, where the probability of occurrence for these two variables simultaneously in its scenarios is 0.024, while the probability for the first variable occurring in its scenario because the second one has been presented in its scenario is 0.129. In addition, the previous data indicates that it is possible for the company to obtain high *Productivity benefits* even when the *PM implementation* levels are low, and this is due to the fact that these benefits can be originated by different sources.
- Finally, the most pessimistic scenario for the maintenance manager is when there are low values simultaneously in the *Productivity benefits* and in the *PM implementation* process, which has a probability of 0.065, and the probability that the first variable occurs in its scenario, since it has occurred in the second is 0.343. The previous information indicates that low *PM implementation* levels will result in low levels in the *Productivity benefits*.

12.2.5.5 Sensitivity Analysis: *Warehouse Management* and Productivity Benefits (H_5)—Complex Model 6

Productivity benefits can have diverse sources, this time the Sensitivity analysis is where those benefits with *Warehouse management* are related. In addition, Table 12.17 illustrates the Sensitivity analysis for the relationship between these two variables, where it is observed that the probability of obtaining *Productivity benefits* at its high level is 0.190, while at its low levels is 0.155, while the probability of having the *Warehouse management* at its high levels is 0.188 and the probability that its levels are low is 0.141.

According to the data in Table 12.7, the following can be concluded regarding the combined scenarios in the two latent variables:

Table 12.17 Sensitivity analysis: *Warehouse management* and Productivity benefits (H_5)—Complex Model 6

Warehouse management	*Productivity benefits*	
	High	Low
High	Productivity benefits+ = 0.190 Warehouse management+ = 0.188 Productivity benefits+ *&* Warehouse management+ = 0.073 Productivity benefits+ *if* Warehouse management+ = 0.391	Productivity benefits− = 0.155 Warehouse management+ = 0.188 Productivity benefits− *&* Warehouse management+ = 0.000 Productivity benefits− *if* Warehouse management+ = 0.000
Low	Productivity benefits+ = 0.190 Warehouse management− = 0.141 Productivity benefits+ *&* Warehouse management− = 0.008 Productivity benefits+ *if* Warehouse management− = 0.058	Productivity benefits− = 0.155 Warehouse management− = 0.141 Productivity benefits− *&* Warehouse management− = 0.076 Productivity benefits− *if* Warehouse management− = 0.538

- In the most optimistic scenario, when the *Productivity benefits* and the *Warehouse management* are at their high levels, the probability of occurrence for those two variables simultaneously is 0.073, which can be considered a moderate value. In the same way, the probability that the first variable will occur in its scenario, because the second one has been presented is 0.391, which indicates that high *Warehouse management* levels are associated with the *Productivity benefits* that are acquired when implementing TPM.
- The previous information is clearly demonstrated when analyzing the scenario where the *Productivity benefits* have low levels and the *Warehouse management* high levels, since the probability of occurrence for that scenario simultaneously is zero. Likewise, the probability that the first variable will occur because the second one has been presented is zero. The previous data indicates that high *Warehouse management* levels are not related to low levels in the *Productivity benefits*.
- The previous conclusion is also demonstrated when analyzing the scenario where the *Productivity benefits* have high levels and the *Warehouse management* low levels, which have a probability of occurrence of 0.008 simultaneously, an almost invalid value that indicates that is unlikely to happen. Also, the probability of presenting the first variable in its scenario, because the second has occurred, is 0.0058, which indicates that there is little probability that the *Warehouse management* at its low levels will result in *Productivity benefits* at high levels.

- Finally, the worst scenario in the analysis for these two variables is when the *Productivity benefits* and *Warehouse management* have low values in a simultaneous manner, which has a probability of occurrence of 0.076 and is considered a moderate risk for the maintenance manager. Likewise, the probability that the first variable is presented in its scenario, because the second has occurred, is zero, which indicates that low levels in the *Warehouse management* are related to low levels in the *Productivity benefits*, therefore, the manager must focus on having a clean and organized warehouse in order to be able to guarantee TPM benefits.

In this case, the Sensitivity analysis between the *Benefits for the organization* and the *Productivity benefits* is not performed, since they have been reported in another model.

References

Aga DA, Noorderhaven N, Vallejo B (2016) Transformational leadership and project success: the mediating role of team-building. Int J Project Manag 34(5):806–818. https://doi.org/10.1016/j.ijproman.2016.02.012

Anand A, Singhal S, Panwar S, Singh O (2018) Optimal price and warranty length for profit determination: an evaluation based on preventive maintenance. In: Kapur PK, Kumar U, Verma AK (eds) Quality, IT and business operations: modeling and optimization. Springer Singapore, pp 265–277. https://doi.org/10.1007/978-981-10-5577-5_21

Amik G, Deshmukh SG (2006) Maintenance management: literature review and directions. J Qual Maint Eng 12(3):205–238. https://doi.org/10.1108/13552510610685075

Arnaiz A, Konde E, Alarcón J (2013) Continuous improvement on information and on-line maintenance technologies for increased cost-effectiveness. Procedia CIRP 11:193–198. https://doi.org/10.1016/j.procir.2013.07.038

Bourke J, Roper S (2016) AMT adoption and innovation: an investigation of dynamic and complementary effects. Technovation 55–56:42–55. https://doi.org/10.1016/j.technovation.2016.05.003

Campbell JD, Jardine AK, McGlynn J (2016) Asset management excellence: optimizing equipment life-cycle decisions. CRC Press.

Cardoso RdR, Pinheiro de Lima E, Gouvea da Costa SE (2012) Identifying organizational requirements for the implementation of advanced manufacturing technologies (AMT). J Manuf Syst 31(3):367–378. https://doi.org/10.1016/j.jmsy.2012.04.003

Chand G, Shirvani B (2000) Implementation of TPM in cellular manufacture. J Mater Process Technol 103(1):149–154. https://doi.org/10.1016/S0924-0136(00)00407-6

Choe J-M (2004) Impact of management accounting information and AMT on organizational performance. J Inf Technol 19(3):203–214. https://doi.org/10.1057/palgrave.jit.2000013

Christine C (2008) Warehouse management technologies. Sensor Rev 28(2):108–114. https://doi.org/10.1108/02602280810856660

Colledani M, Tolio T (2012) Integrated quality, production logistics and maintenance analysis of multi-stage asynchronous manufacturing systems with degrading machines. CIRP Annals 61 (1):455–458. https://doi.org/10.1016/j.cirp.2012.03.072

Das K, Lashkari RS, Sengupta S (2007) Machine reliability and preventive maintenance planning for cellular manufacturing systems. Eur J Oper Res 183(1):162–180. https://doi.org/10.1016/j.ejor.2006.09.079

Davila A, Elvira MM (2012) Humanistic leadership: lessons from Latin America. J World Bus 47 (4):548–554. https://doi.org/10.1016/j.jwb.2012.01.008

Eti MC, Ogaji SOT, Probert SD (2006) Reducing the cost of preventive maintenance (PM) through adopting a proactive reliability-focused culture. Appl Energy 83(11):1235–1248. https://doi.org/10.1016/j.apenergy.2006.01.002

Fulton M, Hon B (2010) Managing advanced manufacturing technology (AMT) implementation in manufacturing SMEs. Int J Prod Perform Manag 59(4):351–371. https://doi.org/10.1108/17410401011038900

Fumagalli L, Macchi M, Giacomin A (2017) Orchestration of preventive maintenance interventions. IFAC-PapersOnLine 50(1):13976–13981. https://doi.org/10.1016/j.ifacol.2017.08.2417

Gao L, Yu F, Gao L, Xiong N, Yang G (2016) Consistency maintenance of compound operations in real-time collaborative environments. Comput Electr Eng 50:217–235. https://doi.org/10.1016/j.compeleceng.2015.06.021

Gouiaa-Mtibaa A, Dellagi S, Achour Z, Erray W (2018) Integrated maintenance-quality policy with rework process under improved imperfect preventive maintenance. Reliab Eng Syst Saf 173:1–11. https://doi.org/10.1016/j.ress.2017.12.020

Hu Q, Boylan JE, Chen H, Labib A (2018) OR in spare parts management: A review. Eur J Oper Res 266 (2):395–414. https://doi.org/10.1016/j.ejor.2017.07.058

Ilgin MA, Tunali S (2007) Joint optimization of spare parts inventory and maintenance policies using genetic algorithms. Int J Adv Manuf Technol 34(5):594–604. https://doi.org/10.1007/s00170-006-0618-z

Jin M, Tang R, Ji Y, Liu F, Gao L, Huisingh D (2017) Impact of advanced manufacturing on sustainability: an overview of the special volume on advanced manufacturing for sustainability and low fossil carbon emissions. J Clean Prod 161:69–74. https://doi.org/10.1016/j.jclepro.2017.05.101

Kamran SM, John SU (2010) Optimal preventive maintenance and replacement schedules with variable improvement factor. J Qual Maint Eng 16(3):271–287. https://doi.org/10.1108/13552511011072916

Kang K, Subramaniam V (2018) Integrated control policy of production and preventive maintenance for a deteriorating manufacturing system. Comput Ind Eng. https://doi.org/10.1016/j.cie.2018.02.026

Lewis MW, Boyer KK (2002) Factors impacting AMT implementation: an integrative and controlled study. J Eng Tech Manag 19(2):111–130. https://doi.org/10.1016/S0923-4748(02)00005-X

Medina-Herrera N, Jiménez-Gutiérrez A, Grossmann IE (2014) A mathematical programming model for optimal layout considering quantitative risk analysis. Comput Chem Eng 68:165–181. https://doi.org/10.1016/j.compchemeng.2014.05.019

Mohsen, Hassan (2002) A framework for the design of warehouse layout. Facilities 20(13/14):432–440. https://doi.org/10.1108/02632770210454377

Mwanza BG, Mbohwa C (2015) Design of a total productive maintenance model for effective implementation: case study of a chemical manufacturing company. Procedia Manuf 4:461–470. https://doi.org/10.1016/j.promfg.2015.11.063

Ni J, Jin X (2012) Decision support systems for effective maintenance operations. CIRP Annals 61 (1):411–414. https://doi.org/10.1016/j.cirp.2012.03.065

Pan E-s, Liao W-z, Zhuo M-l (2010) Periodic preventive maintenance policy with infinite time and limit of reliability based on health index. J Shanghai Jiaotong Univ (Science) 15(2):231–235. https://doi.org/10.1007/s12204-010-9512-9

Pang S, Jia Y, Liu X, Deng Y (2016) Study on simulation modeling and evaluation of equipment maintenance. J Shanghai Jiaotong Univ (Sci) 21(5):594–599. https://doi.org/10.1007/s12204-016-1768-2

Pintelon L, Muchiri PN (2009) Safety and maintenance. In: Ben-Daya M, Duffuaa SO, Raouf A, Knezevic J, Ait-Kadi D (eds) Handbook of maintenance management and engineering. Springer, London, pp 613–648. https://doi.org/10.1007/978-1-84882-472-0_22

Shrivastava D, Kulkarni MS, Vrat P (2016) Integrated design of preventive maintenance and quality control policy parameters with CUSUM chart. Int J Adv Manuf Technol 82(9):2101–2112. https://doi.org/10.1007/s00170-015-7502-7

Singh R, Gohil AM, Shah DB, Desai S (2013) Total productive maintenance (TPM) implementation in a machine shop: a case study. Procedia Eng 51:592–599. https://doi.org/10.1016/j.proeng.2013.01.084

Smith JD (1998) The warehouse management handbook. Tompkins Press.

Spanos YE, Voudouris I (2009) Antecedents and trajectories of AMT adoption: the case of Greek manufacturing SMEs. Res Policy 38(1):144–155. https://doi.org/10.1016/j.respol.2008.09.006

Sutton I (2017) Maintenance. In: Plant design and operations, 2nd edn. Gulf Professional Publishing, pp 265–292. https://doi.org/10.1016/B978-0-12-812883-1.00009-7

Szőke G, Antal G, Nagy C, Ferenc R, Gyimóthy T (2017) Empirical study on refactoring large-scale industrial systems and its effects on maintainability. J Syst Softw 129:107–126. https://doi.org/10.1016/j.jss.2016.08.071

Tu PYL, Yam R, Tse P, Sun AO (2001) An integrated maintenance management system for an advanced manufacturing company. Int J Adv Manuf Technol 17(9):692–703. https://doi.org/10.1007/s001700170135

Turrini L, Meissner J (2017) Spare parts inventory management: new evidence from distribution fitting. Eur J Oper Res. https://doi.org/10.1016/j.ejor.2017.09.039

Vilarinho S, Lopes I, Oliveira JA (2017) Preventive maintenance decisions through maintenance optimization models: a case study. Procedia Manuf 11:1170–1177. https://doi.org/10.1016/j.promfg.2017.07.241

Vilarinho S, Lopes I, Sousa S (2018) Developing dashboards for SMEs to improve performance of productive equipment and processes. J Ind Inf Integr. https://doi.org/10.1016/j.jii.2018.02.003

Yang L, Ma X, Zhao Y (2017) A condition-based maintenance model for a three-state system subject to degradation and environmental shocks. Comput Ind Eng 105:210–222. https://doi.org/10.1016/j.cie.2017.01.012

Yi W, Chi H-L, Wang S (2018) Mathematical programming models for construction site layout problems. Autom Constr 85:241–248. https://doi.org/10.1016/j.autcon.2017.10.031

Zhang Y, Chen J (2018) Supply chain coordination of incomplete preventive maintenance service based on multimedia remote monitoring. Multimedia Tools and Applications. https://doi.org/10.1007/s11042-018-5977-6

Chapter 13
Structural Equation Models—Methodological Factors

Abstract In this chapter, two structural equation models are presented. The first one refers to a model that integrates two technical variables and two benefits, while the second model is an integrator, since it integrates the *Human factor* independent latent variable where the independent *Operating factor* and *Benefits* latent variables intervene in the total productive maintenance programs implementing process. In addition, each of the latent variables is validated and the model is evaluated through the partial least squares technique. Also, the direct effects, indirect effect, and total effects between the variables are measured. Similarly, a series of industrial implications from the results obtained are displayed.

13.1 Complex Model 7

The Complex Model 7 integrates four variables, two of them refer to technological or operational aspects while two of them are obtained *Benefits*, which are the *Technological status*, the *TPM implementation*, the *Benefits for the organization*, and the *Safety benefits*. These variables are related to generate six hypotheses, which are justified in the section below.

13.1.1 Hypotheses

The equipment/machines importance is increasing and raising with fast-growing technology (Maggard and Rhyne 1992). In addition, the technology usage in maintenance has become a key success factor in its management (Liliane et al. 1999), that's the reason why it has become an essential function in a manufacturing environment (Shamsuddin et al. 2005) and integral quality, plants productivity, and the production strategy (Kumar et al. 2004). Also, the maintenance function effective integration and quality practices will help companies to save time, money,

J. R. Díaz-Reza et al., *Impact Analysis of Total Productive Maintenance*,
https://doi.org/10.1007/978-3-030-01725-5_13

and other resources when productive systems have high technological levels (Sachin and Sanjay 2016).

Furthermore, organizations invest in maintenance activities, because it involves the technology usage, training, and people, and as a result, they adopt different strategies as a preventive measure to avoid machinery and equipment failures (Sachin and Sanjay 2016). Additionally, it is assumed that teams with a high technological level are easier to keep in adequate conditions, because they have many interfaces that notify by different means of their operational status, either through visual or sound aids (Lewis and Boyer 2002), which allows the quick maintenance specialists intervention, preventing major damages. In this way, the following hypothesis can be established:

H_1: The machinery and equipment *Technological status* in the productive system has a direct and positive effect on the *TPM implementation*.

Investment in maintenance is one of the basic functions in a company, which returns better quality, safety, reliability, flexibility, and delivery schedules (Teresko 1992; Willmott and McCarthy 2001) and consequently it has become a strategic tool to increase competitiveness, instead of simply being a general cost that must be controlled (Waeyenbergh and Pintelon 2009). In the same way, the technology team and the development capabilities have become relevant factors that demonstrate the strength in an organization and differentiate it from others (Marcello et al. 2006), in that case, TPM strongly affects technology management through well-maintained equipment/machines; therefore, well-maintained equipment will have a control in repeatability and reproducibility terms to assemble flawless products (Sachin and Sanjay 2016).

In addition, the TPM implementation leads to the product innovation through well-maintained processes and equipment along with the technology management (Ahuja and Khamba 2008b), which allows to obtain a better operational performance (Sachin and Sanjay 2016). Since Sohal et al. (1991) it is stated that the new technologies implementation in the production systems represented a benefit for the company if it could carry out its adequate administration. According to Chung (1991) it is claimed that there are many advantages because of new technologies; however, for many companies, the problem was administrative and that is why Zammuto and O'Connor (1992) remark the important role of senior management and cultural aspects in order to guarantee the *Benefits*. Therefore, the following hypothesis presented:

H_2: The machinery and equipment *Technological status* in the production system has a direct and positive effect on the *Benefits for the organization* that are obtained.

Any TPM program objective is to improve productivity and quality along with a higher employee morale and job satisfaction, as well as being an innovative approach to maintenance that optimizes the equipment effectiveness, promotes the autonomous maintenance of the equipment operator through the daily activities that involve the total workforce (Singh et al. 2013). The TPM implementation will lead industries toward manufacturing excellence (Shinde and Prasad 2017). Also, its approach optimizes the equipment effectiveness, eliminates breakdowns, and promotes maintenance autonomy with a total workforce, since it is a system that covers the maintenance prevention, preventive maintenance, and maintenance related to improvements, with the ultimate goal of preventing losses and waste (Lai Wan and Tat Yuen 2017); therefore, the following hypothesis can be presented:

H_3: The *TPM implementation* in a productive system has a direct and positive impact on the *Benefits for the organization*.

TPM represents a system for the effective process of technology use, as it is designed to efficiently manage fixed assets such as machines, equipment, and properties throughout its life cycle (McKone and Weiss 1998). In addition, these poorly maintained or not maintained equipment implies a great danger to the plant (Pintelon and Muchiri 2009) and despite the increase in security knowledge that is currently available, accidents are still reported. In order to avoid these risks, there has been an increasement in the production system, automation because both the operative and security-related teams are more complex to understand and maintain properly (Pintelon and Muchiri 2009).

In order to reduce risks, current machinery designs have to be more ergonomic and safe for operators (Dalkilic 2017). The maintenance function can affect the plant safeness by improving unsafe conditions, avoiding unsafe tasks, and improving the safety management performance (Pintelon and Muchiri 2009). According to the previous information, the following hypothesis can be proposed:

H_4 The machinery and equipment *Technological status* in a production system has a direct and positive effect on the *Safety benefits* obtained

Moreover, modern manufacturing requires that organizations that want to be successful and achieve world-class manufacturing must have effective and efficient maintenance, and TPM is considered as an effective strategic improvement initiative to improve the maintenance engineering activities quality (Ahuja and Pankaj 2009). In addition, the role of maintenance to keep and improve the plant and equipment availability, the product quality, safety requirement, and the plants cost-effectiveness level, establish an important part from the operating budget in the companies (Al-Najjar and Alsyouf 2003). Currently, its importance has increased due to a higher level of automation, and the maintenance function is now recognized as a strategic tool to increase competitiveness, since it has an impact on capacity, quality, costs, the environment, and safety (Alsyouf 2007). Also, systematic TPM interventions in the organization contribute significantly to improving the productivity, quality, and manufacturing system safeness as well as the workforce morale (Ahuja and Pankaj 2009). Analyzing the *Benefits* that are obtained from the TPM implementation, the following hypothesis is proposed:

H_5 The *TPM implementation* process in a production system has a direct and positive effect on the *Safety benefits* obtained in the company

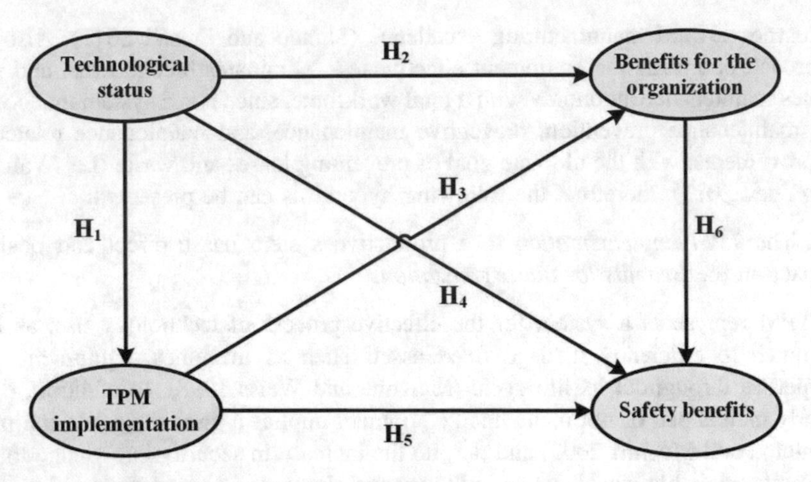

Fig. 13.1 Proposed hypotheses from Complex Model 7

In this case, hypothesis H_6 presented in this model has already been validated and justified in a previous model; therefore, it is displayed in the next manner:

H_6 The *Benefits for the organization* that are acquired by implementing TPM in a productive system has a direct and positive effect on the *Safety benefits* obtained

Figure 13.1 illustrates the relationships among the variables as hypotheses, which are represented by arrows.

The variables analyzed in this model have already been validated in another chapter; all of them have achieved the requirements expected, consequently, they are integrated into the model, which is discussed in the following sections.

13.1.2 Efficiency Indexes

The model is evaluated to acknowledge the efficiency indexes that it has, as well as to make improvements in case it requires some. In addition, the efficiency indexes are mentioned below in the following list:

- Average path coefficient (APC) = 0.380, $p < 0.001$
- Average R-squared (ARS) = 0.531, $p < 0.001$
- Average adjusted R-squared (AARS) = 0.528, $p < 0.001$
- Average block VIF (AVIF) = 1.944, acceptable if $\leq$ 5, ideally $\leq$ 3.3
- Average full collinearity VIF (AFVIF) = 2.750, acceptable if $\leq$ 5, ideally $\leq$ 3.3
- Tenenhaus GoF (GoF) = 0.600, small $\geq$ 0.1, medium $\geq$ 0.25, large $\geq$ 0.36
- Sympson's paradox ratio (SPR) = 1.000, acceptable if $\geq$ 0.7, ideally = 1

- R-squared contribution ratio (RSCR) = 1.000, acceptable if $\geq$ 0.9, ideally =1
- Statistical suppression ratio (SSR) = 1.000, acceptable if $\geq$ 0.7
- Nonlinear bivariate causality direction ratio (NLBCDR) = 1.000, acceptable if $\geq$ 0.7

According to the efficiency indexes in the Complex Model 7, the following is concluded, and therefore the model is interpreted:

- The APC index indicates a good relationship between the variables, the associated p-value is under 0.05, and it means that there is predictive validity.
- The ARS and AARS indexes indicate that the latent dependent variables have enough parametric predictive validity.
- There are no collinearity problems among the latent variables, since the AVIF and AFVIF indexes are under 3.3; the maximum accepted value.
- It is observed that the data obtained from the maquiladora industry has an adequate adjustment to the model, since the Tenenhaus index is over 0.36.
- There are no directionality problems in the latent variables, since the SPR and other indexes are all equal to one.

13.1.3 Results—Complex Model 7

Because the efficiency indexes from the model have shown that it is valid and that all of them are achieved, its results interpretation are proceeding, which is portrayed in Fig. 13.2, where for each of the hypotheses or relationships among the variables,

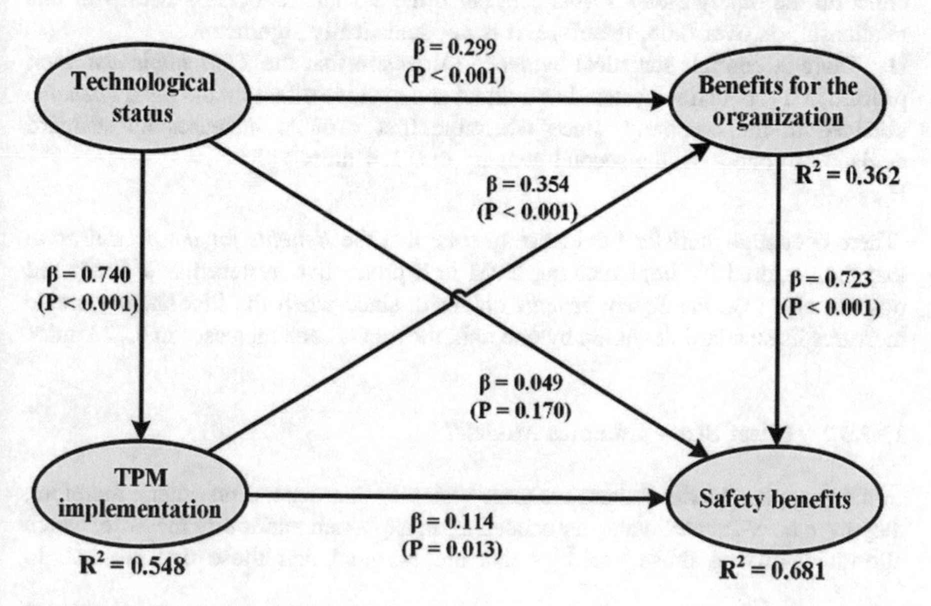

Fig. 13.2 Evaluated Complex Model 7

the β value, the p-value for the statistical significance test, and the R-squared value for each independent latent variable is illustrated.

13.1.3.1 Direct Effects—Complex Model 7

The direct effects support to validate the hypotheses that have been proposed in the model from Fig. 13.1. In addition, this conclusion is developed based on the values that are held in the β and the associated p-value. Also, the conclusions related to the hypotheses are the following:

H_1: There is enough statistical evidence to declare that the machinery and equipment *Technological status* in the productive system has a direct and positive effect on the *TPM implementation*, since when the first latent variable increases its standard deviation in one unit, the second one increases in 0.740 units
H_2: There is enough statistical evidence to state that the machinery and equipment *Technological status* in the production system has a direct and positive effect on the *Benefits for the organization* that are obtained, since when the first latent variable increases its standard deviation in one unit, the second one goes up in 0.299 units
H_3: There is enough statistical evidence to state that the *TPM implementation* in a productive system has a direct and positive impact on the *Benefits for the organization*, since when the first latent variable increases its standard deviation in one unit, the second one increases in 0.354 units
H_4: There is not enough statistical evidence to state that the machinery and equipment *Technological status* in a production system has a direct and positive effect on the *Safety benefits* obtained, since the p-value associated with β in this relationship is over 0.05, therefore, it is not statistically significant
H_5: There is enough statistical evidence to declare that the *TPM implementation* process in a production system has a direct and positive effect on the *Safety benefits* obtained in the company, since when the first variable increases its standard deviation in one unit, the second goes up in 0.114 units
H_6:

There is enough statistical evidence to state that the *Benefits for the organization* that are acquired by implementing TPM in a productive system has a direct and positive effect on the *Safety benefits* obtained, since when the first latent variable increases its standard deviation by one unit, the second one increases in 0.723 units.

13.1.3.2 Effect Size—Complex Model 7

In the Complex Model 7 there are three variables that depend on others; therefore, they have an R-squared value associated as an explained variability measure, which allows identifying those variables that are essential and those that are not. In

Table 13.1 Effects size—Complex Model 7

Dependent variable	Independent variable			R^2
	Benefits for the organization	*TPM implementation*	*Technological status*	
Benefits for the organization		0.199	0.163	0.362
TPM implementation			0.548	0.548
Safety benefits	0.59	0.065	0.026	0.681

addition, the variance decomposition is presented in Table 13.1, based on this, the following is concluded:

- The success in the *TPM implementation* is explained in 54.8% by the *Technological status* variable, which indicates that managers must focus on looking for productive systems with a high technological level to guarantee the TPM success.
- The *Benefits for the organization* variable are explained in 36.2% by two variables, the first is the *Technological status* that explains 16.3% while the second is the *TPM implementation* that can explain 19.9%, which indicates that the second variable is more important because of its greater explanatory power.
- The *Safety benefits* variable is explained by three variables in 68.1%, the first variable is *Benefits for the organization* that can explain 59% from the variance, the second is *TPM implementation* that explains only 0.065, finally, the *Technological status* explains 2.6%. According to the previous information, it is concluded that the most important variable in this model to obtain *Safety benefits* is to obtain the *Benefits for the organization.*

13.1.3.3 Total Indirect Effects—Complex Model 7

The indirect effects often help to understand relationships that are not visible or have no common sense in the industrial praxis. In addition, Table 13.2 shows the total indirect effects between the latent variables in the Complex Model 7, based on the data analysis, the following can be concluded:

- Although the direct effect between the *Technological status* and *Safety benefits* variable was statistically not significant, it is observed that the indirect effect between these two variables is the highest in Table 13.2 and it is statistically significant, with a 0.491 value. This indicates that the *Technological status* does influence on the *Safety benefits*, but it uses the *TPM implementation* and the *Benefits for the organization* as mediating variables.
- The *Technological status* variable has an indirect effect on the *Benefits for the organization* variable with a 0.262 value, which is because the *TPM*

Table 13.2 Total indirect effects, effect size, and *p*-values

Dependent variable	Independent variable	
	TPM implementation	*Technological status*
Benefits for the organization		0.262 ($p < 0.001$) ES = 0.143
Safety benefits	0.256 ($p < 0.001$) ES = 0.145	0.491 ($p < 0.001$) ES = 0.257

implementation mediating variable. This indicates that the *Technological status* facilitates the TPM programs implementation in the production system, which results in *Benefits for the organization*.

- Finally, it is observed that the *TPM implementation* variable has an indirect effect on the *Safety benefits* variable, which is possible through the *Benefits for the organization* mediating variable, which has a 0.256 value. In addition, this indirect effect is significant, since the direct effect can be only 0.114 units, that is, the indirect effect is greater than the direct effect, which demonstrates the importance in that relationship of focusing first on obtaining *Benefits for the organization*.

13.1.3.4 Total Effects—Complex Model 7

Table 13.3 illustrates the total effects between the variables involved in the Complex Model 7, where the direct and indirect effects are already included. In addition, the *p*-values are also shown as a statistical significance test for these effects, as well as the effect size as an explained variance measure through that relationship.

According to the data in Table 13.3, the following conclusions can be obtained about the relationships in the latent variables from the Complex Model 7, such as the following:

Table 13.3 Total effects, effect size, and *p*-values

	Benefits of the organization	*TPM implementation*	*Technological status*
Benefits for the organization		0.354 ($p < 0.001$) ES = 0.199	0.561 ($p < 0.001$) ES = 0.307
TPM implementation			0.740 ($p < 0.001$) ES = 0.548
Safety benefits	0.723 ($p < 0.001$) ES = 0.590	0.371 ($p < 0.001$) ES = 0.209	0.540 ($p < 0.001$) ES = 0.283

- All the total effects are statistically significant, since the associated *p*-value is under 0.05, including the relationship between the *Technological status* and the *Safety benefits*, where the direct effect was statistically nonsignificant.
- Due to the total effects, the relationship between the *Benefits for the organization* and the *Safety benefits*, although it is a direct effect only, it is a relationship between the latent variables analyzed in this model, since it has a 0.723 value. Also, the same happens in the relationship between the *Technological status* and the *TPM implementation* variable, where the beta value is 0.740 standard deviation units.
- The *Technological status* also has a strong relationship with the *Benefits for the organization*; however, this total effect includes a direct and an indirect effect that occurs through the mediating *TPM implementation* variable. This indicates that managers must focus on performing an adequate *TPM implementation,* and always seek to have a high *Technological status* on the installed machines and equipment in the production system.

13.1.4 Conclusions and Industrial Implications—Complex Model 7

The results obtained from the Complex Model 7 analysis allow to make a series of general conclusions and recommendations, as well as a series of industrial implications, such as the following:

- Maintenance managers should always try to maintain a high *Technological status* in the machines and equipment that they have installed in the production systems, since it depends on the *TPM implementation* success process, which demonstrates easily due to the high beta value that relates them.
- In the same way, it is important that managers inquire to obtain *Benefits for the organization* to be able to guarantee *Safety benefits*. For example, they should focus on motivating employees, providing them a clean and healthy environment in order that they can carry out their work, create an equipment preservation culture, promote lifelong learning among the company's employees, and generate and promote effective communication networks.
- Although the relationship between *Technological status* and *Safety benefits* was statistically not significant when analyzing the indirect effect, which does not make sense in the industrial praxis, it is observed that the total effect is statistically significant. However, in order to that relationship exists, the *Technological status* is not enough, because it is required that these machines are adequately maintained, and that they should focus on obtaining *Benefits for the organization*, since in the end they will be transformed into *Safety benefits*.

- According to the beta value size in the relationship between *Benefits for the organization* and *Safety benefits*, it is recommended that maintenance managers focus on achieving the first variable beforehand.

13.1.5 Sensitivity Analysis—Complex Model 7

The sensitivity analysis in a model helps to know the behavior that variables will have when they acquire certain behavior. In this section, this analysis is performed for the Complex Model 7, where each of the hypotheses or relationships between the variables, their possible scenarios, and probabilities of occurrence are discussed.

13.1.5.1 Sensitivity Analysis: *Technological Status* Con *TPM Implementation* (H_1)—Complex Model 7

Undoubtedly, there is a strong relationship between the *Technological status* and the *TPM implementation*, since the beta value was the largest found in the analyzed model, with a 0.740 value. In addition, Table 13.4 shows the results obtained from the Sensitivity analysis performed on these two variables, where it can be observed that the *TPM implementation* has a probability of occurrence at its high level of 0.193, but a probability of occurrence at its low level of 0.168. In the same way, the *Technological status* has a probability of occurrence at its high level of 0.163 while a probability of occurrence of 0.182 at its low level.

From the analysis of the combined scenarios that these two variables may have, the following is discussed:

Table 13.4 Sensitivity analysis: *Technological status* and *TPM implementation* (H_1)—Complex Model 7

Technological status	*TPM implementation*	
	High	Low
High	TPM implementation+ = 0.193 Technological status+ = 0.163 TPM implementation+ & Technological status+ = 0.103 TPM implementation+ *if* Technological status+ = 0.633	TPM implementation− = 0.168 Technological status+ = 0.163 TPM implementation− & Technological status+ = 0.003 TPM implementation− *if* Technological status+ = 0.017
Low	TPM implementation+ = 0193 Technological status− = 0.182 TPM implementation+ & Technological status− = 0.000 TPM implementation+ *if* Technological status− = 0.000	TPM implementation− = 0.168 Technological status− = 0.182 TPM implementation− & Technological status− = 0.114 *TPM implementation− if* Technological status− = 0.627

- In the most optimistic scenario, when the *TPM implementation* and the *Technological status* have high levels, the probability of occurrence for these two variables simultaneously in those scenarios is 0.103, which can be considered a low value, since the maintenance manager would like to have these two variables in their high levels almost always. Likewise, it is observed that the probability of occurrence for the first variable in its scenario, because the second has occurred, is 0.633, which indicates that high levels in the *Technological status* always result in *TPM implementation* high levels.
- The previous assertion is easily demonstrated when analyzing the scenario where the *TPM implementation* has low levels and the *Technological status* has high levels, where the probability of occurrence for these two variable scenarios simultaneously is 0.003, which indicates that it will almost never be presented. Similarly, the probability that the first latent variable happens in its scenario because the second has happened is 0.017. It indicates that high levels in the machines and equipment *Technological status* is not related to low levels in the *TPM implementation*.
- Also, when analyzing the scenario where the *TPM implementation* has high levels and the *Technological status* low levels, it is observed that the probability of simultaneous occurrence for these two variables in their scenarios is zero. In the same way, the probability of presenting the first latent variable in its scenario, because the second has occurred, is zero. The previous statement indicates that the low levels in the *Technological status* are not related to high levels in the *TPM implementation*.
- Finally, the worst scenario that can be presented for a maintenance engineer is that the *TPM implementation* and the *Technological status* have low levels, which has a probability of occurrence of 0.114, which is a high value that can be considered as a critical risk. Likewise, the probability that the first variable occurs in its scenario because the second variable has occurred is 0.627, which indicates that the low levels in the *Technological status* are strongly related to the low levels in the *TPM implementation*.

13.1.5.2 Sensitivity Analysis: *Technological Status* and *Benefits for the Organization* (H_2)—Complex Model 7

Companies invest in advanced technologies for manufacturing in order to obtain some kind of *Benefits*, and the relationship between these two variables has been widely investigated. In this subsection, the Sensitivity analysis is performed between the *Technological status* that the company has and the *Benefits for the organization* that are obtained. In addition, Table 13.5 shows that the probability of occurrence independently for the *Technological status* at its high level is 0.163, but at its low level is 0.182. In the same way, the probability of having *Benefits for the organization* at its high level is 0.163 while the probability at its low level is 0.166.

Table 13.5 Sensitivity analysis: *Technological status* and *Benefits for the organization* (H_2)—Complex Model 7

Technological status	*Benefits for the organization*	
	High	Low
High	Benefits for the organization + = 0.163 Technological status+ = 0.163 Benefits for the organization+ & Technological status+ = 0.065 Benefits for the organization+ *if* Technological status+ = 0.400	Benefits for the organization − = 0.166 Technological status+ = 0.163 Benefits for the organization− & Technological status+ = 0.000 Benefits for the organization−*if* Technological status+ = 0.000
Low	Benefits for the organization + = 0.163 Technological status− = 0.182 Benefits for the organization+ & Technological status− = 0.022 Benefits for the organization+ *if* Technological status− = 0.119	Benefits for the organization − = 0.166 Technological status− = 0.182 Benefits for the organization− & Technological status− = 0.073 Benefits for the organization− *if* Technological status− = 0.403

According to the data analysis in Table 13.5, specifically from the cross-level relationships, it can be concluded by recommending the following:

- In the optimistic scenario, where the *Benefits for the organization* and the *Technological status* have a high level, there is a probability of occurrence of 0.065, which can be considered as a low value, while the probability that the first variable happens in its scenario because the second one has happened is 0.400. The previous data indicates that high levels in the machinery and equipment *Technological status* are related to the *Benefits for the organization* at its high levels.
- The previous data is verified by analyzing the scenario where the *Benefits for the organization* have low levels but the *Technological status* has high levels, and the probability of occurrence for those two variables happening simultaneously in those scenarios is zero, also it is observed that, the probability of obtaining the first variable in its scenario because the second variable has occurred is zero, which indicates that the *Technological status* at its high levels is not related to the *Benefits for the organization* at its low levels.
- Likewise, when analyzing the scenario where the *Benefits for the organization* have high levels and the *Technological status* low levels, it is observed that the probability of occurrence for both variables happening simultaneously in those scenarios is 0.022; a low value. Also, the probability that the first variable occurring in its scenario because the second one has occurred is 0.119, which indicates that it is possible to obtain *Benefits for the organization* when there is a low *Technological status*, because these *Benefits* can have another source and they do not only depend on the technological level in the company.

- Finally, the worst scenario occurs when the *Benefits for the organization* and the *Technological status* have simultaneously low levels, which has a probability of occurrence of 0.073, which can be considered as a moderate risk. Similarly, the probability of occurrence for the first variable in its scenario because the second has occurred is 0.403, which indicates that low levels in the *Technological status* are associated with low levels in the *Benefits for the organization;* therefore, managers must focus on raising the machinery and equipment technological level that they have installed in the production systems.

13.1.5.3 Sensitivity Analysis: *TPM Implementation* and *Benefits for the Organization* (H_3)—Complex Model 7

If the *TPM implementation* did not offer *Benefits for the organization*, then companies would not look to implement it, since it would only represent a cost. In this subsection, the Sensitivity analysis is performed to relate these two variables to their respective high and low levels, which is illustrated in Table 13.6. In addition, it is observed that the probability of occurrence for the *Benefits for the organization* variable at its high level is 0.163, but the probability at its low levels is 0.166, while the probability of occurrence for the *TPM implementation* variable at its high levels is 0.193 while at its low levels is 0.168.

From the data analysis in Table 13.6 to combine the scenarios where the previous variables can be presented, the following is concluded and recommended:

- In the most optimistic scenario, where the *Benefits for the organization* and the *TPM implementation* are simultaneously at their high levels, the probability of

Table 13.6 Sensitivity analysis: *TPM implementation* and *Benefits for the organization* (H_3)—Complex Model 7

TPM implementation	*Benefits for the organization*	
	High	Low
High	Benefits for the organization + = 0.163 TPM implementation+ = 0.193 Benefits for the organization+ & TPM implementation+ = 0.068 Benefits for the organization+ *if* TPM implementation+ = 0.352	Benefits for the organization − = 0.166 TPM implementation+ = 0.193 Benefits for the organization− & TPM implementation+ = 0.003 Benefits for the organization− *if* TPM implementation+ = 0.014
Low	Benefits for the organization + = 0.163 TPM implementation− = 0.168 Benefits for the organization+ & TPM implementation− = 0.019 Benefits for the organization+ *if* TPM implementation− = 0.113	Benefits for the organization − = 0.166 TPM implementation− = 0.168 Benefits for the organization− & TPM implementation− = 0.082 Benefits for the organization− *if* TPM implementation− = 0.484

occurrence is 0.068, while the probability for the first variable occurring in their scenario, because the second has happened is 0.352, which indicates that high levels in the *TPM implementation* are associated with high levels in *the Benefits for the organization;* therefore, the manager must focus on planning and executing adequately the maintenance programs.

- The previous statement is easily demonstrated when analyzing the scenario where there are *Benefits for the organization* at its low level and *TPM implementation* at its high level, where the probability of occurrence for these variables simultaneously in their scenarios is only 0.003, and the probability for the first variable occurring in its scenario because the second has occurred is 0.014; they represent low values. In addition, it indicates that high levels in the *TPM implementation* are not associated with low levels in the *Benefits for the organization.*
- A similar conclusion to the previous one is obtained when analyzing the scenario where the *Benefits for the organization* have high levels but the *TPM implementation* low levels, where the probability of simultaneous occurrence for those two variables in its scenarios is 0.019, and the probability for the first variable occurring in its scenario because the second has occurred is 0.113. The previous data indicates that it is possible to obtain *Benefits for the organization* moderately, even when there are low levels in the *TPM implementation.*
- Finally, the worst scenario that can arise is when the *Benefits for the organization* and the *TPM implementation* have low levels simultaneously, which has a probability of occurrence of 0.082; it represents a moderate risk. However, it is observed that the probability of occurrence for the first variable in its scenario because the second has occurred is 0.484, which indicates that low levels in the *TPM implementation* are associated with low levels in the *Benefits for the organization*, as a result, maintenance managers should focus on fulfilling the established plans and programs.

13.1.5.4 Sensitivity Analysis: *Technological Status* and *Safety Benefits* (H_4)—Complex Model 7

One of the advantages, from modifying and improving the machinery and equipment *Technological status* in a production process, is that *Safety benefits* can be offered to operators. In this subsection, the Sensitivity analysis is carried out for the relationship between these two variables in the results obtained, which are summarized in Table 13.7, where it is observed that the *Safety benefits* have a probability of occurrence, at its high level of 0.220, but a probability of 0.166 at its low level. In the same way, it is observed that the *Technological status* has a probability of occurrence of 0.163 at its high level while a probability of 0.182 at its low level.

From the data analysis in Table 13.7, the following can be concluded and inferred:

Table 13.7 Sensitivity analysis: *Technological status* and *Safety benefits* (H_4)—Complex Model 7

Technological status	*Safety benefits*	
	High	Low
High	Safety benefits+ = 0.220 Technological status+ = 0.163 Safety benefits+ & Technological status+ = 0.090 Safety benefits+ *if* Technological status+ = 0.550	Safety benefits− = 0.166 Technological status+ = 0.163 Safety benefits− & Technological status+ = 0.003 Safety benefits− *if* Technological status+ = 0.017
Low	Safety benefits+ = 0.220 Technological status− = 0.182 Safety benefits+ & Technological status− = 0.033 Safety benefits+ *if* PM implementation− = 0.179	Safety benefits− = 0.166 Technological status− = 0.182 Safety benefits− & Technological status− = 0.082 Safety benefits− *if* Technological status− = 0.448

- In the most optimistic scenario, when the *Safety benefits* and the *Technological status* have high levels simultaneously, the probability of occurrence for these two variables is 0.090, while the probability for the first variable occurring because the second variable has occurred is 0.550, which indicates that high levels in the *Technological status* guarantee *Safety benefits* for the machines and tools operators.
- The previous statement is easily verified when analyzing the scenario where the *Safety benefits* have low levels and the *Technological status* high levels, where the probability of occurrence for those two variables in their scenario is 0.003, while the probability of occurrence for the first variable because the second variable has occurred is 0.017, which indicates that the high *Technological status* is not associated with *Safety benefits* at its low level.
- Another scenario may occur when there are high levels in the *Safety benefits*, but low levels in the *Technological status*, where the probability of occurrence for these two variables simultaneously in their scenarios is 0.033 while the probability for the first variable happening in its scenario, because the second one has occurred is 0.179, which indicates that it is possible to obtain *Safety benefits* moderately, even when the machines and equipment *Technological status* have low levels.
- Finally, the worst scenario that a maintenance manager can have is when the *Safety benefits* and the *Technological status* have low levels, where the probability of occurrence for those two variables simultaneously in their scenarios and levels is 0.082, which represents a moderate risk. Also, the probability that the first variable occurs because the second variable has occurred is 0.448, which indicates that low levels in the *Technological status* will also represent low levels in the *Safety benefits*.

13.1.5.5 Sensitivity Analysis: *TPM Implementation* and *Safety Benefits* (H_5)—Complex Model 7

In this subsection, the Sensitivity analysis about the relationship between the *TPM implementation* and the *Safety benefits* is reported, which is illustrated in summary in Table 13.8. In addition, it is observed that the first variable has a probability of occurrence at its high level of 0.193 and a probability at its low level of 0.168, while the second variable has a probability of occurrence at its high level of 0.220 and a probability at its low level of 0.166.

From the data analysis in Table 13.8, the following can be summarized and concluded:

- The best scenario for the maintenance manager occurs when the *Safety benefits* and the *TPM implementation* have high levels simultaneously, which has a probability of occurrence of 0.092, which is a low value, because the ideal would be to have high levels in both variables. Likewise, the probability that the first variable occurs in its scenario, because the second has occurred is 0.479. The previous information indicates that there is an adequate *TPM implementation*, *Safety benefits* are guaranteed.
- The previous statement is easily verified when analyzing the scenario where the *Safety benefits* have a low level and the *TPM implementation* has a high level, which has a probability of occurrence of 0.005. Also, the probability that the first variable happens in its scenario, because the second variable has occurred is 0.028, which indicates that the high level in the *TPM implementation* is not associated with a low level from *Safety benefits*.
- When analyzing the scenario where the *Safety benefits* have a high level and the *TPM implementation* a low level, it is observed that the probability of occurrence for these two variables simultaneously is 0.045; however, the probability

Table 13.8 Sensitivity analysis: *TPM implementation* and *Safety benefits* (H4)—Complex Model 7

TPM implementation	*Safety benefits*	
	High	Low
High	Safety benefits+ = 0.220 TPM implementation+ = 0.193 Safety benefits+ & TPM implementation+ = 0.092 Safety benefits+ *if* TPM implementation+ = 0.479	Safety benefits− = 0.166 TPM implementation+ = 0.193 Safety benefits− & TPM implementation+ = 0.005 Safety benefits− *if* TPM implementation+ = 0.028
Low	Safety benefits+ = 0.220 TPM implementation− = 0.168 Safety benefits+ & TPM implementation− = 0.024 Safety benefits+ *if* PM implementation− = 0.145	Safety benefits− = 0.166 TPM implementation− = 0.168 Safety benefits− & TPM implementation− = 0.076 Safety benefits− *if* TPM implementation− = 0.452

that the first variable occurring in its scenario, because the second variable has occurred, is 0.145, which indicates that it is possible to have moderate *Safety benefits* even when the *TPM implementation* levels are low, since these may have other sources.

- Finally, the most pessimistic scenario for the maintenance engineer is presented when the *Safety benefits* and the *TPM implementation* process have low levels simultaneously, which has a probability of occurrence of 0.076, which can be considered as a moderate risk. In the same way, the probability that the first variable is presented at its low level, because the second has a low level as well, is 0.452, which indicates that a low level in the *TPM implementation* would also result in a low level for the *Safety benefits*.

13.1.5.6 Sensitivity Analysis: *Benefits for the Organization* and *Safety Benefits* (H_5)—Complex Model 7

In a previous model, the Sensitivity analysis has been performed for the relationship between *Benefits for the organization* and *Safety benefits;* therefore, they are not reported in this section.

13.2 Integrator Model

In this model, three latent variables that intervene in the *TPM implementation* process (Total Productive Maintenance) are related: *Human factor* which is integrated by variables such as Work Culture, Suppliers, Clients, and Managerial Commitment, the *Operating factor* variable that is integrated by the *TPM implementation*, PM implementation, *Technological status*, Layout, and Warehouse management variables, finally, the *Benefits* variable, which is integrated by the *Benefits for the organization*, Productivity benefits, and *Safety benefits*. In addition, in this model, the *Human factor* variable is assumed as an independent variable, since the TPM is a holistic approach whose implementation depends on the human resources commitment in the company and the activities that each of them performs in order to achieve it, but most important is that the benefits from TPM after its implementation are acquired. Consequently, the dependent variables are the *Operating factor* and those *Benefits*.

Also, a set of three hypotheses that relate the variables that integrate this model are generated, which are discussed below.

13.2.1 Hypotheses—Integrator Model

The manufacturing industry in Mexico has experienced an unprecedented degree of changes in the last three decades, which imply strong alterations in management approaches, products and processes technologies, clients' expectations, suppliers' attitudes, as well as competitive behavior (Ahuja et al. 2006). Most industries have the intention to become world-class organizations (Norddin and Saman 2012) and TPM contributes effectively to improve industries' competitiveness and effectiveness in the maintenance field (Sharma et al. 2012), since it is a production guided by improvement methodology that is designed to optimize the equipment reliability and ensure the plant's efficient use through the employee participation and empowerment, linking the manufacturing, maintenance, and engineering, although the commitment and leadership from the senior management is essential for the TPM success (Jitendra et al. 2014).

Moreover, TPM embraces empowerment through production operators that establish an ownership sense in their daily operating team (Albert and Chan 2000). This ownership sense is an important factor that supports the continued success of TPM, as each operator is responsible for ensuring that machines remain in good conditions for maximum utilization (Jitendra et al. 2014).

Also, TPM is a team-based asset management strategy that emphasizes cooperation between operations and maintenance departments with the goal of zero defects, zero failures, and effective design in the workplace (Gulati and Smith 2009). In addition, the *TPM implementation* methodology provides organizations with a guide to fundamentally transform their plant production by integrating culture, the process, and technology (Moore 1997).

Considering that the senior management is responsible for establishing the maintenance plans, the maintenance managers to follow up these plans, the maintenance technicians to execute the activities, and the operator to keep the equipment in good condition, undoubtedly, the *Human factor* affects the operative performance indexes; therefore, the following hypothesis is proposed:

H_1 The *Human factor* during the *TPM implementation* process has a direct and positive effect on the *Operating factor* and its indexes.

Furthermore, TPM is a preventive and productive equipment-based maintenance, both abbreviated as "PM" that involves all levels and all functions in the organization, from senior executives to operators in the production floor (Sun et al. 2003). In addition, maintenance includes activities related to the spare parts stock maintenance, human resources, and risk management, and in a broader sense, it includes all decisions at all levels from the organization related to the acquisition and maintenance for the high availability and reliability level of its assets (Al-Turki et al. 2014).

The management structure is usually organized into many decision layers and parallel functional areas; marketing, purchasing, production, engineering, and maintenance are common functional areas that generally share the manufacturing facilities' benefit in different objectives, which in many cases may have a conflict with each other, and that proper synchronization is essential in order to success globally on business (Al-Turki et al. 2014). Hence, the TPM successful implementation is a team effort where individuals from each and every department must work together, focusing on the employees, because due to the experience, the operator knows more about a machine and its process (Shinde and Prasad 2017). In addition, managers and the maintenance team according to their experience can judge the tasks to be performed or the processes to be eliminated (Shinde and Prasad 2017), and they can try different alternatives to make the team as well as the process more effective in achieving the *Benefits* offered by TPM (Ahuja and Khamba 2008c).

Also, TPM is an improvement process and problem-solving approach to improve the organization's response capacity to meet the customer needs, as well as affect the cost optimization as part of the management strategy to increase market incomes and maximize the *Benefits* (Ahuja and Khamba 2008a). Thus, for instance, compliance with maintenance plans and programs depends on their operators and technicians, as well as managers capabilities; therefore, it is considered that these *benefits* are linked to the skills and commitment from those who perform these activities, and as a result, the following hypothesis is proposed:

H_2 The *Human factor* associated with the *TPM implementation* has a direct and positive effect on the *Benefits* that are obtained.

Additionally, in order that a production system works without waste and to be profitable, the maintenance system should work effectively, this is necessary because the large investments by organizations must generate profits, and the best way to maintain the equipment operation is through the administration of its maintenance (Teonas et al. 2014). Also, effective integration of the maintenance function with engineering functions and other manufacturing functions can help to save large amounts of time, capital, and other useful resources dealing with reliability, availability, maintenance, and performance issues (Moubray 2003).

Also, an effective maintenance program can contribute significantly to the business performance, since in order to support production, maintenance must guarantee the equipment availability to produce products in the required quantity and quality levels (Ahuja and Khamba 2008b). In financial terms, maintenance can be represented from 20 to 40% according to the value added in a product as it moves through the plant (Eti et al. 2004).

Moreover, maintenance influences workers safety and health in two ways. First, regular maintenance that is planned and carried out correctly is essential to keep both machines and the work environment safe and reliable (Jasiulewicz-Kaczmarek 2014). Second, the maintenance itself must be carried out safely, with adequate protection for maintenance traders and others people in the workplace (Butlewski 2012). Also, TPM integrates tasks, transforming traditional management models through the continuous search for: waste elimination, people improvement, production processes improvement, quality, and service, and with this transformation an evolution will come to the company looking for a greater competitiveness

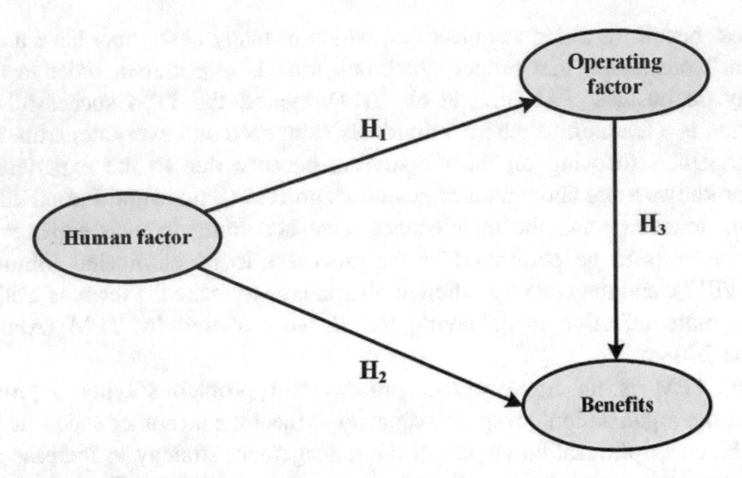

Fig. 13.3 Hypotheses—Integrator Model

(Teonas et al. 2014). According to the previous data, the following hypothesis can be presented:

H_3 The *Operating factor* during the *TPM implementation* process has a direct and positive effect on the *Benefits* obtained.

Figure 13.3 graphically shows the relationships between the latent variables, where each arrow represents a hypothesis.

13.2.2 Validation of Variables—Integrator Model

As a matter of fact, many latent variables that have been analyzed in previous models were validated in an earlier chapter. However, this model is a second-order integrator and the latent variables analyzed are integrated by latent variables, and a validation process has not been carried out; therefore, in this section, the validation process is performed in the same order to continue with the model analysis and interpretation. In addition, Table 13.9 portrays the indexes obtained from the validation process of these variables as well as their analysis.

As a result, the following is concluded:

- The *R*-squared and adjusted *R*-squared values indicate an adequate predictive validity from a parametric point of view, since the values are over 0.02; however, the *Q*-square indicates that there is an adequate nonparametric predictive validity, since it is greater than zero and similar to *R*-square.
- There is enough internal and content validity, since the Cronbach's Alpha values and the compound reliability are over 0.7 in each of the analyzed variables.

Table 13.9 Validity indexes—Integrator Model

Coefficient	*Human factor*	*Operating factor*	*Benefits*
R-squared coefficients		0.756	0.467
Adjusted R-squared coefficients		0.755	0.464
Composite reliability coefficients	0.877	0.924	0.956
Cronbach's alpha coefficients	0.812	0.897	0.931
Average variances extracted	0.643	0.709	0.878
Full collinearity VIFs	4.335	4.217	1.815
Q-squared coefficients		0.756	0.469

- There is enough convergent validity in the latent variables, since the variance average extracted values over 0.5 in all variables.
- Since the inflation variance rates are under five, it is considered that collinearity is not a severe problem.

13.2.3 Efficiency Indexes

Since the latent variables have shown the required validity indexes, they are joined into the Integrator Model. In addition, in the section below there are ten model efficiency indexes lists that were used to validate it, which are illustrated in Fig. 13.3.

- Average path coefficient (APC) = 0.527, $p < 0.001$;
- Average R-squared (ARS) = 0.611, $p < 0.001$;
- Average adjusted R-squared (AARS) = 0.610, $p < 0.001$;
- Average block VIF (AVIF) = 3.558, acceptable if $\leq$ 5, ideally $\leq$ 3.3;
- Average full collinearity VIF (AFVIF) = 3.456, acceptable if $\leq$ 5, ideally $\leq$ 3.3;
- Tenenhaus GoF (GoF) = 0.674, small $\geq$ 0.1, medium $\geq$ 0.25, large $\geq$ 0.36;
- Sympson's paradox ratio (SPR) = 1.000, acceptable if $\geq$ 0.7, ideally = 1;
- R-squared contribution ratio (RSCR) = 1.000, acceptable if $\geq$ 0.9, ideally = 1;
- Statistical suppression ratio (SSR) = 1.000, acceptable if $\geq$ 0.7; and
- Nonlinear bivariate causality direction ratio (NLBCDR) = 1.000, acceptable if $\geq$ 0.7.

According to the previous list, regarding the Integrator Model the following can be concluded:

- According to the APC index or β average, it is concluded that their average is statistically significant, since the p-associated value of p is under 0.05, and therefore the model is predictive.

- According to the ARS and AARS indexes, it is concluded that there is enough predictive validity, since the associated *p*-value is under 0.05, and it can be declared with a 95% of reliability that the relationships are different from zero, and as a result, the dependent variables are explained in an average of 61.1% by the independent variables.
- According to the AVIF and AFVIF indexes, it can be concluded that there are no collinearity and multicollinearity problems within the model, since the values are under five.
- Regarding the GoF index, it can be concluded that the model has a very high explanatory power, since the value is 0.674 and it is over the 0.36 suggested as large, which indicates that the data fit the model properly.
- The other indexes are equal to one, which indicates that there are no problems regarding how the hypotheses have been presented.

In addition, since the values and conclusions previously described for each efficiency index in the model have indicated that they are ideal, it is proceeding to analyze the results from the Integrator Model.

13.2.4 Results—Integrator Model

Figure 13.4 presents the results from the evaluated structural equation model; the β values that represent each hypothesis in Fig. 13.3 can be observed, as well as its *p*-value as the statistical significance evidence for each of the betas and their

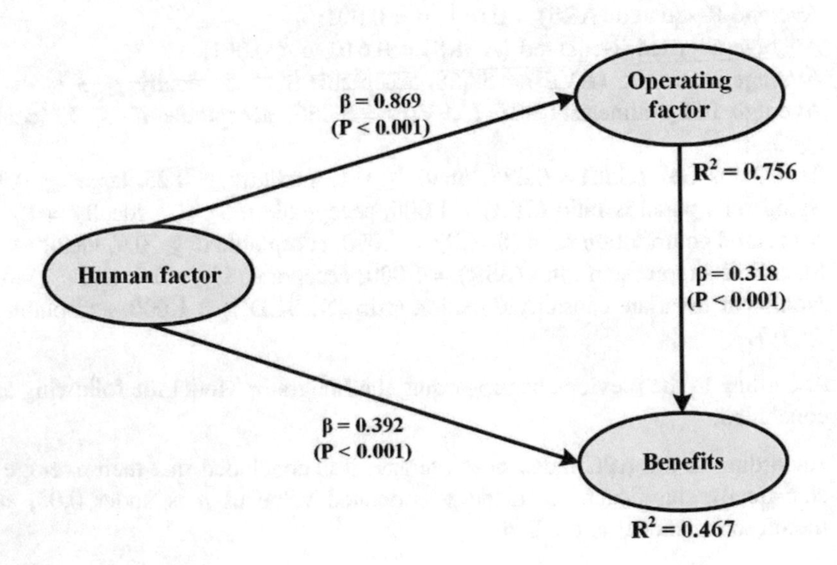

Fig. 13.4 Evaluated Integrator Model

dependency address. Also, the *R*-squared value is observed, which measures the explained variance by each of the independent variables on the dependent variables.

13.2.4.1 Direct Effects—Integrator Model

According to the betas (β) coefficients and the *p*-associated values, the following can be concluded:

H_1 There is enough statistical evidence to declare that the *Human factor* during the *TPM implementation* process has a direct and positive effect on the *Operating factor* and its indexes, since when the first latent variable increases its standard deviation by one unit, the second increases in 0.869 units.

H_2 There is enough statistical evidencc to declare that the *Human factor* associated with the *TPM implementation* has a direct and positive effect on the *Benefits* that are obtained, since when the first latent variable increases its standard deviation by one, the second increases in 0.392 units.

H_3 There is enough statistical evidence to state that the *Operating factor* during the *TPM implementation* process has a direct and positive effect on the *Benefits* obtained, since when the first latent variable increases its standard deviation by one unit, the second increases in 0.392 units

In addition, Table 13.10 shows the variance contributions (the *R*-squared shown in Fig. 13.4) explained by each of the independent latent variables on the latent dependent variables, and in this sense, it is determined which activities are within the *TPM implementation* that influence more within the relationship between variables. Also, in this Integrator Model only two dependent latent variables are available.

- The *Operating factor* latent variable is explained only by the *Human factor* latent variable, and this explains 75.60% from the first variable variance, in such a way that in order to the operative factors to be performed, in other words, in order that the *TPM implementation* is executed, all members participation from the company is required, from the senior management to the machines operators. Similarly, capital investment is needed for the modern machinery acquisition to facilitate manufacturing and to be updated, technologically speaking. In the same way, a good production floor distribution is important to facilitate movement in the company. Finally, it is significant to have an appropriate *Warehouse management*, which has enough spare parts to carry out the

Table 13.10 *R*-squared contribution—Integrator Model

	Independent variable		R^2
Dependent variable	*Human factor*	*Operating factor*	
Operating factor	0.756		0.756
Benefits	0.260	0.207	0.467

machines preventive maintenance in time, as well as to identify them to avoid delays in the maintenance.

- Regarding the variance decomposition from the dependent latent variable called *Benefits*, which is explained in a 26.00% by the *Human factor* latent variable, and in a 20.70% by the *Operating factor* variable, as a total of 46.70%. In addition, it means that there are TPM *Benefits* that will be obtained by people, that is, workers permanent learning, increased morale, as well as *Safety benefits* such as prevention and elimination of potential accidents causes, but most important, *Benefits* related to machinery will be obtained, which will result in an improvement in the final product quality, which in turn will create manufacturing industries competitive capacities.

13.2.4.2 Total Indirect Effects—Integrator Model

As it was already mentioned in the methodology section, as well as it has been reported in the previous complex models, within the structural equation models there are indirect effects as result of mediating variables; in this case, there is only an indirect effect (therefore, the result of this effect is not presented in a table), which is due to the *Human factor* latent variable toward the *Benefits* variable through the *Operating factor* mediating variable. In addition, the effect between these variables is $\beta = 0.277$, which means that when the *Human factor* variable increases its standard deviation by one unit, the *Benefits* variable increases in 0.277 units, and hence the indirect effect is 0.183.

13.2.4.3 Total Effects—Integrator Model

Table 13.11 presents the total effects, which are the total direct and indirect effects, where the β value can be observed, as well as the p-value for each effect; therefore, in all cases, this value is under 0.05, which makes them statistically significant.

According to the previous data analysis, the following can be concluded:

- The most significant variable in this model is the *Human factor*, since it has the total effect value equal to 0.756, which many authors support, because TPM is a

Table 13.11 Total effects—Integrator Model

Dependent variable	Independent variable	
	Human factor	*Operating factor*
Operating factor	0.869 $p < 0.001$ ES = 0.756	
Benefits	0.669 $p < 0.001$ ES = 0.443	0.318 $p < 0.001$ ES = 0.207

tool that must be implemented in a holistic way, and one of the most critical success factors is the management and operators commitment.

- The *Human factor* latent variable explains a 0.443 from the *Benefits* variable, and from this value, the 0.183 is explained indirectly by the *Operating factor* mediating variable.
- Finally, as it was previously mentioned, the *Operating factor* variable explains a 20.70% from the *Benefits* that are obtained from a correct *TPM implementation*.

13.2.5 Conclusions and Industrial Implications—Integrator Model

In this model, three latent variables and three hypotheses were related, which turned out to be statistically significant, therefore, and according to the values obtained from the relationships between the variables, the following is concluded:

- In order that the *Operating factor* is developed within the companies, the participation of all the members in the company is vital, which is facilitated by activities in small groups (Nakajima 1988), since TPM is a continuous improvement process focused on structured teams that seek to optimize production efficiency by identifying and eliminating equipment losses, and production efficiency throughout the production system life cycle through the employees active participation in all the levels of the operational hierarchy (Ahuja and Khamba 2008b). Consequently, there must be an understanding and belief from the management regarding the TPM concept (Lycke 2003).
- It is crucial to have an appropriate *Layout* within the companies, since there can be a waste due to the movement economy violation, which can be the differences in the ability to walk and waste because of an inefficient *Layout* (Ahuja and Khamba 2008b).
- It is important that the warehouse management is performed correctly and that the parts are well identified, recorded entries (when new parts arrive) and outcomes (when the machines are maintained). Also, it is important that there is a proper stock inside the warehouse, since the stock parts are significant, since it represents a high storage cost when it is presented, and when it is not present it can cause high costs due to the high availability, and therefore it is important to estimate the appropriate availability level (Barberá et al. 2012).
- The companies' managers must be aware of the technological advances in production terms, and in that case, they must be in constant communication with their equipment suppliers, since the team's technology and development capabilities have become relevant factors that demonstrate the strength of an organization and differentiate it from others (Marcello et al. 2006).
- Throughout the *TPM implementation* in the companies, certain benefits are obtained, since TPM requires its traders training in order that they are able to

carry out several activities which make them experts in multiple skills, lead to work, and greater workers flexibility enrichment; the operators participation in daily maintenance creates a responsibility sense, pride, and ownership, where delay times are reduced and productivity is increased; besides, teamwork is promoted between operations and maintenance (Duffuaa and Raouf 2015).

13.2.6 Sensitivity Analysis—Integrator Model

In this model, three hypotheses were analyzed among the three latent variables, where the probabilities of the possible high and low scenarios for each one of the variables that intervene in the relationships are estimated.

13.2.6.1 Sensitivity Analysis: *Human Factor* and *Operating Factor* (H_1) —Integrator Model

As it was observed in the results section in this chapter, the first hypothesis H_1 is the most important, since the *Human factor* variable explains 75.60% of the *Operating factor* variable; therefore, in order that all the activities involve in the total productive maintenance program implementation within the companies are performed, the people participation is imperative. In addition, Table 13.12 shows the possible scenarios where there are low and high participation in the *Human factor*, as well as in the *Operating factor*.

- There is a probability of 0.185 that the *Operating factor* variable is presented in its high scenario, and a probability of 0.196 that the variable is presented in the same way; these values may seem to be very low, because there is less than a

Table 13.12 Sensitivity analysis: *Human factor* and *Operating factor*—Integrator Model

Human factor	*Operating factor*	
	High	Low
High	Operating factor+ = 0.185 Human factor+ = 0.196 Operating factor+ & Human factor + = 0.122 Operating factor+ *if* Human factor + = 0.625	Operating factor− = 0.168 Human factor+ = 0.196 Operating factor− & Human factor + = 0.000 Operating factor− *if* Human factor + = 0.000
Low	Operating factor+ = 0.185 Human factor− = 0.163 Operating factor+ & Human factor − = 0.000 Operating factor+ *if* Human factor − = 0.000	Operating factor− = 0.168 Human factor− = 0.163 Operating factor− & Human factor − = 0.111 Operating factor− *if* Human factor − = 0.683

fifth of the probability for these scenarios being presented independently. In addition, if it is observed that the probability for these two variables being presented at their high levels simultaneously is 0.122, even lower, which suggests whether or not to invest in a tool as TPM. However, while observing that the probability when the *Operating factor* variable is present in its high scenario because there is a *Human factor* variable at its high level as well, it is 0.625. The previous information leads to conclude that in these types of tools the most significant aspect is the human resources commitment at all levels within the companies, since it will guide them toward the right path during the *TPM implementation*, which guarantees the *Operating factor* presence.

- Moreover, the scenario where the probabilities that the *Operating factor* variable is presented at its high level and the *Human factor* variable is at its low level is analyzed; the probabilities are 0.185 and 0.163, respectively. In addition, the value from the last variable was slightly reduced. Then, what would happen if the *Human factor* variable is presented at its low level? As it was mentioned in the previous section, in order that the TPM activities are implemented, everybody's participation within the company is crucial; therefore, the probabilities of having a high *Operating factor* and a low *Human factor* are practically impossible, and as a result, it is difficult to carry out the activities that the *Operating factor* requires, since there is a little or nonparticipation from the *Human factor*.
- The probabilities that the two variables are presented at their low and high levels individually have already been analyzed. In this case, the probability of simultaneous presentation for the scenario where the *Operating factor* is at its low level and the *Human factor* is high, can be seen to be 0.000, which can be concluded that as long as the human resources perform their activities within the *TPM implementation*, there will always be operational metrics that are improved. Therefore, there is a probability that the *Operating factor* will not be developed because there is high human resources participation.
- Finally, in the scenario where the variable *Operating factor* is presented in its low level as well as the *Human factor* variable is 0.111, that is, there is very little chance that it would be presented.

13.2.6.2 Sensitivity Analysis: *Human Factor* and Benefits (H_2)—Integrator Model

In this section, the Sensitivity analysis between the *Human factor* and the *Benefits* obtained latent variables when implementing TPM is described, which is represented by the second hypothesis about the high and low scenarios for each variable. In addition, the probability that the *Human factor* variable is presented at its high level is 0.196 while at its low level is 0.163. In the same way, the probabilities that the *Benefits* variable is presented at its high and low levels are 0.177 and 0.152,

Table 13.13 Sensitivity analysis: *Human factor* and Benefits—Integrator Model

Human factor	Benefits	
	High	Low
High	Benefits+ = 0.177 Human factor+ = 0.196 Benefits+ & Human factor+ = 0.090 Benefits+ *if* Human factor+ = 0.458	Benefits− = 0.152 Human factor+ = 0.196 Benefits− & Human factor+ = 0.000 Benefits− *if* Human factor+ = 0.000
Low	Benefits+ = 0.177 Human factor− = 0.163 Benefits+ & Human factor − = 0.011 Benefits+ *if* Human factor− = 0.067	Benefits− = 0.152 Human factor− = 0.163 Benefits− & Human factor − = 0.082 Benefits− *if* Human factor− = 0.500

respectively. Also, these and other results from the probabilities of these variables can be observed in Table 13.13. In the section below, the combination of these scenarios is analyzed:

- The probability that the *Benefits* variable is presented at its high level as well as the *Human factor* variable is 0.090, and it is a very low probability because it is expected from operators, managers, and people in charge of TPM to have an influence on the *Benefits* that this tool provides. The previous information can be observed when the conditional probability is analyzed, that is, there is a probability of 0.458 that there will be a high probability of having the *Human factor* at its high level. In addition, similar to the analysis from the previous hypothesis, the *Human factor* participation is the most important in these scenarios, because it is about the people who perform each of the tasks within the company.
- In the same way, the probability that there will be high *Benefits* and the *Human factor* at its low level is 0.011, which makes the probability of this scenario almost null. Then, the probability that there will be high *Benefits* and low *Human factor* participation is like the previous scenario, almost null. Therefore, *Benefits* are not expected without the *Human factor* participation.
- In the scenario, where the *Benefits* variable is presented at its low level and the *Human factor* variable at its high level is zero, as well as if there are low *Benefits* since there is a high *Human factor* participation, it is also null. In addition, this is logical because the consequence of a high *Human factor* participation is the *Benefits* that are achieved because of it.
- Finally, the scenario where both variables are presented at their low level is 0.082, which is a kind of low value, but there is a probability that this will happen. Also, what is the probability that there are low *Benefits* since there is a low *Human factor* participation? A probability of 0.500 is obtained, and therefore it can be concluded that it is very likely that *Benefits* will not be obtained because the *Human factor* performs few activities within the *TPM implementation*.

13.2.6.3 Sensitivity Analysis: *Operating Factor* and Benefits (H_3)—Integrator Model

Table 13.14 shows the values of the probabilities from the combinations about the high and low scenarios for the *Benefits* and *Operating factor* variables. In addition, the probabilities to have high and low scenarios of obtaining *Benefits* are 0.177 and 0.152, respectively. Also, the probability to have high and low *Operating factor* scenarios are 0.185 and 0.168, respectively; therefore, the combination of these scenarios is analyzed:

- The probability of presenting the two variables in their high level simultaneously is 0.092 but the probability of obtaining high *Benefits* since the *Operating factor* is high, which is 0.500; it makes sense, because if the *Human factor* performs operative activities, consequently there will be *Benefits*.
- The probability that the *Benefits* latent variable is presented at its high level and the latent variable *Operating factor* is presented at its low level is 0.011, which makes this scenario almost impossible and indicates that the *Operating factor* is essential to generate *Benefits*. Likewise, the probability of obtaining high *Benefits,* since there is a low *Operating factor* participation, which is 0.065; it makes it very unlikely, and in other words, it would not be expected to have *Benefits* if the *Operating factor* activities are not carried out.
- In the scenario, where the *Benefits* latent variable is presented at its low level and the *Operating factor* latent variable at its high level, it is 0.000 while the *Benefits* that are obtained, since there is a high *Operating factor* participation; in other words, the activities that are carried out as planned is 0.000; it is expected that if these activities are performed as planned, consequently such *Benefits* will be obtained.

Table 13.14 Sensitivity analysis: *Operating factor* and Benefits—Integrator Model

Operating factor	Benefits	
	High	Low
High	Benefits+ = 0.177 Operating factor+ = 0.185 Benefits+ & Operating factor + = 0.092 Benefits+ *if* Operating factor + = 0.500	Benefits− = 0.152 Operating factor+ = 0.185 Benefits− & Operating factor + = 0.000 Benefits− *if* Operating factor + = 0.000
Low	Benefits+ = 0.177 Operating factor− = 0.168 Benefits+ & Operating factor − = 0.011 Benefits+ *if* Operating factor − = 0.065	Benefits− = 0.152 Human factor− = 0.168 Benefits− & Operating factor − = 0.082 Benefits− *if* Operating factor − = 0.484

- In the pessimistic scenario, where both latent variables are presented simultaneously and in their low levels, it is 0.000, while the *Benefits* latent variable is at its low level the *Operating factor* latent variable is at its low level as well, it is 0.484, which leads to the conclusion that it is possible that *Benefits* will not be obtained if the activities are not carried out in the operational phase. The previous information indicates that managers should not wait for that situation to occur, and they must focus on the operational aspects, since these guarantee *Benefits*.

13.2.6.4 Sensitivity Analysis: *Human Factor* and *Operating Factor* Along with *Benefits*—Integrator Model

In the Sensitivity analysis previously performed only the relationship between two latent variables was analyzed, and in this case, the three variables will be analyzed; in other words, what happens when the *Human factor* variable is presented at its high or low level along with the *Operating factor* variable at its high and low level, having the objective of analyzing the high and low levels scenarios from the *Benefits* that are obtained.

In addition, Table 13.15 illustrates the arrangement of the possible scenarios that can be presented during the *TPM implementation*, for instance, it may be the case that the *Human factor* variable is presented at its high level, the *Operating factor* variable is also presented at its high level and high *Benefits* are obtained, which is denoted by +++. In addition, another scenario that can be presented is where the *Human factor* variable is presented at its low level and the *Operating factor* variable at its high level, and high *Benefits* are obtained, which is denoted by −++. Also, the other scenarios can be identified in Table 13.15. Additionally, Table 13.16 shows the Sensitivity analysis for the possible scenarios of these combinations among the variables.

Once the sensitivity analysis is completed, the following can be concluded:

- The probabilities that the *Benefits* variable is presented at its high and low level are 0.177 and 0.152, respectively. It can be seen that they are low probabilities and they are not very encouraging for the managers in the companies.

Table 13.15 Possible combined scenarios arrangement

		Human factor			
		+		−	
		Operating factor		*Operating factor*	
		+	−	+	−
Benefits	+	+++	+−+	−++	−+
	−	++−	+−	−+−	−

Table 13.16 Sensitivity analysis: *Human factor* and *Operating factor* along with *Benefits*—Integrator Model

		Human factor			
		+		−	
		Operating factor		*Operating factor*	
		+	−	+	−
Benefits	+	Ben+ = 0.177 H. Fact+ & Op. Fact + = 0.122 Ben+ and H. Fact+ & Op. Fact + = 0.076 Ben+ if H. Fact + & Op. Fact + = 0.622	Ben+ = 0.177 H. Fact+ & Op. Fact − = 0.000 Ben+ and H. Fact+ & Op. Fact − = 0.000 Ben+ if H. Fact + & Op. Fact − = 0.000	Ben+ = 0.177 H. Fact− & Op. Fact + = 0.000 Ben+ and H. Fact− & Op. Fact + = 0.000 Ben+ if H. Fact − & Op. Fact + = Undefined	Ben+ = 0.177 H. Fact− & Op. Fact− = 0.111 Ben+ and H. Fact− & Op. Fact − = 0.005 Ben+ if H. Fact − & Op. Fact − = 0.049
	−	Ben− = 0.152 H. Fact+ & Op. Fact + = 0.122 Ben− and H. Fact+ & Op. Fact + = 0.000 Ben− if H. Fact+ & Op. Fact + = 0.000	Ben− = 0.152 H. Fact+ & Op. Fact − = 0.000 Ben− and H. Fact+ & Op. Fact − = 0.000 Ben− if H. Fact + & Op. Fact − = Undefined	Ben− = 0.152 H. Fact− & Op. Fact + = 0.000 Ben− and H. Fact− & Op. Fact + = 0.000 Ben− if H. Fact − & Op. Fact + = Undefined	Ben− = 0.152 H. Fact− & Op. Fact − = 0.111 Ben− and H. Fact− & Op. Fact = 0.068 Ben− if H. Fact − & Op. Fact − = 0610

Operating factor = Op. Fact; Human factor = H. Fact; Benefits = Ben

- The probability that the *Human factor* and *Operating factor* variables are presented at their high and low level are 0.122 and 0.111, respectively.
- The probability of presenting the three variables at their high simultaneous level is 0.076, but what happens to the *Benefits* variable if the *Human factor* and *Operating factor* variables are presented at their high level? The probability of having these *Benefits* in these conditions is 0.622, which makes it necessary to pay full attention to the operational activities that are carried out by the people who work within the company, since this way it ensures that those *Benefits* are going to be obtained.
- Contrary to the previous point, it is when the *Benefits* variable is at its low level because the *Human factor* and *Operating factor* variables are at their high levels; in other words, it means that the activities regarding the *TPM implementation* are being executed; the probability of that scenario is 0.0.
- Some operations or probabilities of occurrence are undefined, since it is divided by a probability equal to zero.

References

Ahuja IPS, Khamba JS (2008a) Strategies and success factors for overcoming challenges in TPM implementation in Indian manufacturing industry. J Qual Maintenance Eng 14(2):123–147. https://doi.org/10.1108/13552510810877647

Ahuja IPS, Khamba JS (2008b) Total productive maintenance: literature review and directions. Int J Qual Reliab Manag 25(7):709–756. https://doi.org/10.1108/02656710810890890

Ahuja IPS, Khamba JS (2008c) Strategies and success factors for overcoming challenges in TPM implementation in Indian manufacturing industry. J Qual Maintenance Eng 14(2):123–147. https://doi.org/10.1108/13552510810877647

Ahuja IPS, Pankaj K (2009) A case study of total productive maintenance implementation at precision tube mills. J Qual Maintenance Eng 15(3):241–258. https://doi.org/10.1108/13552510910983198

Ahuja IS, Khamba JS, Choudhary R (2006) Improved organizational behavior through strategic total productive maintenance implementation. Am Soc Mech Eng 47748:91–98. https://doi.org/10.1115/imece2006-15783

Albert HCT, Chan PK (2000) TPM implementation in China: a case study. Int J Qual Reliab Manag 17(2):144–157. https://doi.org/10.1108/02656710010304555

Al-Najjar B, Alsyouf I (2003) Selecting the most efficient maintenance approach using fuzzy multiple criteria decision making. Int J Prod Econ 84(1):85–100. https://doi.org/10.1016/S0925-5273(02)00380-8

Alsyouf I (2007) The role of maintenance in improving companies' productivity and profitability. Int J Prod Econ 105(1):70–78. https://doi.org/10.1016/j.ijpe.2004.06.057

Al-Turki UM, Ayar T, Yilbas BS, Sahin AZ (2014) Maintenance in manufacturing environment: an overview. In: Al-Turki UM, Ayar T, Yilbas BS, Sahin AZ (eds) Integrated maintenance planning in manufacturing systems. Springer, Cham, pp 5–23. https://doi.org/10.1007/978-3-319-06290-7_2

Barberá L, Crespo A, Viveros P, Stegmaier R (2012) Advanced model for maintenance management in a continuous improvement cycle: integration into the business strategy. Int J Syst Assur Eng Manag 3(1):47–63. https://doi.org/10.1007/s13198-012-0092-y

Butlewski M (2012) The issue of product safety in contemporary design. Safety of the system, Technical, organizational and human work safety determinants Red Szymon Salamon Wyd PCzęst Częstochowa, pp 1428–1600

Chung KB (1991) Deriving advantages from advanced manufacturing technology—an organizing paradigm. Int J Prod Econ 25(1):13–21. https://doi.org/10.1016/0925-5273(91)90126-E

Dalkilic S (2017) Improving aircraft safety and reliability by aircraft maintenance technician training. Eng Fail Anal 82:687–694. https://doi.org/10.1016/j.engfailanal.2017.06.008

Duffuaa SO, Raouf A (2015) Total productive maintenance. In: Duffuaa SO, Raouf A (eds) Planning and control of maintenance systems: modelling and analysis. Springer, Cham, pp 261–270. https://doi.org/10.1007/978-3-319-19803-3_12

Eti MC, Ogaji SOT, Probert SD (2004) Implementing total productive maintenance in Nigerian manufacturing industries. Appl Energy 79(4):385–401. https://doi.org/10.1016/j.apenergy.2004.01.007

Gulati R, Smith R (2009) Maintenance and reliability best practices. Industrial Press Inc., New York

Jasiulewicz-Kaczmarek M (2014) Integrating safety, health and environment (SHE) into the autonomous maintenance activities. In: Stephanidis C (ed) HCI International 2014—posters' extended abstracts. Springer, Cham, pp 467–472

Jitendra K, Vimlesh Kumar S, Geeta A (2014) Impact of TPM implementation on Indian manufacturing industry. Int J Prod Perform Manag 63(1):44–56. https://doi.org/10.1108/IJPPM-06-2012-0051

Kumar P, Wadood A, Ahuja IPS, Singh TP, Sushil M (2004) Total productive maintenance implementation in Indian manufacturing industry for sustained competitiveness. Paper presented at the 34th International Conference on Computers and Industrial Engineering, San Francisco, CA, 14–16 November

Lai Wan H, Tat Yuen L (2017) Total productive maintenance and manufacturing performance improvement. J Qual Maintenance Eng 23(1):2–21. https://doi.org/10.1108/JQME-07-2015-0033

Lewis MW, Boyer KK (2002) Factors impacting AMT implementation: an integrative and controlled study. J Eng Tech Manage 19(2):111–130. https://doi.org/10.1016/S0923-4748(02)00005-X

Liliane P, Du Niek P, Van Frank P (1999) Information technology: opportunities for maintenance management. J Qual Maintenance Eng 5(1):9–24. https://doi.org/10.1108/13552519910257032

Lycke L (2003) Team development when implementing TPM. Total Qual Manag Bus Excellence 14(2):205–213. https://doi.org/10.1080/1478336032000051395

Maggard BN, Rhyne DM (1992) Total productive maintenance: a timely integration of production and maintenance. Prod Inventory Manag J 33(4):6

Marcello B, Gionata C, Marco F, Andrea G (2006) AHP-based evaluation of CMMS software. J Manufact Technol Manag 17(5):585–602. https://doi.org/10.1108/17410380610668531

McKone KE, Weiss EN (1998) TPM: planned and autonomous maintenance: bridging the gap between practice and research. Prod Oper Manag 7(4):335–351

Moore R (1997) Combining TPM and reliability-focused maintenance. Plant Eng 51(6):88–90

Moubray J (2003) Twenty-first century maintenance organization: part I—the asset management model, maintenance technology. Applied Technology Publications, Barrington, IL

Nakajima S (1988) Introduction to TPM: total productive maintenance. (Translation). Productivity Press, Inc, p 129

Norddin K, Saman MZM (2012) Implementation of total productive maintenance concept in a fertilizer process plant. Jurnal Makanikal 32:66–82

Pintelon L, Muchiri PN (2009) Safety and Maintenance. In: Ben-Daya M, Duffuaa SO, Raouf A, Knezevic J, Ait-Kadi D (eds) Handbook of Maintenance Management and Engineering. Springer, London, pp 613–648. https://doi.org/10.1007/978-1-84882-472-0_22

Sachin M, Sanjay S (2016) Total productive maintenance, total quality management and operational performance: an empirical study of Indian pharmaceutical industry. J Qual Maintenance Eng 22(4):353–377. https://doi.org/10.1108/JQME-10-2015-0048

Shamsuddin A, Masjuki Hj H, Zahari T (2005) TPM can go beyond maintenance: excerpt from a case implementation. J Qual Maintenance Eng 11(1):19–42. https://doi.org/10.1108/13552510510589352

Sharma K, Gera G, Kumar R, Chaudhary H, Gupta S (2012) An empirical study approach on TPM implementation in manufacturing industry. Int J Emerg Technol 3(1):18–23

Shinde DD, Prasad R (2017) Application of AHP for ranking of total productive maintenance pillars. Wirel Pers Commun. https://doi.org/10.1007/s11277-017-5084-4

Singh R, Gohil AM, Shah DB, Desai S (2013) Total productive maintenance (TPM) implementation in a machine shop: a case study. Procedia Eng 51:592–599. https://doi.org/10.1016/j.proeng.2013.01.084

Sohal A, Samson D, Weill P (1991) Manufacturing and technology strategy: a survey of planning for AMT. Comput Integr Manuf Syst 4(2):71–79. https://doi.org/10.1016/0951-5240(91)90023-R

Sun H, Yam R, Wai-Keung N (2003) The implementation and evaluation of Total Productive Maintenance (TPM)—an action case study in a Hong Kong manufacturing company. Int J Adv Manuf Technol 22(3):224–228. https://doi.org/10.1007/s00170-002-1463-3

Teonas B, Julio Cezar Mairesse S, Ana Paula Barth B (2014) Improvement of industrial performance with TPM implementation. J Qual Maintenance Eng 20(1):2–19. https://doi.org/10.1108/JQME-07-2012-0025

Teresko J (1992) Time bomb or profit center? Ind Week 241(3):6

Waeyenbergh G, Pintelon L (2009) CIBOCOF: A framework for industrial maintenance concept development. Int J Prod Econ 121(2):633–640. https://doi.org/10.1016/j.ijpe.2006.10.012

Willmott P, McCarthy D (2001) 2—Assessing the true costs and benefits of TPM. In: Total productivity maintenance, 2nd edn. Butterworth-Heinemann, Oxford, pp 17–22. https://doi.org/10.1016/B978-075064447-1/50005-9

Zammuto RF, O'Connor EJ (1992) Gaining advanced manufacturing technologies' benefits: the roles of organization design and culture. Acad Manag Rev 17(4):701–728. https://doi.org/10.2307/258805

Printed in the United States
By Bookmasters